Geometry and Theoretical Physics

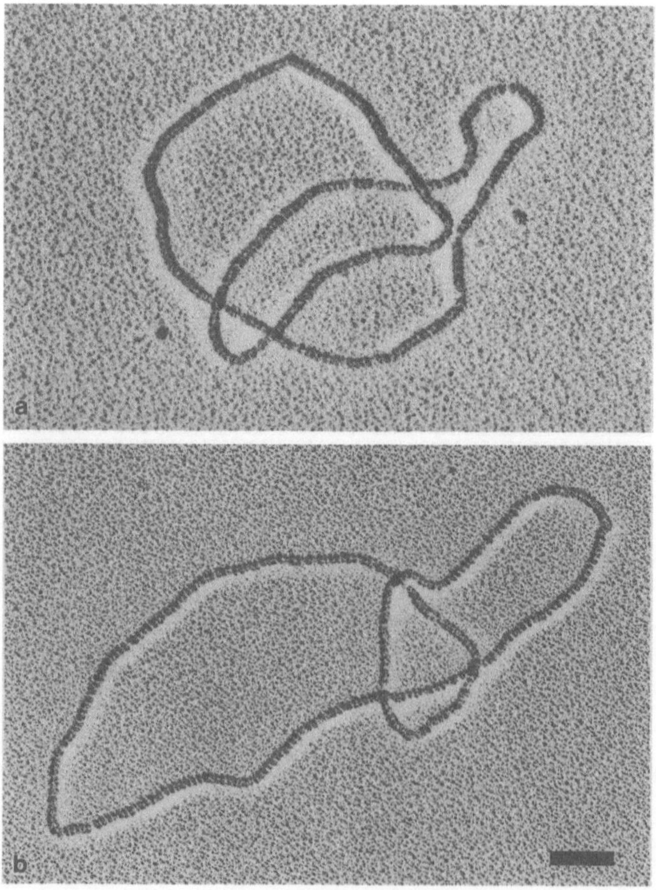

Knots tied by Nature: The positive and negative trefoil of DNA under the electron microscope. The scale bar measures 10^{-7} m.

J. Debrus A. C. Hirshfeld (Eds.)

Geometry and Theoretical Physics

Edited in Cooperation with the Deutsche Physikalische Gesellschaft

With 32 Figures

Springer-Verlag
Berlin Heidelberg New York
London Paris Tokyo
Hong Kong Barcelona
Budapest

Dr. Joachim Debrus

Physikzentrum Bad Honnef, Hauptstraße 5
W-5340 Bad Honnef, Fed. Rep. of Germany

Priv.-Doz. Dr. Allen C. Hirshfeld

Lehrstuhl für Theoretische Physik III, Universität Dortmund
Postfach 500500, W-4600 Dortmund 1, Fed. Rep. of Germany

ISBN-13:978-3-642-76355-7 e-ISBN-13:978-3-642-76353-3
DOI: 10.1007/978-3-642-76353-3

Softcover reprint the hardcover 1st edition 1991

Typesetting: Camera ready by editors

56/3140-543210 – Printed on acid-free paper

Preface

The interaction between geometry and theoretical physics has often been very fruitful. A highlight in this century was Einstein's creation of the theory of general relativity. Equally impressive was the recognition, starting from the work of Yang and Mills and culminating in the Weinberg–Salam theory of the electroweak interaction and quantum chromodynamics, that the fundamental interactions of elementary particles are governed by gauge fields, which in mathematical terms are connections in principal fibre bundles. Theoretical physicists became increasingly aware of the fact that the use of modern mathematical methods may be necessary in the treatment of problems of physical interest. Since some of these topics are covered at most summarily in the usual curriculum, there is a need for extra-curricular efforts to provide an opportunity for learning these techniques and their physical applications.

In this context we arranged a meeting at the Physikzentrum Bad Honnef 12–16 February 1990 on the subject "Geometry and Theoretical Physics", in the series of physics schools organized by the German Physical Society. The participants were graduate students from German universities and research institutes. Since the meeting occurred only a short time after freedom of travel between East and West Germany became a reality, this was for many from the East the first opportunity to attend a scientific meeting in the West, and for many from the West the first chance to become personally acquainted with colleagues from the East. Indeed, nearly one-third of the approximately 60 participants came from the region which was then still the German Democratic Republic.

The lectures at the school were devoted to various topics within this general framework. A mathematical introduction to the relevant differential geometric methods was given by E. Binz. Since these lectures were not written up for publication we are happy to be able to include in this volume lecture notes on fibre bundles and modern differential geometry written by U. Kasper. This material is important because it provides some of the mathematical background necessary for the understanding of the following articles, and as a consequence contributes to making this volume in some measure self-contained. E. Binz went on to discuss the application of the above methods to the physical description of smoothly deformable media.

In two lectures F.W. Hehl presented a very detailed and complete discussion of fermions in a gravitational field, largely within the framework of special relativity. This material, which presents some of the more recent developments for the first time in a coherent and systematic fashion, should prove an important reference for this subject. Professor Hehl also spoke on a new continuum-theoretical approach to spacetime physics; he is reserving the publication of this material for a future opportunity. K.-H. Rehren spoke on new applications of the methods of algebraic field theory in low-dimensional spacetime, which explain the occurrence of braid group statistics and link into investigations of 2-dimensional conformal field theories. J. Hoppe discussed infinite-dimensional algebras and $(2+1)$-dimensional field theories.

M. Forger presented recent developments in conformal field theory, and W. Nahm spoke on solitons and integrable theories, but neither of these contributions could be included in this publication.

A series of articles in this volume is concerned with a special topic which has attracted the attention of many mathematical physicists: anomalies in quantum field theory. The reason for this interest is the fact that we encounter here a particularly impressive example of the application of powerful methods of contemporary mathematics to a problem of eminent physical significance. A. Hirshfeld provided a general introduction to this subject. N.A. Papadopoulos discussed the role of stratification in anomalies. H. Römer spoke in a wider context on field theoretic applications of the index theorem. F. Brandt discussed a complete classification of anomalies in general gauge theories. M. Reuter illustrated how new mathematical methods can yield interesting insights into relationships in physics in his talk on modular invariance, causality, and the PCT-theorem.

T. Schücker gave an introduction to new developments linking statistical mechanics, field theory, and the mathematical theory of knots. There are even applications beyond physics, namely in biology. In his abstract he writes: "The discovery of a link between apparently unrelated fields is always a particular highlight in the development of natural science." In this spirit we decided to choose for the frontispiece of this volume a photograph of "Knots tied by Nature": positive and negative trefoil knots in DNA molecules. This figure was generously provided by Prof. T. Koller from the ETH in Zürich. We thank him and his colleagues, as well as the editors of *Nature*, for their permission to reproduce this figure.

We were especially privileged to hear a talk of K. Bleuler about his mentor Wolfgang Pauli, discussing both his scientific work and, beyond that, his ideas on the foundations of physics and the sources of scientific inquiry. We have included here both a version of this talk in the original German and an English translation, accomplished with Professor Bleuler's kind assistance.

We wish to thank all the lecturers at the school, who provided so unstintingly of their knowledge and their time, as well as the participants, for their interest and their encouragement. The staff of the Physikzentrum Bad Honnef furnished their customary excellent service, establishing the ambience which is the necessary condition for the success of any scientific meeting. Stefan Groote, University of Dortmund, provided invaluable technical and editorial assistance in the preparation of this volume for publication. We gratefully acknowledge the financial support of the Volkswagen Foundation.

Bad Honnef
Dortmund
February 1991

J.Debrus
A.C. Hirshfeld

Contents

Fibre Bundles: An Introduction to Concepts of Modern Differential Geometry[1]

U. Kasper

Zentralinstitut für Astrophysik
D(O)-1561 Potsdam, Federal Republic of Germany

Abstract

An introduction to concepts of modern differential geometry is given, emphasizing in particular differentiable manifolds, vector fields, principal fibre bundles and connections. The relevance of these concepts for gauge theories and for theories of gravitation is indicated.

1. Introduction

The physicist meets fibre bundles in the form of gauge theories. They prove especially useful when the spacetime structure is no longer so simple as in the case, say, of Minkowski space. But also for local considerations they are often worthwhile, since they make it easier to understand what is really done.

The following description attempts to make the reader familiar with the most important concepts in this area. Proofs of theorems are rarely given. These can be found in the relevant literature [1, 2]. For German speaking people, we refer to [3], note, however, that the notations here deviate slightly from those used in [1, 2]. For a detailed study of all the matter presented here, the reader also interested in physical applications should consult [4].

In order to give a first general idea of the concepts to be treated, the definitions of a connection in a principal fibre bundle and the principal fibre bundle itself are given below, without presuming the reader to understand them at this point: $P(M, G)$ is a principal fibre bundle over the manifold M with the structure group G, $T_u(P)$ is the tangent space to P at $u \in P$ and $vT_u(P)$ is a subspace of $T_u(P)$ consisting of vectors which touch the fibre through u at u.

[1] The material presented here was the basis of four lectures given at the university of Konstanz in 1986. Dr. A. Hirshfeld has structured the notes and translated them into English. The author wishes to thank him for doing this tremendous work.

1.1 Definition:

A *connection* Γ in P assignes to every $u \in P$ a subspace $hT_u(P)$ of $T_u(P)$ with the following properties:

1. $T_u(P) = vT_u(P) \oplus hT_u(P)$ ($\oplus$ denotes the "direct sum" of the two sub-vector spaces).
2. The action R_a of an element $a \in G$ on P induces a mapping R_{a*} of the tangent spaces of P. For every $u \in P$ we have: $hT_{ua}(P) = R_{a*}hT_u(P)$.
3. $hT_u(P)$ depends smoothly on u.

Now we define the principal bundle itself. Let P and M be differentiable manifolds and π a differentiable mapping of P onto M. The Lie group G acts smoothly on P from the right.

1.2 Definition:

(P, M, π, G) is called a *principal fibre bundle over M with structure group G*, when

1. G acts *freely* on P, i.e., $u \cdot a = u$ for some $u \in P$ implies $a = e \in G$.
2. Let $u_1, u_2 \in P$ with $\pi(u_1) = \pi(u_2)$. This holds if and only if an $a \in G$ exists, such that $u_1 \cdot a = u_2$.
3. P is locally trivial over M, i.e., for arbitrary $x \in M$ there is a neighborhood U and a diffeomorphism $\psi : \pi^{-1}(U) \to U \times G$, such that $\psi(u) = (\pi(u), \eta(u))$ with $\eta(u) \in G$ and $\psi(u \cdot a) = (\pi(u), \eta(u)a)$.

We see, therefore, that fibre bundles are quite highly structured mathematical objects. One has to know what *differentiable manifolds* and *Lie groups* are. One needs further structures on the manifolds, such as *vector fields* and *differentiable forms*. Also a firm grasp of the concept of a *mapping* is of the utmost importance. This point is already a stumbling block for the physicist attempting to penetrate the mysteries of modern differential geometry. He thinks more in terms of input/output and often does not pay so much attention to the intermediate object. But it is precisely this intermediate object which is here of outstanding importance. In addition, the physicist often writes down things which are different from what he really means.

For example, $A^k(x^m)$ is written to denote a vector field. But what it really specifies are the components of a vector at a point in relation to some basis which is not explicitly specified. Often, however, the notation $A^k(x^m)$ signifies that a "multiplicity" of vector fields exist over a space, and a specific prescription chooses a particular representative out of this multiplicity. We shall later recognize a vector field as a section of a vector bundle. This is a mapping A of a manifold M in the set of all tangent spaces to M at every point of M. This set is denoted by $T(M)$, and we denote the vector field by $A : M \to T(M)$. The vector at the point x is then the image of x under the mapping $A : x \mapsto A(x) \in T_x(M)$, which is an element of the tangent space to M at x.

2. Differentiable Manifolds and Their Structures

We now wish to begin with a more detailed discussion of the concepts appearing in the definitions above. First of all, what is a differentiable manifold?

2.1 Definition:

A set M becomes a *topological space*, when we identify in M a family of subsets $\mathcal{O}$ with the following properties:

1. The union of any non-empty family of sets from $\mathcal{O}$ belongs to $\mathcal{O}$.
2. $\emptyset \in \mathcal{O}$ ($\emptyset$ denotes the empty set).
3. The intersection of any two sets of $\mathcal{O}$ belongs to $\mathcal{O}$.
4. $M \in \mathcal{O}$.

$\mathcal{O}$ is called a *topology* on M. The elements of $\mathcal{O}$ are called *open sets*. A subset $U \subset M$ is called a *neighborhood* of $x \in M$, if there is some $O \in \mathcal{O}$ with $x \in O$ and $O \subset U$. One says that $\mathcal{O}$ *is finer than* $\mathcal{O}'$ when $\mathcal{O}' \subset \mathcal{O}$. A family of sets Ω is called a *basis* of $\mathcal{O}$, if $\Omega \subset \mathcal{O}$, and any set in $\mathcal{O}$ can be represented as a union of sets in Ω. A family of subsets Ω of the topological space M is called *locally finite*, if every point of M has a neighborhood which intersects only with a finite number of sets of Ω. A topological space is *paracompact*, if for any open covering $\mathcal{U}$ of M there is a locally finite covering which is finer than $\mathcal{U}$.

In the theory of fibre bundles paracompact manifolds are used as basis manifolds, because then, e.g., the existence of connections is ensured.

Finally we need the concept of a Hausdorff space: A topological space M is called a *Hausdorff space*, if for $x, y \in M$ with $x \neq y$, there are open sets U,V with $x \in U, y \in V$ and $U \cap V = \emptyset$ (Roughly speaking, we are told here whether two elements of M are considered different).

2.2 Definition:

A Hausdorff space with a countable basis is called an *n-dimensional manifold* if it is locally homeomorphic to the Euclidean space $\mathbb{R}^n$. Here *locally homeomorphic* means that every point $x \in M$ possesses a neighborhood which is homeomorphic to an open subset of $\mathbb{R}^n$.

A *homeomorphism* is a one-to-one continuous mapping, whose inverse is also continuous (We should proceed here to explain, what *continuous* means. However, the reader is familiar with this concept in special contexts. This should suffice to give an intuitive feeling for continuity, and in the following we shall in any case have no occasion to use the general topological definition).

An important aspect of the concepts still to be treated is already here apparent. A manifold is locally similar to the Euclidean space, and so one tries to transfer all one has learned about Euclidean spaces and the structures defined on them to the more general context of manifolds.

2.3 Definition:

For every $x \in M$ there is a neighborhood U_ι and a homeomorphism φ_ι of the type discussed above. The pair (U_ι, φ_ι) is called a *chart of M*. A family ϕ of charts is called an *atlas of M*, when the domains U_ι of the mappings φ_ι form a covering of M.

Let a_i, $i = 1, \ldots, n$ be a basis of $\mathbb{R}^n$, which shall be fixed for the purposes of the following discussion. The representation of a chart $\varphi : U \to \mathbb{R}^n$ with respect to this basis is (summation over i, ranging from 1 to n)

$$\varphi : x \in U \mapsto \varphi(x) = x^i(x) a_i \in \mathbb{R}^n. \tag{2.1}$$

For $x^i(x)a_i$ we sometimes also write shortly (x^i). The $x^i(x)$ are called *coordinates of* $x \in M$ with respect to the chart (U, φ). Let (U_ι, φ_ι), $(U_\kappa, \varphi_\kappa)$ be two charts with $U_\iota \cap U_\kappa \neq \emptyset$. A point $x \in U_\iota \cap U_\kappa$ has the coordinates $x^i(x)$ with respect to the chart (U_ι, φ_ι) and the coordinates $x^{i'}(x)$ with respect to the chart $(U_\kappa, \varphi_\kappa)$.

The coordinates with respect to the different charts are related by the *coordinate transformation*

$$\varphi_{\kappa\iota} : x^i \in \varphi_\iota(U_\iota \cap U_\kappa) \mapsto x^{i'} = \varphi_\kappa \circ \varphi_\iota^{-1}(x^i) \in \varphi_\kappa(U_\iota \cap U_\kappa). \tag{2.2}$$

$\varphi_{\kappa\iota}$ is a homeomorphism of open sets of $\mathbb{R}^n$. An atlas φ is said to be *of class* C^k, if all the transformations $\varphi_{\kappa\iota}$ are continuous and differentiable to the k-th order.

2.4 Definition:

An n-dimensional *differentiable manifold* is an n-dimensional manifold with a maximal atlas φ of class C^k.

Why is it important to start from a Hausdorff space with countable basis? The condition for M to be Hausdorff excludes certain "pathological" cases. For example:

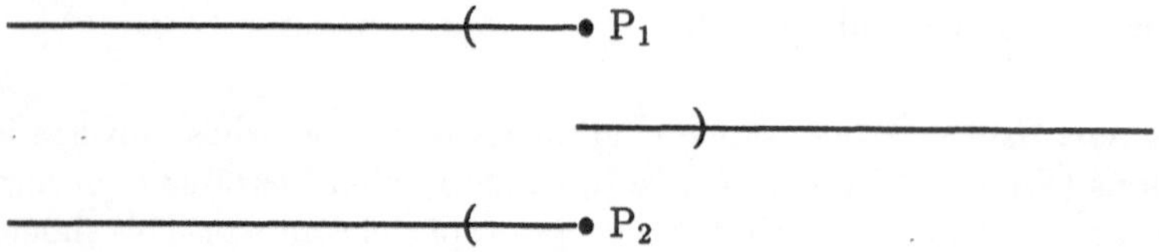

Fig. 1. P_1 and P_2 have no neighborhoods U and V with $P_1 \in U, P_2 \in V$ and $U \cap V = \emptyset$.

The condition that the basis be countable prevents, for example, giving $\mathbb{R}^2$ the structure of $\mathbb{R}$.

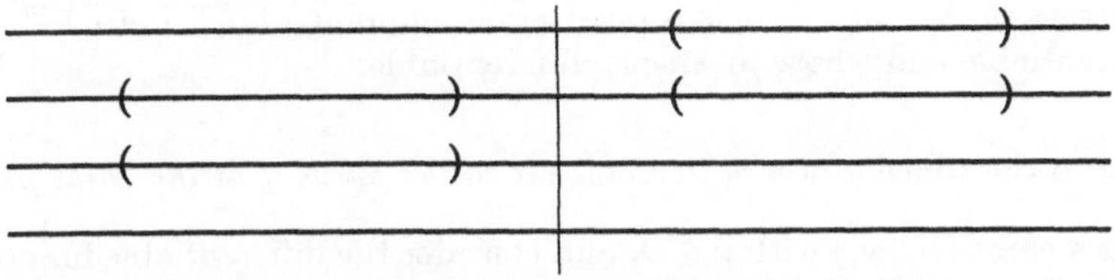

Fig. 2. A covering of $\mathbb{R}^2$ with one-dimensional open sets does not have a countable basis.

Given two manifolds X and Y with their respective charts $(U_\iota, \varphi_\iota)_{\iota\in I}$ and $(V_\kappa, \psi_\kappa)_{\kappa\in J}$, one can construct the *product manifold* $X \times Y$ with the charts $(U_\iota \times V_\kappa, \varphi_{\iota,\kappa})_{(\iota,\kappa)\in I\times J}$, where the homeomorphisms $\varphi_{\iota\kappa}$ are defined by

$$\varphi_{\iota,\kappa} : (x, y) \in U_\iota \times V_\kappa \mapsto (\varphi_\iota(x), \psi_\kappa(y)) \in \mathbb{R}^{n+m}. \tag{2.3}$$

An *open submanifold* Y of the n-dimensional mainfold X^n is an open subset of X^n, whose differentiable structure follows from a restriction of the atlas of X^n to Y.

3. Differentiable Maps and Vector Fields

We now discuss differentiable maps between differentiable manifolds.

Let $f : X^n \to Y^m$ be a mapping of X^n into Y^m, where X^n, Y^m are of class C^k. Let (U_ι, φ_ι), respectively (V_κ, ψ_κ), be charts of X^n, respectively Y^m, with $U_\iota \cap f^{-1}(V_\kappa) \neq \emptyset$. The *representation of* f relative to the given charts is

$$f_{\kappa\iota}(x^i) = \psi_\kappa \circ f \circ \varphi_\iota^{-1}(x^i), \quad (x^i) \in \varphi_\iota(U_\iota \cap f^{-1}(V_\kappa)). \tag{3.1}$$

$f_{\kappa\iota}$ is a mapping from $\mathbb{R}^n$ into $\mathbb{R}^m$. Now let $\varphi_{\iota\lambda}$, respectively $\psi_{\mu\kappa}$, be coordinate transformations of X^n, respectively Y^m. If $x \in U_\iota \cap U_\lambda \cap f^{-1}(V_\mu \cap V_\kappa)$, we have

$$f_{\mu\lambda} = \psi_{\mu\kappa} \circ f_{\kappa\iota} \circ \varphi_{\iota\lambda} = \psi_\mu \circ f \circ \varphi_\lambda^{-1}. \tag{3.2}$$

The mapping f is said to be *continuously differentiable to order s at the point* $x \in X^n$, if for two charts $(U_\iota, \varphi_\iota), (V_\kappa, \psi_\kappa)$ with $x \in U_\iota$ and $y = f(x) \in V_\kappa$ the representation $f_{\kappa\iota}$ is continuously differentiable of order s at the point $(x^i(x))$. It follows from the formula for the transition from $f_{\kappa\iota}$ to $f_{\mu\lambda}$ that $s \leq r$. This definition is independent of the choice of the charts.

A *curve* γ *of class* C^r is an example of a differentiable mapping from $I \in \mathbb{R}^1$ into the manifold X^n:

$$\gamma : t \in I \mapsto \gamma(t) \in X. \tag{3.3}$$

3.1 Definition:

A *diffeomorphism* of X^n onto Y^m is a one-to-one mapping of X^n onto Y^m, which is differentiable and whose inverse is differentiable.

We now come to the concept of a *vector tangent to the curve* γ *at the point* p.

We choose a chart (U_ι, φ_ι) with $p \in U_\iota$ and consider the differentiable function f on U_ι and a curve $\gamma(t)$ with $\gamma(t_0) = p$. With the help of the coordinate representations of f and γ we can construct

$$\frac{\partial f_\iota}{\partial x^i} \frac{dx^i}{dt}\bigg|_{t=t_0}. \tag{3.4}$$

This value is independent of the choice of charts, and we write it as

$$Xf = \frac{df(x(t))}{dt}\bigg|_{t=t_0} = \xi^i \frac{\partial}{\partial x^i} f. \tag{3.5}$$

X is called the *tangent vector to* γ *at* p. $\partial/\partial x^i$ are the tangent vectors to those curves whose coordinate representations are given by

$$\begin{aligned} x^j &= \text{constant} \qquad \text{for } j \neq i, \\ x^i &= x_0^i + (t - t_0). \end{aligned} \tag{3.6}$$

3.2 Definition:

The set of all vectors tangent to curves through p constitute the *tangent space* $T_p(X^n)$ to X^n at p, and the disjoint union of all the $T_p(X^n)$ is called the *tangent bundle of* X^n:

$$T(X^n) = \bigcup_{p \in X^n} T_p(X^n). \tag{3.7}$$

To every chart (U_ι, φ_ι) with $x \in U_\iota$ corresponds a bijective mapping

$$\varphi_{\iota x} : T_x(X^n) \to \mathbb{R}^n, \tag{3.8}$$

which assigns to the vector $X \in T_x(X^n)$ its coordinate representation $X_\iota \in \mathbb{R}^n$. If X_κ is the coordinate representation of X with respect to the chart $(U_\kappa, \varphi_\kappa)$, the relation between X_ι and X_κ is given by the well-known law for the transformation of the components of a vector.

The tangent bundle $T(X^n)$ is our first example of a structure which we shall later call a *fibre bundle*. $T(X^n)$ itself can be given the structure of a differentiable manifold, which is locally a product manifold. Namely, we may locally represent its elements as pairs (x, X), where $x \in X^n$ and $X \in T_x(X^n)$. $T(X^n)$ is called a *bundle manifold*. Any element of $T_x(X^n)$ may be related to any other element through a linear transformation, i.e., an element of the transformation group $GL(n, \mathbb{R})$, here called the *structure group*, which acts on the bundle manifold. Finally, there exists a *projection* from $T(X^n)$ onto the *base manifold* X^n which maps the *fibre* over x, namely $T_x(X^n)$, onto x. A *vector field* is the assignment of a tangent vector $X_p \in T_p(X^n)$ to every point $p \in X^n$.

We cannot enter here into questions of differentiability.

4. Differential Forms

4.1 Definition:

A map $f : Y^m \to X^n$ induces a map of the tangent bundle of Y^m, $T(Y^m)$, into $T(X^n)$, called the *differential of f* and denoted by *df*:

$$df : T(Y^m) \to T(X^n). \tag{4.1}$$

This mapping will again be defined by using specific charts. The independence of the result from the choice of charts then guarantees that the result represents a real "geometric object". Take (U_ι, φ_ι), with $U_\iota \subset Y^m$ and $y \in U_\iota$. Take, further, (V_κ, ψ_κ) with $f(U_\iota) \subset V_\kappa \subset X^n$ and $f(y) = x \in V_\kappa$. The coordinate representation of f is $f_{\kappa\iota} = \psi_\kappa \circ f \circ \varphi_\iota^{-1} : \varphi_\iota(U_\iota) \subset \mathbb{R}^m \to \psi_\kappa(V_\kappa) \subset \mathbb{R}^n$. $df_{\kappa\iota} : \mathbb{R}^m \to \mathbb{R}^n$ is the well-known differential of a mapping from $\mathbb{R}^m$ into $\mathbb{R}^n$ (given by $(\partial x^i/\partial y^\alpha)$). However, combined with φ_ι and ψ_κ we also have the mappings $\varphi_{\iota y} : T_y(Y^m) \to \mathbb{R}^m$ and $\psi_{\kappa x} : T_x(X^n) \to \mathbb{R}^n$, which associate to each tangent vector its coordinate representation.

Now take $s \in T_y(Y^m)$ and define

$$s \in T_y(Y^m) \mapsto df_y(s) = \psi_{\kappa x}^{-1} \circ df_{\kappa\iota} \circ \varphi_{\iota y}(s) \in T_x(X^n). \tag{4.2}$$

This mapping is independent of the choice of charts.

4.2 Definition:

In particular, if $X^n = \mathbb{R}$ and we identify $\mathbb{R}^1$ with $\mathbb{R}$, then df_y is called a *linear form* on $T_y(Y^m)$. We have

$$df_y(s) = \left.\frac{\partial f_\iota}{\partial y^i}\right|_y s^i. \tag{4.3}$$

Take, for example, $f = y^j(y)$ and $s = \partial/\partial y^i$. We then have

$$< dy^i, \frac{\partial}{\partial y^i} >:= dy^j(\frac{\partial}{\partial y^i}) = \delta^j_i \quad \text{(Kronecker Symbol).} \tag{4.4}$$

The space of linear forms on $T_y(Y^m)$ is denoted by $T^*_y(Y^m)$, and the disjoint union

$$T^*(Y^m) = \bigcup_{y \in Y^m} T^*_y(Y^m) \tag{4.5}$$

is called the *cotangent bundle* of Y^m (and is another example of a fibre bundle).

4.3 Definition:

A mapping $f : Y^m \to X^n$ induces a *pullback mapping* $f^* : T^*(X^n) \to T^*(Y^m)$ of the cotangent bundle on X^n into the cotangent bundle on Y^m:

$$f^* : \omega \in T^*_x(X^n) \mapsto f^*\omega \in T^*_y(Y^m) \tag{4.6}$$

with

$$(f^*\omega)_y(s) = \omega(df_y(s)), \tag{4.7}$$

where $x = f(y)$ and $s \in T_y(Y^m)$.

Besides the linear forms (also called one-forms) we need the higher-order *r-forms*. These are antisymmetric r-linear mappings defined over the algebra of functions on Y^m, $\mathcal{F}(Y^m)$, from the r-fold product of the set $\mathcal{X}(Y^m)$ of vector fields,

$$\underbrace{\mathcal{X}(Y^m) \times \cdots \times \mathcal{X}(Y^m)}_{r \text{ times}}, \quad \text{into} \quad \mathcal{F}(Y^m). \tag{4.8}$$

From one-forms $\omega_1, \ldots, \omega_r$ we can build up a special r-form, written as

$$\omega = \omega_1 \wedge \omega_2 \wedge \ldots \wedge \omega_r. \tag{4.9}$$

It is defined as follows:
If $X_1, \ldots, X_r$ are r vector fields, then the result of the r-linear mapping ω is:

$$(\omega_1 \wedge \ldots \wedge \omega_r)(X_1, \ldots, X_r) = \frac{1}{r!}\det(\omega_j(X_k)). \tag{4.10}$$

The values of the individual matrix elements $\omega_j(X_k)$ are obtained after choosing a specific chart:

$$\begin{aligned} \omega_j &= f_{(j)i} dy^i \qquad (i = 1, \ldots, m), \\ X_k &= X^p_{(k)} \frac{\partial}{\partial y^p} \qquad (p = 1, \ldots, m), \end{aligned} \tag{4.11}$$

and

$$\omega_j(X_k) = f_{(j)i} X^p_{(k)} dy^i(\frac{\partial}{\partial y^p}) = f_{(j)i} X^i. \tag{4.12}$$

This is a function on Y^m. A general r-form is a linear combination of special r-forms over the algebra $\mathcal{F}(Y^m)$. And with ω_r being a r-form, ω_s a s-form, one can define a $(r+s)$-form $\omega_r \wedge \omega_s$, called the *wedge product* of ω_r and ω_s.

Just as the mapping $f : Y^m \to X^n$ induces the pullback mapping on linear forms given above, it also induces a pull-back mapping on the multilinear forms, and we get a relation analogous to Eq. (4.7).

4.4 Definition:

d is the *operator of the exterior derivative.* It acts on the r-form ω to yield the $(r+1)$-form $d\omega$, and is defined by the following rules:

1. For 0-forms (functions) the application of d yields the total differential df of the function f.
2. For the wedge product $\omega \wedge \pi$ of the r-form ω and the s-form π we get

$$d(\omega \wedge \pi) = d\omega \wedge \pi + (-1)^r \omega \wedge d\pi. \tag{4.13}$$

3. We have the Poincarè Lemma

$$d^2 = 0. \tag{4.14}$$

In a given chart the r-form ω is given by

$$\omega = \sum_{i_1<\ldots<i_r} f_{i_1\ldots i_r} du^{i_1} \wedge \ldots \wedge du^{i_r}. \tag{4.15}$$

$d\omega$ is then

$$d\omega = \sum_{i_1<\ldots<i_r} df_{i_1\ldots i_r} \wedge du^{i_1} \wedge \ldots \wedge du^{i_r}. \tag{4.16}$$

The operator d commutes with the pull-back operation f^* induced by the function f:

$$d(f^*\omega) = f^*(d\omega). \tag{4.17}$$

Both $T(X^n)$ and $T^*(X^n)$ are differential manifolds. Their charts can be obtained from those of X^n: for $t \in T_x(X^n)$,

$$\pi : t \mapsto \pi(t) = x \in X^n \tag{4.18}$$

is the projection from $T(X^n)$ to X^n. Further, set $\bar{U}_\iota = \pi^{-1}(U_\iota)$, where $U_\iota \subset X^n$ is a coordinate-neighborhood of X^n. The chart $(\bar{U}_\iota, \bar{\varphi}_\iota)$ of $T(X^n)$ is then defined by

$$\bar{\varphi}_\iota : t \in \bar{U}_\iota \mapsto (\varphi_\iota(\pi(t)), t_\iota) \in \varphi_\iota(U_\iota) \times \mathbb{R}^n \subset \mathbb{R}^{2n}. \tag{4.19}$$

(for the meaning of t_ι, see below.) Correspondingly, for $w \in T^*(X^n)$ let

$$\tilde{\pi} : w \mapsto \tilde{\pi}(w) = x \in X^n \tag{4.20}$$

be the projection from $T^*(X^n)$ onto X^n and $\tilde{U}_\iota = \tilde{\pi}^{-1}(U_\iota)$. The chart $(\tilde{U}_\iota, \tilde{\varphi}_\iota)$ of $T^*(X^n)$ is defined by

$$\tilde{\varphi}_\iota : w \in \tilde{U}_\iota \mapsto (\varphi_\iota(\tilde{\pi}(w)), w_\iota) \in \varphi_\iota(U_\iota) \times \mathbb{R}^{n*}. \tag{4.21}$$

Here t_ι, w_ι are elements of $\mathbb{R}^n, \mathbb{R}^{n*}$, respectively, which are obtained from t, w by the assignment of coordinates ($\mathbb{R}^n$ and $\mathbb{R}^{n*}$ are the tangent space and the space of one-forms to $\mathbb{R}^n$, respectively. $\mathbb{R}^n$ is a special manifold because of the fact that $\mathbb{R}^n$, its tangent space and its space of one-forms at a point can be identified in a natural way).

Example:

Take a given map $f : X^m \to Y^n$.
The coordinate representation of f is $y^\alpha = f^\alpha(x^i)$. Let

$$s^i \frac{\partial}{\partial x^i} = s \in T(X^m), \qquad g_\beta dy^\beta = \omega \in T^*(Y^n) \tag{4.22}$$

be representations of s and ω in a given basis. We know that

$$\omega(df(s)) = (f^*\omega)(s). \tag{4.23}$$

We wish to determine $f^*\omega \in T^*(X^m)$.

$$df(s) = \frac{\partial f^\alpha}{\partial x^i} s^i \frac{\partial}{\partial y^\alpha}, \qquad \omega(df(s)) = g_\beta \frac{\partial f^\beta}{\partial x^i} s^i. \tag{4.24}$$

Let the representation of $f^*\omega$ in the given basis be $h_i dx^i$.

$$(f^*\omega)(s) = h_i s^i, \qquad \text{so} \quad h_i s^i = g_\beta \frac{\partial f^\beta}{\partial x^i} s^i. \tag{4.25}$$

This must hold for all s^i. Hence

$$f^*\omega = g_\beta(f(x)) \frac{\partial f^\beta}{\partial x^i} dx^i. \tag{4.26}$$

Furthermore,

$$d\omega = dg_\beta \wedge dy^\beta \quad \text{and} \quad f^* d\omega = dg_\beta \wedge \frac{\partial f^\beta}{\partial x^i} dx^i, \tag{4.27}$$

as well as

$$d(f^*\omega) = dg_\beta \wedge \frac{\partial f^\beta}{\partial x^i} dx^i, \tag{4.28}$$

which implies that

$$f^*(d\omega) = d(f^*\omega). \tag{4.29}$$

With this we close our discussion of differentiable manifolds and the structures which live on them, and turn to a discussion of Lie groups and of Lie groups of transformations.

5. Lie Groups and Lie Transformation Groups

5.1 Definition:

A manifold G^r which is at the same time a group is called a *Lie group*, if the mapping

$$(g,h) \in G^r \times G^r \mapsto g \cdot h^{-1} \in G^r \tag{5.1}$$

is differentiable. Details concerning the degree of differentiability will again be suppressed. The concepts we have developed in the previous section concerning differentiable manifolds are here applied to the theory of Lie groups.

The group elements generate the following important diffeomorphisms of G^r onto itself:

1. Left translation $L_g : x \in G^r \mapsto L_g x = gx \in G^r$.
2. Right translation $R_g : x \in G^r \mapsto R_g x = xg \in G^r$.
3. Inner automorphisms $\alpha_g = L_g \circ R_{g^{-1}}$.

α_g is also written as $\operatorname{ad} g$.

The following relations hold:

$$\begin{aligned} L_{gh} &= L_g \circ L_h, \quad L_{g^{-1}} = L_g^{-1} \\ R_{gh} &= R_h \circ R_g, \quad R_{g^{-1}} = R_g^{-1} \\ L_g \circ R_h &= R_h \circ L_g, \quad \alpha_{gh} = \alpha_g \circ \alpha_h, \quad \alpha_{g^{-1}} = \alpha_g^{-1}. \end{aligned} \tag{5.2}$$

In order to deal with *Lie subgroups* we must return to the concept of *submanifolds*, because in the previous section we treated only the concept of *open* submanifolds. Let $f : M \to M'$ be a bijective mapping, whose differential df is also bijective at every point of M. We then say that M, or also $f(M)$, is a *submanifold* of M'.

5.2 Definition:

A *Lie subgroup* H of a Lie group G is a subgroup of G, which is at the same time a submanifold of G, and which is itself, with this manifold structure, a Lie group.

Among all *vector fields* on a Lie group the *left invariant*, repectively *right invariant*, vector fields play a special rôle. Let X be a vector field on G, where $X_a \in T_a(G)$, and let dL_g be the differential of the mapping $L_g : L_g a = ga$. Then $dL_g X_a$ is a vector at $ga : dL_g X_a \in T_{ga}(G)$.

5.3 Definition:

The vector field X is *left invariant*, if $X_{ga} = dL_g X_a$ for all elements g and a of G. The analogue holds for right invariant vector fields.

Using the known rules for the addition of vector fields, the multiplication with functions, and taking a commutator of two vector fields as "multiplication", the left invariant vector fields become a Lie algebra, the *Lie algebra of the Lie group*, denoted by $\mathcal{G}$. We speak of the *left invariant differential forms* on G, when $dL_g\omega = \omega$ for all $g \in G$ holds. The set of all left invariant one-forms $\mathcal{G}^*$ yields a vector space which is dual to the set of all the left invariant vector fields.

Every automorphism φ of a Lie group induces an automorphism $d\varphi$ of the Lie algebra $\mathcal{G}$ of G. This is true in particular for the inner automorphisms $\operatorname{ad} g (g \in G)$. Instead of $d(\operatorname{ad} g)$ we shall just write in the following $\operatorname{ad} g$. A similar notation is often used for all mappings induced by a mapping f, such as df, f^*, etc. It should be possible to recognize from the context which mapping is meant.

5.4 Definition:

The representation $a \mapsto \operatorname{ad} a$, $a \in G$, is called the *adjoint representation of G in $\mathcal{G}$*. For $A \in \mathcal{G}$, we have

$$(\operatorname{ad} a)A = R_{a^{-1}}A, \tag{5.3}$$

which really means

$$d(\operatorname{ad} a)(A) = dR_{a^{-1}}(A). \tag{5.4}$$

To see that this is correct, note that $\operatorname{ad} a(x) = axa^{-1}, x \in G$. dL_a acts on A_x to yield A at the point ax. $dR_{a^{-1}}$ now acts on the result to give $dR_{a^{-1}}A_{ax}$, which is again a vector at $a^{-1} \circ ax = x$ which is, in general, different from $A_{axa^{-1}}$. These considerations illustrate how relations between fields can be used to obtain relations between the relevant objects at points of the manifold.

The space $\mathcal{G}$ may be identified with the tangent space $T_e(G)$ at the unit element e of G, because every left invariant vector field may be generated from an element of T_eG by left translation, and vice versa it can be "pushed back" to e.

Up to now we have considered "simple" differential forms on differentiable manifolds. They map vector fields into real functions. At every point of the manifold, therefore, a real number is assigned to every vector. We shall, however, have need of a generalization of this concept of differential forms, so that the values are not real numbers but elements of certain vector spaces. We then speak of *differential forms with values in the vector space V^N*. An example of this kind is the *canonical one-form* θ on the Lie group G. This is a left invariant one-form, which assigns to every element $A \in \mathcal{G}$ the same element: $\theta(A) = A$.

We now consider the action of Lie groups on differentiable manifolds. These differentiable manifolds may themselves be Lie groups.

5.5 Definition:

A pair $[G^r, X^n]$ is called a *transformation group with right action*, when

1. G^r is a Lie group,
2. X^n is a differentiable manifold,
3. $(x, g) \in X^n \times G^r \mapsto xg \in X^n$ is a differentiable mapping,
4. For every $g \in G^r$ the mapping $r_g : x \in X^n \mapsto r_g(x) = xg \in X^n$ is a diffeomorphism of X^n onto itself, and $r_{gh} = r_h \circ r_g$.

If instead of (3) we have the mapping $(g, x) \in G^r \times X^n \mapsto gx \in X^n$, and instead of (4) the diffeomorphism $\ell_g : x \in X^n \mapsto \ell_g(x) = gx \in X^n$ and $\ell_{gh} = \ell_g \circ \ell_h$, we speak of a *transformation group with left action.*

5.6 Definition:

We say that G *acts freely (effectively)* on X, when $r_g x = x$ for some x (for all $x \in X$) implies $g = e$.

The free action of G is important in the theory of fibre bundles, because it has as a consequence that the tangent spaces to the fibres of a principal bundle are isomorphic to the Lie algebra of the structure group.

5.7 Definition:

We speak of a (local) *one-parameter transformation group* in X, when there is a mapping from $I \times X \to X$, $I \subset \mathbb{R}^1$,

$$(t, p) \in I \times X \mapsto \varphi_t(p) \in X, \tag{5.5}$$

such that the following conditions are fulfilled:

1. For every $t \in I$, $\quad \varphi_t : p \mapsto \varphi_t(p)$ is a diffeomorphism of X onto itself,
2. For every $t, s \in I$ and $p \in X$, $\quad \varphi_{t+s}(p) = \varphi_t(\varphi_s(p))$.

For $p \in X$, every one-parameter transformation group generates a curve in X, which is called the *orbit* through p. It also induces on X a vector field, which consists of the tangent vectors to the orbit through p. Vice-versa, a vector field yields a one-parameter transformation group, when the vectors of the vector field at every point are interpreted as tangent vectors to an orbit. In particular, the elements of a Lie algebra generate a one-parameter transformation group of the Lie group itself. The one-parameter subgroups of a Lie transformation group, which are generated by the elements of $\mathcal{G}$, produce orbits on X and thereby vector fields; the vector field on X associated with $A \in \mathcal{G}$ is denoted by A^*.

6. Fibre Bundles

We are now prepared to take up again the theme introduced at the beginning of this article, namely principal fibre bundles. Utilizing the material presented involving differentiable manifolds and Lie groups we are now in a position to understand the relevant definitions.

Locally, principal fibre bundles look like the direct product of a coordinate neighborhood of the basis manifold with a Lie group. Only when we consider global aspects we may notice that the structure is not the product of two manifolds after all.

Given a principal fibre bundle, it is possible from the local diffeomorphisms $\psi_\alpha : \pi^{-1}(U_\alpha) \to U_\alpha \times G$ to construct the so-called *transition functions* $\psi_{\beta\alpha}(\pi(u))$, for u an element of the total bundle space P, which are mappings of the basis manifold M into the structure group G. The significance of these functions will become apparent from the theorem concerning the construction of principal fibre bundles discussed below.

Thus ψ_α maps $\pi^{-1}(U_\alpha)$ diffeomorphically onto $U_\alpha \times G$ in the following manner: $u \in P$ yields $(\pi(u), \varphi_\alpha(u)) \in U_\alpha \times G$, where $\varphi_\alpha(u) \in G$. φ_α has by definition the property $\varphi_\alpha(ua) = \varphi_\alpha(u)a$ for $a \in G$. For $u \in \pi^{-1}(U_\alpha \cap U_\beta)$ we have

$$\varphi_\beta(ua)(\varphi_\alpha(ua))^{-1} = \varphi_\beta(u)(\varphi_\alpha(u))^{-1}. \tag{6.1}$$

This shows that $\psi_{\beta\alpha} := \varphi_\beta(u)(\varphi_\alpha(u))^{-1}$ really depends only on $\pi(u)$ and not on the element of the fibre $\pi^{-1}(x) = \pi^{-1}(\pi(u))$. The transition functions are thus mappings defined on intersections $U_\alpha \cap U_\beta$, which have the property

$$\psi_{\gamma\alpha}(x) = \psi_{\gamma\beta}(x) \cdot \psi_{\beta\alpha}(x) \tag{6.2}$$

for $x \in U_\alpha \cap U_\beta \cap U_\gamma$.

We now come to the construction of a principal fibre bundle:

Let M be a manifold with a covering (U_α) by open sets U_α. Let G be a Lie group. If for every intersection $U_\alpha \cap U_\beta$ mappings $\psi_{\beta\alpha} : U_\alpha \cap U_\beta \to G$ with the property (6.2) exist, then it is possible to construct a principal bundle $P(M, G)$ with these mappings as transition functions.

This construction is nothing but the pasting together of product manifolds to a bundle space: Let two products $U_\alpha \times G, U_\beta \times G$ be given. $(x, a) \in U_\alpha \times G$ and $(y, b) \in U_\beta \times G$ are equivalent, by definition, when $x = y$ and $b = \psi_{\beta\alpha}a$.

Now let

$$X = \bigcup_\alpha (U_\alpha \times G) \tag{6.3}$$

be the disjoint union of the products $U_\alpha \times G$. The principal bundle P to be constructed is then the set of equivalence classes of equivalent elements. If G were to consist only of the unit element, that is if $\psi_{\beta\alpha}(x) = e$, then we would get, after pasting together the images of the equivalence classes, a global product bundle.

A question of special importance, especially for understanding certain physical situations, is whether a principal fibre bundle can be "reduced", i.e., whether one can go over to a subgroup of the structure group. To formulate this more precisely we need the concept of the homomorphism between principal fibre bundles:

6.1 Definition:

A *homomorphism* f *from the principal fibre bundle* $P'(M', G')$ *into the principal fibre bundle* $P(M, G)$ consists of a mapping $f' : P' \to P$ and a homomorphism $f'' : G' \to G$ with

$$f'(u'a') = f'(u')f''(a') \qquad \text{for all } u' \in P', a' \in G'. \tag{6.4}$$

This bundle homomorphism maps fibres into fibres, so it induces a mapping $M' \to M$, which is also denoted by f.

6.2 Definition:

In the case $M = M'$, f'' an injection of G' into G and the induced mapping $f : M' \to M$ the identity mapping, $f : P'(M', G') \to P(M, G)$ is called *the reduction of the structure group* G *to the subgroup* G'. $P'(M', G')$ is called the *reduced bundle.*

The structure group G is reducible to a subgroup G' if and only if an open covering (U_α) of M exists, together with transition functions $\psi_{\beta\alpha}$ which take values in G'.

In order to make further statements about the reducibility of principal bundles, we need the concepts of associated bundles and sections of bundles. We start from a principal bundle $P(M, G)$ and a manifold F, on which G acts *from the left.* G acts on the product $P \times F$ *from the right*, according to $(u, \xi, a) \in P \times F \times G \mapsto (ua, a^{-1}\xi) \in P \times F$.

6.3 Definition:

Two elements of $P \times F$ are considered equivalent, if an element of G exists, which connects the elements of $P \times F$ according to

$$(u, \xi) \equiv (v, \eta) \Leftrightarrow \exists a \in G : (v, \eta) = (ua, a^{-1}\xi). \tag{6.5}$$

The set of equivalence classes E is denoted by $P \times_G F$, and called the fibre bundle over M with standard fibre F and structure group G, which is *associated to the principal bundle P*.

In order to have a concrete example, think of the principal bundle $L(M)$ of all bases of the tangent spaces to M and the vector spaces over every point of M, which consist of tensors of a certain fixed degree. The structure group of $L(M)$ is $GL(n, \mathbb{R})$ (the general linear group), the standard fibre is the vector space of tensors, and $GL(n, \mathbb{R})$ acts on the tensors through a linear representation of $GL(n, \mathbb{R})$. The tensor fields are then sections (for definition see below) of the tensor bundle associated to $L(M)$.

6.4 Definition:

In general, a *section of a bundle* is a mapping $\sigma : M \to E$, such that $\pi_E \circ \sigma$ is the identity mapping in M. π_E denotes the projection of the associated bundle E onto the basis manifold M. If (v, ξ) is a representative of an element $A \in E$, then $\pi_E(A) = \pi(u)$ for $u \in P(M, G)$.

Besides global sections $\sigma : M \to E$, also local sections $\sigma_\alpha : U_\alpha \to E$ are sometimes considered. If a principal bundle allows a global section it is a direct product bundle. The bundle $E(M, F, G, P)$ associated to the principal bundle $P(M, G)$ allows a global section when M is paracompact and the standard fibre F is diffeomorphic to $\mathbb{R}^n$.

We now return once more to the reduction problem.

The structure group of $P(M, G)$, G, is reducible to a closed subgroup H if and only if the associated bundle $E(M, G/H, G, P)$ allows a section $\sigma : M \to E$.

An example of a physical theory where the reduction problem comes up is the theory of gravitation. We consider the conformal bundle. This is the set of all bases over the points of spacetime which can be connected by a conformal transformation. The group of conformal transformations is isomorphic to the product of the Lorentz group with the multiplicative group $\mathbb{R}_+$ of the positive real numbers. The factorization of the conformal group according to the Lorentz group yields $\mathbb{R}_+$. When the bundle associated to the conformal bundle with standard fibre $\mathbb{R}_+$ allows a global section, which implies the existence of a positive function on spacetime, the conformal bundle is reducible to the Lorentz bundle. This means, however, that a spacetime metric exists.

7. Connections on Principal Fibre Bundles

The material which we have now covered on differentiable manifolds, Lie transformation groups and principal fibre bundles allows us to understand the definition of a connection on a principal fibre bundle referred to in the introduction. One must only take into account that R_{a*} is the differential of the mapping R_a, and that in the chapter on Lie transformation groups we denoted the action of the group on itself by R_a, but the action of the group on a manifold by r_a. In the definition of a connection we should therefore really write r_a instead of R_a. One does not, however, usually make this distinction, since the context makes clear which mappings are meant.

A connection yields a decomposition of the tangent space at any point of the principal fibre bundle. Any tangent vector then has uniquely determined vertical and horizontal components. The space tangent to the fibre, i.e., the vertical component of the "total" tangent space, is isomorphic to the Lie algebra of G. For every tangent vector of the principal fibre bundle there thus exists a uniquely determined element A of the Lie algebra $\mathcal{G}$. In other words: with a connection Γ is associated a *connection one-form* ω on the principal fibre bundle with values in the Lie algebra $\mathcal{G}$. If $X \in T_u(P), u \in P$, then $\omega(X)$ is that element $A \in \mathcal{G}$ whose image A^* under the action of G on P is precisely the vertical component vX of X. ω has the following properties:

1. $\omega(A^*) = A$ for all $A \in \mathcal{G}$,
2. $R_a^* \omega = (\mathrm{ad}\, a^{-1})\omega$.

The latter condition means: ω yields by its action on $X \in T_u(P)$ an $A \in \mathcal{G}$, and G acts on this element according to the adjoint representation of G in $\mathcal{G}$.

Conversely, given a connection one-form on a principal fibre bundle with the properties (1) and (2), then the existence of a connection Γ is assured. The corresponding horizontal subspaces of the tangent spaces at u are

$$hT_u(P) := \{X \in T_u(P) | \omega(X) = 0\}. \tag{7.1}$$

Physicists are usually more familiar with the connection one-form in its local version. So let $\psi_\alpha : \pi^{-1}(U_\alpha) \to U_\alpha \times G$ be a diffeomorphism. A local section $\sigma(x) = \psi_\alpha^{-1}(x, e)$ is thereby given, where $x \in U_\alpha$ and e denotes the unit element of G. Let θ be the canonical one-form on G. All of this is also valid for U_β. Suppose that $U_\alpha \cap U_\beta \neq \emptyset$. On $U_\alpha \cap U_\beta$ we then have the one-form

$$\theta_{\alpha\beta} := \psi_{\alpha\beta}^* \theta \tag{7.2}$$

and the $\mathcal{G}$-valued one-form

$$\omega_\alpha := \sigma_\alpha^* \omega. \tag{7.3}$$

When ω_α is expanded in terms of a basis of $\mathcal{G}$ and U_α we obtain the physicist's familiar connection coefficients. Their transformation law follows from the relation

$$\omega_\beta = \mathrm{ad}(\psi_{\alpha\beta}^{-1})\omega_\alpha + \theta_{\alpha\beta}. \tag{7.4}$$

Example:

To demonstrate this in the case where the structure group is given by the general linear group, we expand ω_α and ω_β with respect to a basis (dx^j) and $(d\bar{x}^j)$ induced by local coordinates (x^j) and $(\bar{x}^j)$ of U_α and U_β in the space of one-forms over U_α and U_β. (X_i^k) are the coordinates of the elements of $GL(n,\mathbb{R})$, and $E_k^i = \delta_k^i$ the coordinates of the unit in $GL(n,\mathbb{R})$. The algebra of $GL(n,\mathbb{R})$ is identified with $T_E(GL(n,\mathbb{R}))$, and $(\partial/\partial E_k^i)$ is a basis of $T_E(GL(n,\mathbb{R}))$. Then

$$\omega_\alpha = \Gamma_{jk}^i dx^j \frac{\partial}{\partial E_k^i} \tag{7.5}$$

$$\text{and} \quad \omega_\beta = \bar{\Gamma}_{jk}^i d\bar{x}^j \frac{\partial}{\partial E_k^i}. \tag{7.6}$$

The canonical one-form θ in $B \in GL(n,\mathbb{R})$ with coordinates B_i^k reads

$$\theta_B = (B^{-1})_p^s dB_k^p \frac{\partial}{\partial E_k^s}. \tag{7.7}$$

For $x \in U_\alpha \cap U_\beta$, we have

$$\frac{\partial}{\partial \bar{x}^k} = B_k^i(x)\frac{\partial}{\partial x^i}, \qquad B_k^i(x) = \frac{\partial x^i}{\partial \bar{x}^k}, \tag{7.8}$$

where $B_k^i(x)$ are the coordinates of the transition function $\psi_{\alpha\beta} : x \to GL(n,\mathbb{R})$. Then

$$\psi_{\alpha\beta}^*\theta = \frac{\partial \bar{x}^s}{\partial x^p}\frac{\partial^2 x^p}{\partial \bar{x}^m \partial \bar{x}^k} d\bar{x}^m \frac{\partial}{\partial E_k^s} \tag{7.9}$$

$$\text{and} \quad \mathrm{ad}(\psi_{\alpha\beta}^{-1})\omega_\alpha = \frac{\partial \bar{x}^i}{\partial x^m}\Gamma_{jn}^m \frac{\partial x^n}{\partial \bar{x}^k} dx^j \frac{\partial}{\partial E_k^i}. \tag{7.10}$$

Eqs. (7.9) and (7.10) together with Eq. (7.6) give the transformation law for the linear connection coefficients:

$$\bar{\Gamma}_{pk}^i \frac{\partial \bar{x}^p}{\partial x^j} = \frac{\partial \bar{x}^i}{\partial x^m}\frac{\partial x^n}{\partial \bar{x}^k}\Gamma_{jn}^m + \frac{\partial \bar{x}^i}{\partial x^p}\frac{\partial \bar{x}^m}{\partial x^j}\frac{\partial^2 x^p}{\partial \bar{x}^m \partial \bar{x}^k}. \tag{7.11}$$

These so-called *natural coordinates* are widely used in the general theory of relativity.

A connection on $P(M,G)$ also allows us to define *the horizontal subspace of a tangent space* at a point of the bundle $E(M,G,P,F)$ *associated to* $P(M,G)$. For this purpose we start from an element $(v,\xi) \in P \times F$, which is projected onto $w \in E$ upon building the factorization. We keep ξ fixed and obtain in this way a mapping $P \to E$. We call the image of $hT_u(P)$ under this mapping a horizontal subspace of $T_w(E)$, $hT_w(E)$. $hT_w(P)$ is independent of the choice of (v,ξ) in the equivalence class.

Given the concept of the horizontal subspace of the tangent space to $P(M,G)$, respectively $E(M,G,P,F)$, one can *lift* a curve $\gamma = x_t$ in M in a unique manner into P, respectively E. The *lift* of γ in P, $\tau = u_t$, is a curve with projection $\pi(u_t) = x_t$, which goes through u_0 with horizontal tangent vector. In the same way, the lift of γ in E is defined by $\tau^* = w_t$, where w_0 satifies $\pi_E(w_0) = x_0$, and generally $\pi_E(w_t) = x_t$.

If u_0, respectively w_0, belong to the fibres $\pi^{-1}(x_0)$, respectively $\pi_E^{-1}(x_0)$, we obtain a mapping of these fibres into the fibres over x_t which is called *parallel translation*. With this we may define:

7.1 Definition:

The *covariant derivative of a section* φ *of a vector bundle in the direction* $\dot{x}_t$ is given by

$$\nabla_{\dot{x}_t}\varphi = \lim_{h\to 0}\frac{1}{h}[\tau_t^{t+h}(\varphi(x_{t+h})) - \varphi(x_t)]. \tag{7.12}$$

Here $\tau_t^{t+h} : \pi_E^{-1}(x_{t+h}) \to \pi_E^{-1}(x_t)$ is the parallel translate from $x_t{+}h$ to x_t along $\gamma = x_t$. The covariant derivative in the direction of the vector $X \in T_x(M)$ is given in terms of $\nabla_{\dot{x}_0}\varphi$ by the choice of a curve whose tangent vector at $x = x_0$ is X. Finally, let X be a vector field on M, then

$$(\nabla_X\varphi)(x) = \nabla_{X_x}\varphi. \tag{7.13}$$

Before going on to present a convenient formula for the calculation of the covariant derivative, we generalize the formalism to consider *forms* on M with values in a vector space. The vector space V is the standard fibre of a bundle $E(P,M,G,V)$ associated to the principal fibre bundle $P(M,G)$. G acts on V according to the representation ρ.

7.2 Definition:

φ is called a *pseudotensorial form* on P of type (ρ, V), if it is V-valued and

$$R_a^*\varphi = \rho(a^{-1})\cdot\varphi \qquad \text{for all } a \in G. \tag{7.14}$$

It is called *tensorial* when in addition $\varphi(X_1,\ldots,X_r) = 0$ holds if at least one of the tangent vectors $X_i \in T_u(P)$ is vertical.

A tensorial form $\hat{\Lambda}$ on P with values in V may be assigned to a form Λ on M with values in E according to

$$\hat{\Lambda}_u(X_1,\ldots,X_r) = u^{-1}\Lambda_{\pi(u)}(d\pi\, X_1,\ldots,d\pi\, X_r). \tag{7.15}$$

Here $u \in P, \pi(u) \in M$, and $X_1,\ldots,X_r \in T_u(P)$. How is u^{-1} to be understood? $\Lambda_{\pi(u)}(d\pi\, X_1,\ldots,d\pi\, X_r)$ is an element of E in $\pi_E^{-1}(x), x = \pi(u)$. This equivalence class has a representative (u,ξ) in $P \times V$. Keeping u fixed, this determines a mapping from V into E. Conversely, for fixed u some $\xi \in V$ is assigned to every $w \in E$. With this understanding of u as a mapping one can assign to every tensorial form over P a form over M with values in E, according to

$$\Lambda_x(t_1,\ldots,t_r) = u\hat{\Lambda}(X_1,\ldots,X_r). \tag{7.16}$$

Here $x = \pi(u)$, $t_1,\ldots,t_r \in T_x(M)$, and $X_1,\ldots,X_r \in T_u(P)$ with $d\pi\, X_i = t_i$.

The connection one-form is an example of a pseudotensorial form of the adjoint type. Finally, we define:

7.3 Definition:

The *local representation* Λ_ι of Λ is a form over M with values in E. Locally, Λ_ι is a form over $U_\iota \subset M$ with values in V, given by

$$\Lambda_\iota = \sigma_\iota^* \hat{\Lambda}, \tag{7.17}$$

where σ_ι is the local section of P, $\sigma_\iota(x) = \psi^{-1}(x,e)$ which is associated with the diffeomorphism $\psi_\iota : \pi^{-1}(U_\iota) \to U_\iota \times G$.

7.4 Definition:

Let $\hat{\Lambda}$ be a pseudotensorial q-form on P. $D\hat{\Lambda} := (d\hat{\Lambda})h$ is called the *exterior covariant derivative of* $\hat{\Lambda}$.

The h in $(d\hat{\Lambda})h$ means that one is to take the horizontal components of the q vectors on which the form $d\hat{\Lambda}$ acts.

$d\hat{\Lambda}$ is a pseudotensorial $(q+1)$-form. The exterior covariant derivative of the connection one-form ω is the *curvature form* $\Omega = D\omega$. It would, of course, be impossible to construct $(d\hat{\Lambda})$ without being given the connection Γ on the principal fibre bundle.

We now set $\widehat{D\Lambda} = D\hat{\Lambda}$. Then to $D\hat{\Lambda}$ there belongs a $(q+1)$-form $D\Lambda$ on M with values in E. We consider in particular 0-forms over M with values in E, and we get $(D\Lambda)(\dot{x}) = \nabla_{\dot{x}}\Lambda$, which is the special case considered above.

Finally, we arrive at the announced computational rules:

Let $\hat{\Lambda}$ be a tensorial p-form on $P(M,G)$ with values in V, $\rho_* : \mathcal{G} \to GL(V)$ the Lie algebra homomorphism induced by the representation ρ of G in $GL(V)$ and ω the connection one-form of $P(M,G)$. Then we have

$$D\hat{\Lambda} = d\hat{\Lambda} + \rho_*(\omega) \wedge \hat{\Lambda}, \tag{7.18}$$

and for every local representation of Λ, which is the p-form over M with values in E associated to $\hat{\Lambda}$, it follows that

$$D\Lambda_\iota = d\Lambda_\iota + \rho_*(\omega_\iota) \wedge \Lambda_\iota. \tag{7.19}$$

This is the form in which the physicist usually becomes acquainted with the covariant derivative.

The specialization of these considerations to the case of generalized affine and linear connections, which is important in the theory of gravitation, can here only be mentioned as connections on the bundles of affine, respectively linear, reference frames over spacetime.

Let us return once more to the reduction of a principal fibre bundle. We considered before a homomorphism of the bundle $P'(M',G')$ into $P(M,G)$. If $P'(M',G')$ has a connection Γ', then there exists a connection Γ on $P(M,G)$, such that under $f : P' \to P$ the horizontal subspaces of Γ' are mapped into the horizontal subspaces of Γ.

If $M' = M$ and $f : M' \to M$ is the identity mapping (see the notations above), then one says that Γ is *reducible to the connection* Γ' in P.

If a connection Γ on $P(M,G)$ is given by the connection one-form ω, and if H is a subgroup of G with the Lie algebra $\mathcal{H}$, then Γ is reducible to a connection on the subbundle $Q(M,H)$ if and only if ω on Q only takes values in the Lie algebra $\mathcal{H}$.

7.5 Definition:

A *parallel section* is a mapping $\sigma : M \to E$ such that the images of the tangent spaces $T_x(M)$ lie in the horizontal subspaces $hT_x(E)$ of the tangent spaces to E.

Let H be a closed subgroup of G, $E(M, G/H, G, P)$ a bundle associated to $P(M,G)$ with a section $\sigma : M \to E$ and $Q(M,H)$ the reduced bundle corresponding to σ. A connection Γ on P is reducible to a connection Γ' in Q if and only if σ *is parallel with respect to* Γ.

References

1. S. Kobayashi, N. Nomizu: ***Foundations of Differential Geometry*** (Wiley, New York 1963)
2. S. Sternberg: ***Lectures on Differential Geometry*** (Prentice Hill, Englewood Cliffs, New Jersey 1964)
3. R. Sulanke, P. Wintgen: ***Differentialgeometrie und Faserbündel*** (VEB Deutscher Verlag der Wissenschaften, Berlin 1972)
4. Y. Choquet-Bruhat, C. DeWitt-Morette, M. Dillard-Bleick: ***Analysis, Manifolds, and Physics*** (North-Holland Publishing Company, Amsterdam, New York, Oxford 1977)

Constitutive Laws of Bounded Smoothly Deformable Media

E. Binz

Lehrstuhl für Mathematik I der Universität
Seminargeb. A5, Schloß, D(W)-6800 Mannheim 1,
Federal Republic of Germany

Contents

1. Introduction

Let us think of a material body moving and deforming in the Euclidean space $\mathbb{R}^n$. We make the geometric assumption that at any time the body is a n-dimensional, compact, connected, oriented and smooth manifold with (oriented) boundary. The boundary shall not necessarily be connected, but the material should be a deformable medium. The deformable medium forming the boundary may differ from the one forming the inside of the body.

During the motion of the body the diffeomorphism type of the manifold with boundary is assumed to be fixed. Hence we can think of a standard body M, which from a geometrical point of view is a manifold diffeomorphic to the one moving and deforming in $\mathbb{R}^n$. Thus a configuration is a smooth embedding from M into $\mathbb{R}^n$. The configuration space is hence the collection $E(M, \mathbb{R}^n)$ of all smooth embeddings of M into $\mathbb{R}^n$. This set, equipped with Whitney's C^∞-topology, is a Fréchet manifold (cf.[1]). A smooth motion of the body in $\mathbb{R}^n$ therefore is described by a smooth curve in $E(M, \mathbb{R}^n)$. The calculus on

Fréchet manifolds used in the sequel is the one presented in [1], which in our setting coincides with the one developed in [2].

The physical quality of the deforming medium certainly enters the work $F(J)(L)$ needed to deform (infinitesimally) the material at any configuration $J \in E(M, \mathbb{R}^n)$ in any direction L. The directions are tangent vectors to $E(M, \mathbb{R}^n)$. Since the ladder space is open in the Fréchet space $C^\infty(M, \mathbb{R}^n)$ of all smooth $\mathbb{R}^n$-valued functions endowed with the C^∞-topology (cf.[3]), a tangent vector is thus nothing else but a function in $C^\infty(M, \mathbb{R}^n)$ and vice versa.

In the following we take F, which is an one-form on $E(M, \mathbb{R}^n)$, as a constitutive law. We do not discuss the question whether F characterizes the material fully or not. Throughout these notes we assume that F is smooth.

To allow only internal physical properties of the material to enter F, we have to specify the constitutive law somewhat more precisely. Basic to this specification is the fact that these constitutive properties should not be affected by the particular location of the body in $\mathbb{R}^n$. Thus F has to be invariant under the operation of the translation group. Moreover, if L is any constant map, we assume that $F(J)(L) = 0$ for all $J \in E(M, \mathbb{R}^n)$, too.

The forms F which have these two properties can be regarded as one-forms on $\{dJ \mid J \in E(M, \mathbb{R}^n)\}$, where dJ is the differential of any J. This set of differentials is equipped with the C^∞-topology as well and is denoted by $E(M, \mathbb{R}^n)/\mathbb{R}^n$. The latter space is a Fréchet manifold, too. It admits a natural weak Riemannian metric of an L_2-type. A smooth one-form on $E(M, \mathbb{R}^n)/\mathbb{R}^n$ will be denoted by $F_{\mathbb{R}^n}$. Hence we deal with one-forms of the type $F = d^* F_{\mathbb{R}^n}$.

To handle this one-form F we assume that $F_{\mathbb{R}^n}$ can be represented via the metric mentioned by an integral, which we call the *Dirichlet integral*, used in the field of partial differential equations on parts of $\mathbb{R}^n$. The integral kernel of F is a differential of some smooth map

$$\mathcal{H} \in C^\infty(E(M, \mathbb{R}^n)/\mathbb{R}^n, C^\infty(M, \mathbb{R}^n)), \tag{1.1}$$

called a *constitutive map*. Hence in our setting we characterize the medium as far as the internal physical properties enter $\mathcal{H}$.

At any $dJ \in E(M, \mathbb{R}^n)/\mathbb{R}^n$ the constitutive function $\mathcal{H}$ determines two smooth force densities $\Phi(dJ)$ and $\phi(dJ)$, linked to $\mathcal{H}$ by

$$\Delta(J)(dJ) = \Phi(dJ) \quad \text{and} \quad d\mathcal{H}(dJ)(N) = \phi(dJ) \tag{1.2}$$

and the integrability condition necessary to solve this Neumann problem. Here $\Delta(J)$ is the Laplacian determined by the Riemannian metric $J^*\langle\, ,\, \rangle$, where $\langle\, ,\, \rangle$ is the fixed scalar product on $\mathbb{R}^n$. N is the positively oriented unit normal of ∂M in M. Vice versa any pair of force densities (Φ, ϕ) satisfying the integrability condition for the Neumann problem determines some constitutive map of the above mentioned type.

Let us point out here that in these notes we neither discuss any dynamics nor do we study equilibrium conditions. We only investigate the notion of a constitutive law in the above sense.

Since F is affected by the material forming the boundary, we treat the boundary material in an analogous way and in analogy to $\mathcal{H}$ exhibit a characteristic constitutive map h. Thus $\Delta(j)h(dJ)$ with $j := J|\partial M$ and $J \in E(M, \mathbb{R}^n)$ describes the force density $\tilde{\phi}(dJ)$ up to a constant force along ∂M. However, $d\mathcal{H}(dJ)(N)$ also determines force densities which cannot be of the form $\tilde{\phi}(dJ)$. Any specific properties of the boundary enter additively in h. An additive part of h is the constitutive map for the boundary material, thought to be detached from the body. Hence the rest of h describes the influence of the body material on the boundary material implemented into the body.

Finally, we show that both $\mathcal{H}$ and h are structured in the following sense: From a mathematical point of view,the work needed to deform volume, area and, respectively, the shape of the body and boundary is encoded generically and naturally in $\mathcal{H}$ and in h. The shape is partly expressed in the unit normal vector field $N(j)$ along the embedding of the boundary. Here $N(j) = dJ(N)$. The procedure to decode the influence mentioned is to use an L_2-splitting of $d\mathcal{H}(dJ)$.

2. The Space of Configurations, the Phase Space and Geometric Preliminaries

Let us think of a material body moving and deforming in the space $\mathbb{R}^n$. We make the geometric assumption that at any time the body is an n-dimensional, compact, connected, oriented and smooth manifold with boundary. The boundary shall be oriented, too, but shall not necessarily be connected. The material should be a deformable medium. The deformable medium forming the boundary may differ from the one forming the inside of the body.

During the motion of the body the diffeomorphism type of the manifold with boundary is assumed to be fixed. Hence we can think of a standard material body M. By this we mean the following: The underlying point set of the body is a smooth, compact, oriented and connected manifold with oriented boundary M. The boundary need not be connected. The dimension of M is assumed to be n. The body constitutes a deformable medium, and we use M to denote both the manifold with boundary and the material body. From this situation we read off what we mean by a configuration:

A *configuration* is a smooth embedding

$$J : M \to \mathbb{R}^n. \tag{2.1}$$

Hence the space of configurations is $E(M, \mathbb{R}^n)$, the collection of all smooth embeddings of M into $\mathbb{R}^n$. Clearly, each $J \in E(M, \mathbb{R}^n)$ induces a smooth embedding

$$J|\partial M : \partial M \to \mathbb{R}^n. \tag{2.2}$$

We call this a *configuration of the boundary of the body.* Let us denote the collection of all smooth embeddings of ∂M into $\mathbb{R}^n$ by $E(\partial M, \mathbb{R}^n)$.

To see what the phase space is, let us first of all observe that the set $E(M, \mathbb{R}^n)$ is obviously a subset of $C^\infty(M, \mathbb{R}^n)$, the collection of all smooth $\mathbb{R}^n$-valued maps of M. Clearly, $C^\infty(M, \mathbb{R}^n)$ is a $\mathbb{R}^n$-vector space. We equip it with the C^∞-topology, also called the *Whitney topology* in [3]. Since M is compact, $C^\infty(M, \mathbb{R}^n)$ is a complete metrizable locally convex space, a so-called *Fréchet space.* $E(M, \mathbb{R}^n)$ is an open set in $C^\infty(M, \mathbb{R}^n)$, which hence inherits the C^∞-topology, too. The phase space is therefore given by

$$TE(M, \mathbb{R}^n) = E(M, \mathbb{R}^n) \times C^\infty(M, \mathbb{R}^n). \tag{2.3}$$

Proceeding for ∂M as for M we obtain $E(\partial M, \mathbb{R}^n)$ as an open subset of the Fréchet space $C^\infty(\partial M, \mathbb{R}^n)$ (cf.[3]). Hence also $E(\partial M, \mathbb{R}^n)$ is a Fréchet manifold with obviously trivial tangent bundle. The phase space for the boundary is then $E(\partial M, \mathbb{R}^n) \times C(\partial M, \mathbb{R}^n)$. The next Lemma shows the relation between the two configuration spaces, i.e. the two spaces of embeddings:

2.1 Lemma:

The restriction map

$$R : C^\infty(M, \mathbb{R}^n) \to C^\infty(\partial M, \mathbb{R}^n), \tag{2.4}$$

assigning to each $J \in C^\infty(M, \mathbb{R}^n)$ the map $J|\partial M$, is surjective. The image $R(E(M, \mathbb{R}^n))$ is open in $E(\partial M, \mathbb{R}^n)$. Hence

$$TR : E(M, \mathbb{R}^n) \times C^\infty(M, \mathbb{R}^n) \to R(E(M, \mathbb{R}^n)) \times C^\infty(\partial M, \mathbb{R}^n) \tag{2.5}$$

has the form $TR = R \times R$ and is surjective.

Proof: Let $j \in C^\infty(\partial M, \mathbb{R}^n)$. By the Collar Theorem (cf.[3], p.113), M admits a collar in M. So there is an open neighborhood $S \subset M$ of ∂M, which is C^∞-diffeomorphic to $\partial M \times [0, \infty)$ via a map ρ, say. For simplicity we identify S with $\partial M \times [0, \infty)$ via ρ.

Given any $l \in C^\infty(\partial M, \mathbb{R}^n)$, let $L \in C^\infty(S, \mathbb{R}^n)$ be defined by

$$L(p,s) = \psi(s) \cdot l(p) \quad \forall s \in [0,\infty) \text{ and } p \in \partial M, \tag{2.6}$$

where $\psi : [0,\infty) \to \mathbb{R}$ is a smooth map being identical to one on $[0,1]$ and vanishing on $[2,\infty)$.

The map L extends l to all of S. The map L itself extends to all of M by putting it identically zero on the complement of S in M. Clearly, $L \in C^\infty(M, \mathbb{R}^n)$ is such that $R(L) = l$. Let us now prove the next assertion:

Let $J \in E(M, \mathbb{R}^n)$ be given, and let us denote $R(J)$ by j. For any $\lambda \in [0,\infty)$ we let $j(\lambda) := J|\partial M \times \lambda$. Clearly, the family $j(\lambda)$ depends smoothly on λ. It obviously defines a smooth curve with $j(0) = j$. Let us choose an open convex neighborhood $O \subset E(\partial M, \mathbb{R}^n)$ of $j \in E(\partial M, \mathbb{R}^n)$, and let $\lambda_0 \in [0,\infty)$ be such that $j(\lambda_0) \in O$. We deviate now from the curve induced by J as follows: We extend the curve $j(\lambda)$ at λ_0 by a straight line along its tangent, up to $j_2 \in O$, say. From here we pass on with a straight line segment to any given $j^1 \in O$. Clearly, we can smooth out this curve at j_2 without affecting $j(\lambda)$ with $\lambda > 2\lambda_0$. Hence we have a smooth curve σ linking $j(\lambda_0)$ with j^1. By construction $\sigma(0) = j$. The smooth embedding

$$\mathcal{J} : \partial M \times [0,\infty) \to \mathbb{R}^n, \tag{2.7}$$

defined by

$$\mathcal{J}(p,\lambda) = \begin{cases} J(p,\lambda)\lambda > 2\lambda_0 & \forall p \in M \\ \sigma(\lambda)(p)\lambda \le 2\lambda_0 & \forall p \in M \end{cases} \tag{2.8}$$

smoothly links j with $J|(M \backslash \partial M \times [0,\infty))$. Thus we have a smooth element $J^1 \in E(M, \mathbb{R}^n)$ such that $J^1|\partial M = j^1$. The remaining assertions are obvious. In the sequel of these notes we write O_∂ instead of $R(E(M, \mathbb{R}^n))$.

On the configuration space we have a natural action a by the translation groups $\mathbb{R}^n$ of $\mathbb{R}^n$, namely

$$a : E(M, \mathbb{R}^n) \times \mathbb{R}^n \to E(M, \mathbb{R}^n), \tag{2.9}$$

assigning to each $J \in E(M, \mathbb{R}^n)$ and each $z \in \mathbb{R}$ the embedding $J + z$. This action extends obviously to $C^\infty(M, \mathbb{R}^n)$. The translation group $\mathbb{R}^n$ acts accordingly on $E(\partial M, \mathbb{R}^n)$. This action restricts to O_∂ and obviously extends also to $C^\infty(\partial M, \mathbb{R}^n)$. The orbit spaces of the respective actions are denoted by

$$\begin{gathered} C^\infty(M, \mathbb{R}^n)/\mathbb{R}^n, \quad C^\infty(\partial M, \mathbb{R}^n)/\mathbb{R}^n, \\ E(M, \mathbb{R}^n)/\mathbb{R}^n, \quad E(\partial M, \mathbb{R}^n)/\mathbb{R}^n \quad \text{and} \quad O_\partial/\mathbb{R}^n. \end{gathered} \tag{2.10}$$

The nature of these spaces is easily understood if for any $L \in C^\infty(M, \mathbb{R}^n)$ we introduce the differential dL, which is locally given by the Fréchet derivative. Hence the tangent map TL of L is (L, dL). The respective notion of $l \in C^\infty(\partial M, \mathbb{R}^n)$ is introduced accordingly. Hence the orbit spaces mentioned above are nothing else than spaces of differentials of the elements of those spaces on which $\mathbb{R}^n$ acts. Via embedding into $\mathbb{R}^n$, M and ∂M inherit some basic geometric structures described below.

Let us fix a scalar product and a normed determinant function $\underline{\triangle}$ (cf.[4]) on M. Then each $j \in E(\partial M, \mathbb{R}^n)$ yields a unit normal vector field

$$N(j) : \partial M \to \mathbb{R}^n \qquad \text{with } \langle N(j), N(j)\rangle = 1, \tag{2.11}$$

for which $j^* i_{N(j)} \underline{\triangle}$ determines the orientation class of ∂M. Here $j^* i_{N(j)} \underline{\triangle}$ denotes the pullback of the $(n-1)$-form $i_{N(j)}$ to ∂M by j. Moreover,

$$i_{N(j)} \underline{\triangle} := \underline{\triangle}(N(j), \ldots). \tag{2.12}$$

Each $J \in E(M, \mathbb{R}^n)$ and each $j \in E(\partial M, \mathbb{R}^n)$ yields Riemannian metrics $m(J)$ and $m(j)$ on M and ∂M respectively. These metrics are defined by

$$m(J)(X, Y) = \langle dJX, dJY\rangle \quad \forall X, Y \in \Gamma TM \tag{2.13}$$

and

$$m(j)(X, Y) = \langle djX, djY\rangle \quad \forall X, Y \in \Gamma T\partial M. \tag{2.14}$$

Here ΓTQ denotes the collection of all smooth vector fields of any smooth manifold Q (with or without boundary). Both $m(J)$ and $m(j)$ depend smoothly on its variables J and j.

For any $J \in E(M, \mathbb{R}^n)$ and any $j \in E(\partial M, \mathbb{R}^n)$, let us denote by $\mu(J)$ the Riemannian volume form determined by $m(J)$ and the orientation of M, and by $\mu(j)$ the volume form determined by $m(j)$ and the orientation of ∂M. Clearly,

$$j^*(i_{N(j)} \underline{\triangle}) = \mu(j). \tag{2.15}$$

Let us point out that there is a unit vector field $N \in \Gamma TM$ normal to $T\partial M$ such that

$$i_N \mu(J) = \mu(j). \tag{2.16}$$

Hence we have

$$dJ(N) = N(j). \tag{2.17}$$

Clearly,

$$\mu(J) = J^* \underline{\triangle}, \tag{2.18}$$

provided J preserves the orientation. These embeddings $J \in E(M, \mathbb{R}^n)$, for which Eq. (2.18) hold, form an open set in $E(M, \mathbb{R}^n)$.

Associated with the metrics $m(J)$ and $m(j)$, we have the respective Levi-Civita connections $\nabla(J)$ on M and $\nabla(j)$ on ∂M. They are determined by

$$dJ\nabla(J)_X Y = d(dJ\, Y)(X) \quad \forall X, Y \in \Gamma TM \tag{2.19}$$

and

$$dj\nabla(j)_X Y = d(dj\, Y)(X) - m(j)(W(j)X, Y) \cdot N(j) \quad \forall X, Y \in \Gamma T\partial M. \tag{2.20}$$

By $W(j)$ we mean the Weingarten map given by

$$dN(j)Z = dj W(j)Z \quad \forall Z \in \Gamma TM. \tag{2.21}$$

3. The Constitutive Law

We characterize the type of the material which constitutes the body M, as far as it affects the work done, by infinitesimally distorting M (cf.[5 – 8]). This idea is formalized by giving a smooth one-form on $E(M, \mathbb{R}^n)$, i.e. a smooth map

$$F : E(M, \mathbb{R}^n) \times C^\infty(M, \mathbb{R}^n) \to \mathbb{R}, \tag{3.1}$$

which varies linearly in the second argument. We interpret $F(J)(L)$ as the work done if M at the configuration $J \in E(M, \mathbb{R}^n)$ is distorted by $L \in C^\infty(M, \mathbb{R}^n)$. We call the medium described by F a *smoothly deformable medium.*

It might be of physical significance that F might depend on further parameters, e.g. in case one wishes to model a visco-elastic material (cf.[8]). However, we restrict ourselves for simplicity to forms of the type given in Eq. (3.1), since complications such as those just mentioned do not affect the basic apparatus.

It is intuitively clear that the work caused by internal physical processes, initiated by a distortion L at a particular configuration J, should not depend on the particular location of $J(M)$ within $\mathbb{R}^n$. That is to say, this work is the same if J is replaced by $J + z$ for any z. Moreover, a distortion by any $z \in \mathbb{R}^n$ should not cause any work due to these processes mentioned above.

These ideas written more formally yield the following equations basic to our further development:

$$F(J + z) = F(J) \quad \forall J \in E(M, \mathbb{R}^n) \text{ and } z \in \mathbb{R}^n \tag{3.2}$$

and

$$F(J)(z) = 0, \quad \forall J \in E(M, \mathbb{R}^n) \text{ and } z \in \mathbb{R}^n. \tag{3.3}$$

A one-form F on $E(M, \mathbb{R}^n)$ satisfying Eqs. (3.2) and (3.3) will in the sequel be called a *constitutive law.*

The Lemma below is obvious:

3.1 Lemma:

A smooth one-form $F : E(M, \mathbb{R}^n) \times C^\infty(M, \mathbb{R}^n)$ is a constitutive law, if it is of the form $F = d^* F_{\mathbb{R}^n}$, that is, if

$$\begin{aligned} F(J)(L) &= F_{\mathbb{R}^n}(dJ)(dL) \\ \forall J \in E(M, \mathbb{R}^n) &\text{ and } L \in C^\infty(M, \mathbb{R}^n), \end{aligned} \tag{3.4}$$

where

$$F_{\mathbb{R}^n} : E(M, \mathbb{R}^n)/\mathbb{R}^n \times C^\infty(M, \mathbb{R}^n)/\mathbb{R}^n \to \mathbb{R} \tag{3.5}$$

is a smooth one-form.

4. Integral Representation of Constitutive Laws; The Dirichlet Integral

The purpose of this section is to define what is meant by an integral representation of a one-form $F_{\mathbb{R}^n}$ on $E(M, \mathbb{R}^n)$. In order to define this representation we shall first introduce a quadric structure on $E(M, \mathbb{R}^n) \times A^1(M, \mathbb{R}^n)$ which is based on the dot product of any two $\mathbb{R}^n$-valued one-forms on M relative to an embedding of M into $\mathbb{R}^n$. We denote the collection of all smooth $\mathbb{R}^m$-valued one-forms of any smooth manifold Q by $A^1(M, \mathbb{R}^m)$.

Let $\gamma \in A^1(M, \mathbb{R}^n)$ and $J \in E(M, \mathbb{R}^n)$ be given. The two-tensor $\langle \gamma, dJ \rangle$ determined by γ and J shall be given by $\langle \gamma X, dJ\, Y \rangle$ for all $X, Y \in \Gamma TM$. This two-tensor $\langle \gamma, dJ \rangle$ yields a unique strong bundle map $A(\gamma, dJ)$ of TM defined by

$$\langle \gamma X, dJ\, Y \rangle = m(J)(A(\gamma, dJ)X, Y) \quad \forall X, Y \in \Gamma TM. \tag{4.1}$$

From this equation we get:

$$\gamma X = dJ\, A(\gamma, dJ)X \quad \forall X \in \Gamma TM. \tag{4.2}$$

For any two one-forms $\gamma_1, \gamma_2 \in A^1(M, \mathbb{R}^n)$ and an embedding $J \in E(M, \mathbb{R}^n)$ we define the above mentioned dot product of γ_1 and γ_2 relative to J by

$$\gamma_1 \cdot \gamma_2 := \operatorname{tr} A(\gamma_1, dJ) \cdot \tilde{A}(\gamma_2, dJ). \tag{4.3}$$

Here $\tilde{A}(\gamma_2, dJ)$ is the adjoint of $A(\gamma_2, dJ)$, formed fibre-wise with respect to $m(J)$. Associated with this product is a weak scalar product $G_{\mathbb{R}^n}(dJ)$ on $A^1(M, \mathbb{R}^n)$, defined by

$$G_{\mathbb{R}^n}(dJ)(\gamma_1, \gamma_2) := \int_M \gamma_1 \cdot \gamma_2\, \mu(J). \tag{4.4}$$

As mentioned before, $\mu(J)$ denotes the Riemannian volume form determined by $m(J)$ and the given orientation of M.

Weak means here that $G_{\mathbb{R}^n}(dJ)$ determines neither the dual space of $C^\infty(M, \mathbb{R}^n)/\mathbb{R}^n$, nor that of $A^1(M, \mathbb{R}^n)$. We equip $A^1(M, \mathbb{R}^n)$ with the C^∞-topology (cf. [1]). The real number $G_{\mathbb{R}^n}(dJ)(\gamma_1, \gamma_2)$ depends smoothly on all of its variables dJ, γ_1 and γ_2. Since $C^\infty(M, \mathbb{R}^n)/\mathbb{R}^n \subset A^1(M, \mathbb{R}^n)$, the quadric structure $G_{\mathbb{R}^n}$ on the space $E(M, \mathbb{R}^n) \times A^1(M, \mathbb{R}^n)$ yields a weak Riemannian structure on $E(M, \mathbb{R}^n)/\mathbb{R}^n$, again denoted by $G_{\mathbb{R}^n}$.

We say that $F_{\mathbb{R}^n}$, a one-form on $E(M, \mathbb{R}^n)/\mathbb{R}^n$, *admits an integral repesentation*, if there exists a smooth map

$$\alpha : E(M, \mathbb{R}^n)/\mathbb{R}^n \to A^1(M, \mathbb{R}^n), \tag{4.5}$$

called the *kernel* of $F_{\mathbb{R}^n}$, such that for any choices of $dJ \in E(M, \mathbb{R}^n)/\mathbb{R}^n$ and $dL \in C^\infty(M, \mathbb{R}^n)/\mathbb{R}^n$,

$$F_{\mathbb{R}^n}(dJ)(dL) = \int_M \alpha(dJ) \cdot dL\mu(J) = G_{\mathbb{R}^n}(J)(\alpha(J), dL). \tag{4.6}$$

We speak of a *constitutive law F with integral kernel α*, if $F = d^* F_{\mathbb{R}^n}$ and $F_{\mathbb{R}^n}$ admit an integral representation with kernel α.

To discuss the uniqueness of α, if it exists at all, we first prove the following:

4.1 Theorem:

Let $\gamma \in A^1(M, \mathbb{R}^n)$ and $J \in E(M, \mathbb{R}^n)$. There exists a uniquely determined differential $d\mathcal{H}$ of some $\mathcal{H} \in C^\infty(M, \mathbb{R}^n)$, called the *exact part of* γ, and a uniquely determined $\beta \in A^1(M, \mathbb{R}^n)$, such that

$$\gamma = d\mathcal{H} + \beta, \tag{4.7}$$

where the exact part of β vanishes. Both $d\mathcal{H}$ and β depend smoothly on J.
If $\mathcal{H}(p_0)$ for some $p_0 \in M$ is kept constant in J, then also $\mathcal{H}$ varies smoothly in J.

Proof: First let us construct $\mathcal{H}$ and β. To this end we fix a basis $e_1, \ldots, e_n$ on $\mathbb{R}^n$, orthonormal with respect to $\langle\ ,\ \rangle$. Then

$$\gamma(X) = \sum_{r=1}^{n} \gamma^r(X) e_r \quad \forall X \in \Gamma TM, \tag{4.8}$$

with $\gamma^r \in A^1(M, \mathbb{R}^n)$ for all $r = 1, \ldots, n$. Since for each r,

$$\gamma^r(X) = m(J)(Y^r, X) \quad \forall X \in \Gamma TM \tag{4.9}$$

for a well defined $Y^r \in \Gamma TM$, due to Hodge's decomposition (cf.[9]), we find a function $\tau^r \in C^\infty(M, \mathbb{R}^n)$ and a uniquely determined vector field $Y_0^r \in \Gamma TM$ such that the following three equations are satisfied:

$$Y^r = \operatorname{grad}_J \tau^r + Y_0^r \quad \text{and} \quad \operatorname{div}_J Y_0^r = 0, \tag{4.10}$$

together with the boundary condition

$$m(J)(Y_0^r, N) = 0 \tag{4.11}$$

along ∂M. Here the indices J in grad_J and div_J mean that the respective operations are formed with respect to $m(J)$.

This decomposition is obtained by solving the Neumann problem

$$\Delta(J)\tau^r = -\operatorname{div}_J Y^r \tag{4.12}$$

with the boundary condition

$$d\tau^r(N) = m(J)(Y^r, N). \tag{4.13}$$

According to [10], this problem has a solution τ^r, unique up to a constant. The desired function $\mathcal{H}$ and the form β are defined by

$$\mathcal{H} := \sum_r \tau^r \cdot e_r \quad \text{and} \quad \beta(X) := \sum_r m(J)(Y_0^r, X)e^r \quad \forall X \in \Gamma TM, \tag{4.14}$$

respectively. It is a matter of routine to show that $d\mathcal{H}$ and β do not depend on the basis chosen. With these notions we immediately deduce

$$\gamma = d\mathcal{H} + \beta. \tag{4.15}$$

To see that the exact part of β vanishes, let us assume that $\psi^r \in C^\infty(M, \mathbb{R}^n)$ is such that for each $r = 1,\ldots,n$,

$$\operatorname{grad}_J \psi^r + Y_{00}^r = Y_0^r \tag{4.16}$$

for some divergence free vector field Y_{00}^r perpendicular to the normal field N. Then

$$\Delta(J)\psi^r = -\operatorname{div}_J \operatorname{grad}_J \psi^r = 0 \quad \text{and} \quad \psi^r = \text{const}. \tag{4.17}$$

Thus the exact part of β vanishes.

To discuss smoothness properties of $\mathcal{H}$ in J let us next show that both $\operatorname{grad}_J \tau^r$ and Y_0^r depend smoothly on $J \in E(M, \mathbb{R}^n)$. To approach our goal, we consider a smoothly parameterized family $J(t) \in E(M, \mathbb{R}^n)$ with t varying in $\mathbb{R}$. We assume that $J(0)$ coincides with a fixed $I \in E(M, \mathbb{R}^n)$. Thus

$$dJ(t) = dI\, A(dJ(t), dI) \quad \forall t \in \mathbb{R} \quad \text{and hence} \tag{4.18}$$

$$\nabla(J(t))_Y X = \nabla(I)_Y X + A(dJ(t), dI)^{-1}\nabla(I)_Y(A(dJ(t), dI))X \tag{4.19}$$

holds for any choice of $X, Y \in \Gamma TM$. Since $\nabla(J(t))$ is torsion-free for any $t \in \mathbb{R}$, the following equation is valid for all $X, Y \in \Gamma TM$:

$$\nabla(I)_Y(A(dJ(t), dI))X = \nabla(I)_X(A(dJ(t), dI))Y. \tag{4.20}$$

With these formulas we deduce immediately

$$\operatorname{grad}_{J(t)} \tau = A(dJ(t), dI)^{-1} \cdot \tilde{A}(dJ(t), dI)^{-1} \operatorname{grad}_J \tau \tag{4.21}$$

for any $\tau \in C^\infty(M, \mathbb{R}^n)$ and

$$\operatorname{div}_{J(t)} X = \operatorname{div}_I X + \operatorname{tr}(A(dJ(t), dI)^{-1}\nabla(I)_X(A(dJ(t), dI))), \tag{4.22}$$

both holding for all t and all $X \in \Gamma TM$.

Let $Y \in \Gamma TM$. First we assume that the following three equations associated with the Hodge decomposition

$$Y = \mathrm{grad}_{J(t)}\, \tau(J(t)) + Y^0(J(t)), \tag{4.23}$$

with

$$\mathrm{div}_{J(t)}\, Y^0(J(t)) = 0 \quad \text{and} \quad d\tau(J(t))(N) = m(J(t))(Y, N) \tag{4.24}$$

all depend smoothly on t. Then Eqs. (4.18), (4.21) and (4.22) yield the next three equations

$$d\dot{J}(0) = dI\, A(d\dot{J}(0), dI), \tag{4.25}$$

$$\begin{aligned} \tfrac{d}{dt}\,\mathrm{grad}\,|_{t=0}\tau &= -2A(d\dot{J}(0), dI)_{\mathrm{sym}}\, \mathrm{grad}_I\, \tau \\ &\forall \text{ fixed } \tau \in C^\infty(M, \mathbb{R}), \end{aligned} \tag{4.26}$$

where $A(d\dot{J}(0)dI)_{\mathrm{sym}}$ denotes the self-adjoint part of $A(d\dot{J}(0), dI)$ formed with respect to $m(I)$ via the polar decomposition (cf.[1]), and finally

$$\frac{d}{dt}\,\mathrm{div}_{J(t)}\,|_{t=0}X = \mathrm{tr}\,\nabla(I)A(d\dot{J}(0), dI)X. \tag{4.27}$$

Using the last three formulas, the derivatives of the expressions in Eqs. (4.23) and (4.24) with respect to t therefore read

$$0 = -2A(d\dot{J}(0), dI)_{\mathrm{sym}}\, \mathrm{grad}_I\, \tau(I) + \mathrm{grad}_I\, \dot{\tau}(I) + \dot{Y}^0(I), \tag{4.28}$$

$$\mathrm{div}_I\, \dot{Y}(I) = -\,\mathrm{tr}\,\nabla(I)A(d\dot{J}(0), dI)Y^0(I) \tag{4.29}$$

and

$$d\dot{\tau}(I)(N) = \dot{m}(I)(Y, N). \tag{4.30}$$

Applying div_I to Eq. (4.28) yields the equation

$$\begin{aligned} \Delta(I)\dot{\tau}(I) = &-2\,\mathrm{div}_I(A(d\dot{J}(0), dI)_{\mathrm{sym}}\, \mathrm{grad}_I\, \tau(I)) \\ &+ \mathrm{tr}\,\nabla(I)A(d\dot{J}(0), dI)Y^0(I), \end{aligned} \tag{4.31}$$

with its boundary condition

$$d\dot{\tau}(I)N = \dot{m}(I)(Y, N). \tag{4.32}$$

Turning back to the problem of showing the smoothness in t of the expressions in Eqs. (4.23) and (4.24), Eqs. (4.31) and (4.32) pose a Neumann problem with $\dot{\tau}(I)$ as the unknown, provided we drop the smoothness assumption in connection with Eqs. (4.23) and (4.24). The right hand sides of both Eqs. (4.31) and (4.32) are smooth. As we already know, such problems have a solution, which is unique up to a constant. Without loss of generality we may assume that for some $p_0 \in M$,

$$\tau(J(t))(p_0) = 0 \quad \forall t \in \mathbb{R}, \tag{4.33}$$

which in turn suggests that $\dot{\tau}(I)(p_0) = 0$.

Eq. (4.31) produces a candidate for $\dot{\tau}(I)$, and if we insert $\dot{\tau}(I)$ into Eq. (4.28) we obtain a candidate for $\dot{Y}^0(I)$. It is now a matter of routine to verify that these candidates in fact do satisfy

$$\lim_{t\to 0}\frac{1}{\tau}(\tau(J(t))-\tau(I))=\dot{\tau}(I) \tag{4.34}$$

$$\text{and}\quad \lim_{t\to 0}\frac{1}{t}(Y^0(J(t))-Y^0(I))=\dot{Y}^0(I), \tag{4.35}$$

respectively. Since $I \in E(M,\mathbb{R}^n)$ was chosen arbitrarily, we obtain $\dot{\tau}(t)$ and $\dot{Y}^0(J(t))$ for all t.

To show the existence of all higher derivatives we have to set up an induction procedure based on Eqs. (4.25), (4.26), (4.27), (4.28), (4.29), (4.30) and (4.31), whose execution is left to the reader. Therefore, both $\tau(J(t))$ and $Y(J(t))$ depend smoothly on $t \in \mathbb{R}$. Since the parameterization in t was arbitrary, by the criterion given in the calculus presented in [2] we conclude that both $\tau(J)$ and $Y(J)$ depend smoothly on $J \in E(M,\mathbb{R}^n)$. This ends the proof.

Some of the calculations in the proof above allow us to look at $G_{\mathbb{R}^n}(J)$ from another point of view. Given $\gamma \in A^1(M,\mathbb{R}^n)$ and $J \in E(M,\mathbb{R}^n)$, according to Eqs. (4.8) and (4.9) we have

$$\gamma(X)=dJ\,A(\gamma,dJ)X=\sum_{r=1}^{n} m(J)(Y^r,X)e_r \quad \forall X \in \Gamma TM. \tag{4.36}$$

Let us denote $(dJ)^{-1}e_r$ by E_r, for all $r=1,\ldots,n$. Then we see from Eq. (4.36) that

$$Y^r=\tilde{A}(\gamma,dJ)E_r \tag{4.37}$$

holds for all $r=1,\ldots,n$. This remark yields the following observation:

4.2 Proposition:

Given $\gamma_1,\gamma_2 \in A^1(M,\mathbb{R}^n)$, $J \in E(M,\mathbb{R}^n)$ and a fixed basis $e_1,\ldots,e_n$ on $\mathbb{R}^n$ orthonormal with respect to $\langle\ ,\ \rangle$, there exist two sets

$$Y_1{}^1,\ldots,Y_1{}^n \quad\text{and}\quad Y_2{}^1,\ldots,Y_2{}^n \tag{4.38}$$

of vector fields in ΓTM, such that

$$\gamma_1\cdot\gamma_2=\sum_{r=1}^{n} m(J)(Y_1{}^r,Y_2{}^r), \quad\text{and hence} \tag{4.39}$$

$$G_{\mathbb{R}^n}(dJ)(\gamma_1,\gamma_2):=\int_M \gamma_1\cdot\gamma_2(J)=\sum_{r=1}^{n}\int_M m(J)(Y_1{}^r,Y_2{}^r)\mu(J). \tag{4.40}$$

If in addition $\gamma_1=d\mathcal{H}$ for some $\mathcal{H}\in C^\infty(M,\mathbb{R}^n)$, then $G_{\mathbb{R}^n}(dJ)(d\mathcal{H},\gamma_2)=0$, provided that the exact part of γ_2 vanishes.

Proof: Let $Y_i{}^r \in \Gamma TM$, $r = 1,\dots,n$ and $i = 1, 2$ be as in Eq. (4.9). Then

$$\begin{aligned}\gamma_1 \cdot \gamma_2 &= \operatorname{tr} A(\gamma_1, dJ) \cdot \tilde{A}(\gamma_2, dJ) \\ &= \sum_{r=1}^{n} m(J)(A(\gamma_1, dJ) \cdot \tilde{A}(\gamma_2, dJ)E_r, E_r) \\ &= \sum_{r=1}^{n} m(J)(Y_1{}^r, Y_2{}^r), \end{aligned} \tag{4.41}$$

establishing Eq. (4.39). To show the last part of the proposition we use Gauss' Theorem as follows:

$$\begin{aligned} G_{\mathbb{R}^n}(dJ)(\gamma_1, \gamma_2) &= \sum_{r=1}^{n} \int_M m(J)(\operatorname{grad}_J \tau^r, Y_0^r) = \sum_{r=1}^{n} \int_M d\tau^r(Y_0^r)\mu(J) \\ &= \sum_{r=1}^{n} \int_M (\operatorname{div}_J(\tau^r Y_0^r) - \tau \cdot \operatorname{div}_J Y_0^r)\mu(J) \\ &= \sum_{r=1}^{n} \int_M m(J)(\tau^r Y_0^r, N)\mu(J) = 0. \end{aligned} \tag{4.42}$$

Here we have $\mathcal{H} = \sum_{r=1}^{n} \tau^r e_r$ and $\operatorname{div}_J Y_0^r = 0$ as well as $m(J)(Y_0^r, N) = 0$.

If γ_1 and γ_2 in the above proposition are exact, the respective vector fields in Eq. (4.39) are gradients. Hence the right- hand-side of the integral in Eq. (4.40) is the classical Dirichlet integral (cf.[11]) for $\mathbb{R}^n$-valued functions.

The integral in the middle part of Eq. (4.40) hence generalizes and reformulates the Dirichlet integral. We therefore call it the Dirichlet integral of any two smooth $\mathbb{R}^n$-valued forms γ_1, γ_2 relative to $J \in E(M, \mathbb{R}^n)$.

Prop. 4.2 also shows that the integral kernel of a constitutive law is not unique at all. To any kernel we may add a map which takes values in the one-forms with vanishing exact part. However, the following theorem guarantees the uniqueness of a very specific type of kernel:

4.3 Theorem:

Let F be a constitutive law with integral kernel. There exists a unique smooth map

$$\alpha : E(M, \mathbb{R}^n)/\mathbb{R}^n \to C^\infty(M, \mathbb{R}^n)/\mathbb{R}^n \subset A^1(M, \mathbb{R}^n), \tag{4.43}$$

such that for any $J \in E(M, \mathbb{R}^n)$ and any $L \in C^\infty(M, \mathbb{R}^n)$,

$$F(J)(L) = \int_M \alpha(dJ) \cdot dL \, \mu(J). \tag{4.44}$$

In fact, there is a unique smooth map

$$\mathcal{H} : E(M, \mathbb{R}^n)/\mathbb{R}^n \to C^\infty(M, \mathbb{R}^n), \tag{4.45}$$

satisfying the equations

$$\alpha(dJ) = d\mathcal{H}(dJ) \quad \forall dJ \in E(M, \mathbb{R}^n)/\mathbb{R}^n \tag{4.46}$$

and

$$\int_{\partial M} \langle \mathcal{H}(dJ), z \rangle \mu(J) = 0 \quad \forall z \in \mathbb{R}^n. \tag{4.47}$$

Proof: The existence of such a kernel is guaranteed by Prop. 4.2; the uniqueness follows easily:

Let α_1 and α_2 be two kernels with values in $C^\infty(M, \mathbb{R}^n)/\mathbb{R}^n$. Then we would have that

$$\int_M (\alpha_1 - \alpha_2)(dJ) \cdot dL \, \mu(J) = 0, \tag{4.48}$$

for all $J \in E(M, \mathbb{R}^n)$ and all $dL \in C^\infty(M, \mathbb{R}^n)$. Since G is positive definite for all $dL = \alpha_1 - \alpha_2$, we conclude that $\alpha_1 = \alpha_2$. To show that $\mathcal{H}$ exists and can be chosen to satisfy Eq. (4.47), we introduce $C^\infty_J(M, \mathbb{R}^n)$, the collection of all $L \in C(M, \mathbb{R}^n)$ satisfying

$$\int_M \langle L, z \rangle \mu(J) = 0 \quad \forall z \in \mathbb{R}^n \tag{4.49}$$

for a given $J \in E(M, \mathbb{R}^n)$. With this space at hand we have the splitting

$$C^\infty(M, \mathbb{R}^n) = \mathbb{R}^n \times C^\infty_J(M, \mathbb{R}^n). \tag{4.50}$$

Equipping $C^\infty_J(M, \mathbb{R}^n)$ with the C^∞-topology yields a Fréchet space, also denoted by $C^\infty_J(M, \mathbb{R}^n)$. Since for any two $I, J \in E(M, \mathbb{R}^n)$

$$m(J)(X, Y) = m(I)(B(dJ, dI)^2 X, Y) \quad \forall X, Y \in \Gamma TM \tag{4.51}$$

for a uniquely determined smooth strong bundle isomorphism $B(dJ, dI)$ of TM, we conclude that

$$C^\infty_J(M, \mathbb{R}^n) = \det(B(dJ, dI)) \cdot C^\infty_I(M, \mathbb{R}^n). \tag{4.52}$$

Clearly,

$$d : C^\infty_J(M, \mathbb{R}^n) \to C^\infty(M, \mathbb{R}^n)/\mathbb{R}^n \tag{4.53}$$

is an isomorphism for each J. Let us denote it by d_J. The desired map $\mathcal{H}$ is given by

$$\mathcal{H}(dJ) := d_J^{-1} \alpha(dJ). \tag{4.54}$$

5. Force Densities Associated with Constitutive Laws Admitting Kernels

The purpose of this section is to associate with any constitutive law admitting integral kernels at any configuration some well defined force densities, one acting upon the whole body and one acting upon the boundary only. Throughout this section, F is a constitutive law admitting a kernel α. By the previous theorem we may assume that α maps into $C^\infty(M, \mathbb{R}^n)/\mathbb{R}^n$.

To construct the force densities just mentioned we use F in the form

$$F(J)(L) = \int_M \operatorname{tr} A(\alpha(dJ), dJ) \cdot \tilde{A}(dL, dJ)\mu(J), \tag{5.1}$$

holding for any of the variables of F. Writing any $L \in C^\infty(M, \mathbb{R}^n)$ relative to a given $J \in E(M, \mathbb{R}^n)$ in the form

$$L = dJ\, X(L, J) \tag{5.2}$$

with a unique $X(L, J) \in \Gamma TM$, we have

$$dL\, X = dJ \nabla(J)_X X(L, J) \quad \forall X \in \Gamma TM, \tag{5.3}$$

and hence derive immediately

$$A(dL, dJ) = \nabla(J)X(L, J) \quad \forall L \in C^\infty(M, \mathbb{R}^n). \tag{5.4}$$

Thus if $e_1, \ldots, e_n$ is a orthonormal basis of $\mathbb{R}^n$, and if we define $E_r \in \Gamma TM$ again by $dJ\, E_r = e_r$ for $r = 1, \ldots, n$, then

$$F(J)(L) = \sum_{r=1}^{n} \int_M m(J)(\tilde{A}(\alpha(dJ), dJ) \cdot \nabla(J)_{E_r} X(L, J), E_r)\mu(J). \tag{5.5}$$

Let us introduce the notion $\operatorname{div}_J T$, the divergence of a strong bundle endomorphism T of TM:

$$\operatorname{div}_J T := \sum_{r=1}^{n} \nabla(J)_{E_r}(T)(E_r). \tag{5.6}$$

This notion does not depend of the basis chosen. Eq. (5.6), together with Eq. (5.5), implies

$$\begin{aligned} F(J)(L) = &\int_M \operatorname{div}_J(\tilde{A}((dJ), dJ)X(L, J))\mu(J) \\ &- \int_M m(J)(\operatorname{div}_J A(\alpha(dJ), dJ), X(L, J))\mu(J). \end{aligned} \tag{5.7}$$

To bring these formulas into a more familiar form, we introduce the notions of a Laplacian $\Delta(J)K$ and $\Delta(J)\gamma$ for any $K \in C^\infty(M, \mathbb{R}^n)$, $\gamma \in A^1(M, \mathbb{R}^n)$. In doing so we follow [12]. We set

$$d^*K = 0. \tag{5.8}$$

If $\gamma \in A^1(M, \mathbb{R}^n)$ for some natural number m, we set

$$d^*\gamma = -\sum_{r=1}^{n} \nabla(J)_{E_r}(\gamma)(E_r). \tag{5.9}$$

Clearly, we have

$$d^*\gamma = -\operatorname{div} Y \quad \text{for } \gamma(X) = m(J)(Y, X) \quad \forall X \in \Gamma TM. \tag{5.10}$$

$\Delta(J)$ is then defined by

$$\Delta(J) := dd^* + d^*d. \tag{5.11}$$

Consequently we have

$$\Delta(J)K = d^*dK = -\sum_{r=1}^{n} \nabla(J)_{E_r}(dK)(E_r). \tag{5.12}$$

Since the two expressions $\Delta(J)_Y(dK)X$ and $\nabla(J)_Y(T)X$, formed for any $K \in C^\infty(M, \mathbb{R}^n)$, any strong bundle map T of TM and any choices of $X, Y \in \Gamma TM$ are by definition $d(dK(X))Y - dK\nabla(J)_X Y$ and $\nabla(J)_Y(TX) - T\nabla(J)_Y X$, we find

$$\begin{aligned} \Delta(J)K &= -(\sum_{r=1}^{n} d(dJ\, A(dK, dJE_r))(E) - dJ\, A(dK, dJ)(\nabla(J)_{E_r} E_r)) \\ &= -\sum_{r=1}^{n} dJ \nabla(J)_{E_r}(A(dK, dJ))E_r \\ &= -dJ \operatorname{div}_J A(dK, dJ). \end{aligned} \tag{5.13}$$

Hence Eq. (5.7) turns out to be

$$\begin{aligned} F(J)(L) = &\int_M \operatorname{div}_J \tilde{A}(\alpha(dJ), dJ) X(L, J)\mu(J) \\ &+ \int_M \langle \Delta(J)\mathcal{H}(dJ), L\rangle \mu(J), \end{aligned} \tag{5.14}$$

with $\alpha(dJ) = d\mathcal{H}(dJ)$ for some $\mathcal{H} \in (C^\infty(E(M, \mathbb{R}^n)/\mathbb{R}^n, C^\infty(M, \mathbb{R}^n))$.

With the help of Gauss' Theorem and Theorem 4.3 we derive:

5.1 Proposition:

Let F be a constitutive law admitting a kernel. Then for each $J \in E(M, \mathbb{R}^n)$ there exists a smooth map

$$\mathcal{H} : E(M, \mathbb{R}^n) \to C^\infty(M, \mathbb{R}^n), \tag{5.15}$$

uniquely determined up to a smooth map from $E(M, \mathbb{R}^n)$ into $\mathbb{R}^n$, for which both

$$F(J)(L) = \int_M \langle \Delta(J)\mathcal{H}(dJ), L\rangle \mu(J) + \int_{\partial M} \langle d\mathcal{H}(dJ)(N), L\rangle i_N \mu(J) \tag{5.16}$$

and hence a Green's equation

$$\int_M \langle \Delta(J)\mathcal{H}(dJ), L\rangle \mu(J) - \int_M \langle \mathcal{H}(dJ)\Delta(J), L\rangle \mu(J) \tag{5.17}$$
$$= \int_{\partial M} \langle dL(N), \mathcal{H}(dJ)\rangle i_N \mu(J) - \int_{\partial M} \langle d\mathcal{H}(dJ)(N), L\rangle i_N \mu(J)$$

hold for all variables of F. Here $i_N \mu(J)$ is the volume element on ∂M, defined by $\mu(J)$ and N, the positively oriented unit normal vector field of $\partial M \subset M$.

The map $\mathcal{H}$ in Prop. 5.1 is called a *constitutive map*, because it fully determines the constitutive law. The proposition motivates us to set

$$\Phi(dJ) := \Delta(J)\mathcal{H}(dJ) \quad \text{and} \quad \varphi(dJ) := d\mathcal{H}(dJ)(N), \tag{5.18}$$

for any $J \in E(M, \mathbb{R}^n)$, with $\mathcal{H}(dJ)$ as in Eq. (5.16).

The maps Φ and φ are called the *force densities associated with* F. These force densities determine F by

$$F(J)(L) = \int_M \langle \Phi(dJ), L\rangle \mu(J) + \int_\partial \langle \varphi(dJ), L\rangle i_N \mu(J) \tag{5.19}$$

for all $J \in E(M, \mathbb{R}^n)$ and all $L \in C^\infty(M, \mathbb{R}^n)$.

Since $\mathcal{H}$ is smooth, both Φ and φ are smooth $C^\infty(M, \mathbb{R}^n)$-valued, respectively $C^\infty(\partial M, \mathbb{R}^n)$-valued functions, on $E(M, \mathbb{R}^n)/\mathbb{R}^n$.

Vice versa, given two smooth maps

$$\begin{aligned} \Phi &: E(M,\mathbb{R}^n)/\mathbb{R}^n \to C^\infty(M,\mathbb{R}^n), \\ \varphi &: E(M,\mathbb{R}^n)/\mathbb{R}^n \to C^\infty(\partial M,\mathbb{R}^n), \end{aligned} \tag{5.20}$$

with integrability condition

$$0 = \int_M \langle \Phi(dJ), z\rangle \mu(J) + \int_{\partial M} \langle \varphi(dJ), z\rangle i_N \mu(J) \quad \forall z \in \mathbb{R}^n, \tag{5.21}$$

there exists for each $dJ \in E(M,\mathbb{R}^n)/\mathbb{R}^n$ a smooth map

$$\bar{\mathcal{H}}(dJ) : M \to \mathbb{R}^n, \tag{5.22}$$

such that the Neumann problem $\Delta(J)\bar{\mathcal{H}}(dJ) = \Phi(dJ)$ with the boundary condition $d\bar{\mathcal{H}}(dJ)(N) = \varphi(dJ)$ is solvable uniquely up to a constant. With these force densities we define a one-form F on $E(M,\mathbb{R}^n)$ by

$$F(J)(L) = \int_M \langle \Delta(J)\bar{\mathcal{H}}(dJ), L\rangle \mu(J) + \int_{\partial M} \langle d\bar{\mathcal{H}}(dJ)(N), L\rangle i_N \mu(J) \tag{5.23}$$

for all $J \in E(M,\mathbb{R}^n)$ and for all $L \in C^\infty(M,\mathbb{R}^n)$. F is a constitutive law due to Eq. (5.21). We now apply Prop. 5.1 to obtain a smooth map

$$\mathcal{H} : E(M,\mathbb{R}^n)/\mathbb{R}^n \to C^\infty(M,\mathbb{R}^n), \tag{5.24}$$

producing

$$\Phi(dJ) = \Delta(J)\mathcal{H}(dJ) \quad \text{and} \quad \varphi(dJ) = d\mathcal{H}(dJ)(N) \tag{5.25}$$

for all $J \in E(M,\mathbb{R}^n)$. Thus we have the following

5.2 Theorem:

Every constitutive law with integral kernel admits a smooth constitutive map

$$\mathcal{H} : E(M,\mathbb{R}^n)/\mathbb{R}^n \to C^\infty(M,\mathbb{R}^n), \tag{5.26}$$

uniquely determined up to a map in $C^\infty(E(M,\mathbb{R}^n)/\mathbb{R}^n,\mathbb{R}^n)$, such that the kernel of F is given by

$$d\mathcal{H} : E(M,\mathbb{R}^n)/\mathbb{R}^n \to C^\infty(M,\mathbb{R}^n)/\mathbb{R}^n \tag{5.27}$$

and which, moreover, satisfies

$$F(J)(L) = \int_M \langle \Delta(J)(dJ), L\rangle \mu(J) + \int_{\partial M} \langle d\mathcal{H}(dJ)(N), L\rangle i_N \mu(J) \tag{5.28}$$

on all of $TE(M,\mathbb{R}^n)$.

The map $\mathcal{H}$ determines two smooth maps

$$\begin{aligned} \Phi &: E(M,\mathbb{R}^n)/\mathbb{R}^n \to C^\infty(M,\mathbb{R}^n) \\ \text{and} \quad \varphi &: E(M,\mathbb{R}^n)/\mathbb{R}^n \to C^\infty(\partial M,\mathbb{R}^n), \end{aligned} \tag{5.29}$$

called the *force densities associated with* F, which for all $J \in E(M,\mathbb{R}^n)$ are given by

$$\Phi(dJ) = \Delta(J)\mathcal{H}(dJ) \quad \text{and} \quad \varphi(dJ) = d\mathcal{H}(dJ)(N) \tag{5.30}$$

satisfying

$$\int_M \langle \Phi(dJ), z\rangle \mu(J) + \int_{\partial M} \langle \varphi(dJ), z\rangle i_N \mu(J) = 0 \quad \forall z \in \mathbb{R}^n. \tag{5.31}$$

Vice versa, given two smooth maps of the form shown in Eqs. (5.20) and satisfying Eq. (5.31), there is a constitutive map $\mathcal{H}$ of the form given in Eq. (5.26), for which Eq. (5.28) holds.

6. Constitutive Laws for the Boundary

The task in this section is to study constitutive laws for the boundary, that is for a deformable medium forming a skin, for which the underlying point set is the manifold M. This skin is thought to be detached from the body. In doing so, we first formulate in analogy to Sects. 2 and 3 what is meant by a constitutive law with integral kernel for the boundary material.

Let us recall that the open set $O_\partial \subset E(\partial M,\mathbb{R}^n)$ is the collection of all $J|\partial M$ with $J \in E(M,\mathbb{R}^n)$. The constitutive laws mentioned above will be given on any open set $O \subset E(M,\mathbb{R}^n)$ and will later be specified on O_∂.

At the very first we introduce the notion corresponding to the Dirichlet integral: Given any $l \in C^\infty(\partial M,\mathbb{R}^n)$ and any $j \in E(\partial M,\mathbb{R}^n)$, then for all $X, Y \in \Gamma T\partial M$,

$$\langle dl\, X, dj\, Y\rangle = m(J)(A(dl,dj)X,Y) \tag{6.1}$$

holds for some smooth strong bundle endomorphism $A(dl,dj)$ of $T\partial M$. Moreover, there is a uniquely defined smooth map

$$c(dl,dj) : \partial M \to \mathbb{R}^n, \tag{6.2}$$

satisfying the following two conditions:

$$c(dl,dj)dj(T_p\partial M) \subset \mathbb{R}\cdot N(j)(p) \quad \forall p \in \partial M \tag{6.3}$$

$$\text{and} \quad c(dl,dj)N(j)(p) \subset dj\, T_p\partial M \quad \forall p \in \partial M, \tag{6.4}$$

such that the equation

$$dl\,X = c(dj, dl)dj\,X + dj\,A(dl, dj) \tag{6.5}$$

holds for any $X \in \Gamma TM$. We refer to [1] or [8] for more details. Based on Eq. (6.4), we introduce $U(dl, dj)$ by

$$c(dl, dj)N(j) = dj\,U(dl, dj). \tag{6.6}$$

This vector field $U(dl, dj) \in \Gamma T\partial M$ is obviously determined uniquely.

Splitting $A(dl, dj)$ into its skewsymmetric, respectively self–adjoint, parts $C(dl, dj)$ and $B(dl, dj)$, formed pointwise with respect to $m(j)$, we end up with

$$dl = c(dl, dj) \cdot dj + dj(C(dl, dj) + B(dl, dj)). \tag{6.7}$$

This decomposition generalizes in the obvious way to any $\gamma \in A^1(\partial M, \mathbb{R}^n)$ and reads as

$$\gamma = c(\gamma, dj) \cdot dj + dj(C(\gamma, dj) + B(\gamma, dj)). \tag{6.8}$$

The metric $G^{\partial}_{\mathbb{R}^n}(dj)$ at $dj \in E(\partial M, \mathbb{R}^n)/\mathbb{R}^n$, applied to any two elements $dl, dk \in C^\infty(M, \mathbb{R}^n)/\mathbb{R}^n$, is defined by integrating the function

$$\begin{aligned} dl \cdot dk :=& -\frac{1}{2}\operatorname{tr} c(dl, dj) \cdot c(dk, dj) \\ &- \operatorname{tr} C(dl, dj) \cdot C(dk, dj) + \operatorname{tr} B(dl, dj) \cdot B(dk, dj) \end{aligned} \tag{6.9}$$

with respect to $\mu(j)$, that is, it is defined by

$$\begin{aligned} G^{\partial}_{\mathbb{R}^n}(dj)(dl, dk) :=& \int_M dl \cdot dk\,\mu(j) \\ =& -\frac{1}{2}\operatorname{tr} c(dl, dj) \cdot c(dk, dj)\mu(j) \\ &- \int_M \operatorname{tr} C(dl, dj) \cdot C(dk, dj)\mu(j) \\ &+ \int_M \operatorname{tr} B(dl, dj) \cdot B(dk, dj)\mu(j). \end{aligned} \tag{6.10}$$

Let $O \subset E(\partial M, \mathbb{R}^n)$ be any open set. We now define a constitutive law F^∂ on O in analogy to Sect. 2, that is we require

$$F_\partial = d^* F^{\partial}_{\mathbb{R}^n} \tag{6.11}$$

for some one-form $F^{\partial}_{\mathbb{R}^n}$ on $O/\mathbb{R}^n$. Accordingly F_∂ is called a constitutive law with kernel α, if for some $\alpha \in C^\infty(O, A^1(\partial M, \mathbb{R}^n))$ the following equation holds:

$$F_\partial(j)(l) = \int_M \alpha(dj) \cdot dl\,\mu(j) \quad \forall l \in C^\infty(\partial M, \mathbb{R}^n) \text{ and } dj \in O/\mathbb{R}^n. \tag{6.12}$$

For any $j \in E(\partial M, \mathbb{R}^n)$ we introduce the Laplacian $\Delta(j)$ according to Eq. (5.11), but require that E_s in this case is a moving frame on M.

With this notion at hand the constitutive laws on O are characterized in detail in

6.1 Theorem:

Let F be a constitutive law on any open set $O \subset E(M, \mathbb{R}^n)$. The following statements are then equivalent:

(i) F_∂ admits a kernel $\alpha \in C^\infty(O/\mathbb{R}^n, A^1(\partial M, \mathbb{R}^n))$.

(ii) There is a smooth map $h \in C^\infty(O/\mathbb{R}^n, C^\infty(\partial M, \mathbb{R}^n))$, uniquely determined up to maps in $C^\infty(O/\mathbb{R}^n)$, such that

$$F_\partial(j)(l) = \int_{\partial M} dh(dj) \cdot dl\, \mu(j) \quad \forall j \in O \text{ and } l \in C^\infty(\partial M, \mathbb{R}^n). \tag{6.13}$$

(iii) There is a unique smooth map $\varphi \in C^\infty(O, C^\infty(\partial M, \mathbb{R}^n))$, such that

$$F_\partial(j)(l) = \int_{\partial M} \langle \varphi(dj), l \rangle \mu(j) \quad \forall j \in O \text{ and } l \in C^\infty(\partial M, \mathbb{R}^n), \tag{6.14}$$

and which satisfies

$$\int_{\partial M} \langle \varphi(dj), z \rangle \mu(j) = 0 \quad \forall j \in O \text{ and } z \in \mathbb{R}^n. \tag{6.15}$$

(iv) There is a smooth map $h \in C^\infty(O/\mathbb{R}^n, C^\infty(\partial M, \mathbb{R}^n))$, uniquely determined up to maps in $C^\infty(\partial M, \mathbb{R}^n)$, such that

$$F_\partial(j)(l) = \int_{\partial M} \langle \Delta(j) h(dj), l \rangle \mu(j) \quad \forall j \in O \text{ and } l \in C^\infty(\partial M, \mathbb{R}^n). \tag{6.16}$$

Proof: The equivalence of (i) with (ii) is the analogy to Theorem 4.3. The proof of this is a sort of reduction theorem and can be found in [7] or [8]. Let us pass next to the equivalence of (ii) with (iv):

Given $j \in O$. For simplicity we write k instead of $h(dj)$. We use the identity

$$\begin{aligned} dl\,Y =& m(j)(\mathrm{grad}_j\, \theta(l,j) - W(j)X(l,j), Y) \cdot N(j) \\ &+ dj(\nabla(j)_Y X(l,j) + \theta(l,j) \cdot W(j))Y \qquad \forall Y \in \Gamma TM, \end{aligned} \tag{6.17}$$

holding for any $l \in C^\infty(\partial M, \mathbb{R}^n)$.

Using any moving frame $E_1, \ldots, E_{n-1}$ on ∂M, orthonormal with respect to $m(j)$, we verify the next set of equations:

$$
\begin{aligned}
dk \cdot dl = &- \frac{1}{2} \operatorname{tr} c(dk, dj) \cdot c(dl, dj) \\
&- \frac{1}{4} \operatorname{tr}(\nabla(j)X(dk, dj) - \tilde{\nabla}(j)X(dk, dj)) \cdot \\
&\quad \cdot (\nabla(j)X(dl, dj) - \tilde{\nabla}(j)X(dl, dj)) \\
&+ \frac{1}{4} \operatorname{tr}(\nabla(j)X(dk, dj) + \tilde{\nabla}(j)X(dk, dj) + \theta(k, j)W(j)) \cdot \\
&\quad \cdot (\nabla(j)X(dl, dj) + \tilde{\nabla}(j)X(dl, dj) + \theta(l, j)W(j)) \\
= &- \frac{1}{2} \operatorname{tr} c(dk, dj) \cdot c(dl, dj) \\
&+ \operatorname{tr}(\nabla(j)X(dk, dj) + \theta(k, j)W(j)) \cdot (\nabla(j)X(dl, dj) + \theta(l, j)W(j)) \\
= &- m(j)(\operatorname{grad}_j \theta(l, j) - W(j)X(l, j)U(k, j)) \\
&+ \textstyle\sum_{i=1}^{n-1} m(j)(((\tilde{\nabla}(j)X(dk, dj) + \theta(k, j)W(j))\nabla(j)_{E_r} X(l, j), E_r) \\
&\qquad + ((\theta(l, j))W(j)(\nabla(j)X(dk, dj) + \theta(k, j) \cdot W(j))E_r, E_r)).
\end{aligned}
\tag{6.18}
$$

The expression for $dk \cdot dl$ is therefore

$$
\begin{aligned}
dk \cdot dl = &m(j)(W(j)U(k, j), X(l, j) - d\theta(l, j)U(k, j)) \\
&+ \operatorname{div}_j((\tilde{\nabla}(j)X(dk, dj) + \theta(k, j)W(j))X(l, j)) \\
&- m(j)(\operatorname{div}_j(\nabla(j)X(dk, dj) + \theta(k, j)W(j)), X(l, j)) \\
&+ \theta(l, j) \operatorname{tr} W(j)(\nabla(j)X(dk, dj) + \theta(k, j)W(j)).
\end{aligned}
\tag{6.19}
$$

On the other hand, we remember that $\Delta(j)$ has been defined as

$$
\Delta(j) = d^* d + d d^* \quad \text{with} \quad d^* l = 0 \quad \text{and} \quad d^* \gamma = - \sum_{r=1}^{n-1} \nabla(j)_{E_r}(\gamma)(E_r) \tag{6.20}
$$

for all $l \in C^\infty(\partial M, \mathbb{R}^n)$, $\gamma \in A^1(M, \mathbb{R}^n)$ and any moving frame $E_1, \ldots, E_{n-1}$ of the above type. Thus the following is also easily verified:

$$
\begin{aligned}
\Delta(j)k = d^* dk = &\, d^*(m(j)(\operatorname{grad}_j \theta(k, j) - W(j)X(k, j), \ldots) \cdot N(j)) \\
&+ d^*(dj \nabla(j)X(k, j) + \theta(k, j) \cdot W(j)) \\
= &-d^*(m(j)(U(dk, dj), \ldots) \cdot N(j)) \\
&+ d^*(dj \nabla(J)X(k, j) + \theta(k, j) \cdot W(j)) \\
= &\, dj\, W(j)U(k, j) + (\operatorname{div}_j U(dk, dj)) \cdot N(j) \\
&- dj(\operatorname{div}_j \nabla(j)X(k, j) + \theta(k, j) \cdot W(j)) \\
&\quad + \operatorname{tr} W(j)(\nabla(j)X(k, j) + \theta(k, j) \cdot W(j)) \cdot N(j).
\end{aligned}
\tag{6.21}
$$

Eqs. (6.19) and (6.21) show the equivalence of (ii) with (iv).

To prove (ii)⇒(iii), we integrate both sides of Eq. (6.19) and pose that the equation

$$\int dk \cdot dl \, \mu(j) = \int \langle \varphi(dj), l \rangle \mu(j) \tag{6.22}$$

yields via Gauss' Theorem the smooth $\varphi(dj)$, given by

$$\begin{aligned}\varphi(dj) = &- dj \operatorname{div}_j(\nabla(j)X(dk, dj) + \theta(k, j) \cdot W(j)) - W(j)U(k, j) \\ &+ (\operatorname{tr} W(j)(\nabla(j)X(dk, dj) + \theta(k, j)W(j) + \operatorname{div}_j U(dk, dj)) \cdot N(j),\end{aligned} \tag{6.23}$$

for each $dj \in E(\partial M, \mathbb{R}^n)/\mathbb{R}^n$. Hence we have

$$\Delta(j)k = \varphi(dj) \quad \forall j \in E(\partial M, \mathbb{R}^n) \tag{6.24}$$

and φ depends smoothly on dj. This implication can be reversed (cf.[10]) due to

$$\int_M \langle \Delta(j)k, z \rangle \nu(j) = 0 \quad \forall z \in \mathbb{R}^n. \tag{6.25}$$

Finally, we concentrate on the equivalence of (iii) and (iv). (iii) yields a map h by solving the equation

$$\Delta(j)h(dj) = \varphi(dj) \tag{6.26}$$

for each $j \in O$, with Eq. (6.15) as integrability condition (cf.[10]). Let us show that $h(dj)$ depends smoothly on j. Without loss of generality we can assume that $h(dj)$ lies in the subspace of $C^\infty(\partial M, \mathbb{R}^n)$ with

$$\int_{\partial M} \langle l, z \rangle \mu(j) = 0 \quad \forall l \in V_j \text{ and } Z \in \mathbb{R}^n. \tag{6.27}$$

This map h, also satisfying Eq. (6.26), is uniquely determined. Since $\Delta(j)$ is self–adjoint with respect to $\int \langle \ , \ \rangle \mu(j)$, we also find

$$F_\partial(j)(l) = \int_{\partial M} \langle h(dj), \Delta(j)l \rangle \mu(j). \tag{6.28}$$

Let $j(t) \in O$ vary smoothly and let $j(t_0) = j$. Since

$$\begin{aligned}&F_\partial(j(t_0 + t))(l) - F_\partial(j)(l) \\ &= \int_{\partial M} \langle h(dj(t_0 + t)), \Delta(j(t_0 + t))l \rangle \mu(j(t_0 + t)) - \int_{\partial M} \langle h(dj), \Delta(j)l \rangle \mu(j) \\ &= \int_{\partial M} \langle h(dj(t_0 + t)) - h(dj), \Delta(j(t_0 + t))l \rangle \mu(j(t_0 + t)) \\ &\quad + \int_{\partial M} \langle h(dj), \Delta(j(t_0 + t))l \rangle \mu(j(t_0 + t)) - \int_{\partial M} \langle h(dj), \Delta(j)l \rangle \mu(j),\end{aligned} \tag{6.29}$$

and since F_∂, $\Delta(j)$ and $\mu(j)$ all vary smoothly in j, we conclude that for all $l \in C^\infty(\partial M, \mathbb{R}^n)$ the limit

$$\lim_{t\to 0} \int_{\partial M} \langle \frac{1}{t}(h(dj(t_0+t)) - h(dj)), \Delta(j(t_0))l\rangle \mu(j(t_0)) \tag{6.30}$$

exists. An induction procedure shows that $h(dj(t))$ varies smoothly in t. Thus by the differentiation theory of [2] not only does $Dh(j)$ exist, we are even ensured that h is smooth. The reverse implication is obvious.

7. The Interplay Between Constitutive Laws of Boundary and Body

The deformable media forming the inside of the body and the boundary may differ, and hence each has to be described, on the one hand, by different constitutive laws. On the other hand, these materials together form one body and should be describable by only one constitutive law holding for the whole body. The qualitative properties of the boundary material attached to the body may be influenced by the deformable material forming the body as a whole.

The purpose of this section is to study the influence of the constitutive properties of the deformable medium forming the body on those forming the boundary of the body. In other words, we will decode the constitutive properties of the boundary material attached to the body from the constitutive law describing the material of the body as a whole.

Let the constitutive law of the deformable medium forming the whole body be again given by F. Moreover, F_∂ denotes the constitutive law of the deformable medium forming the boundary only, and which is thought to be detached from the rest of the body. Thus F_∂ is a one-form on O_∂. Both F and F_∂ are supposed to admit integral representations.

The constitutive law F is, according to Theorem 5.2, determined by a smooth map

$$\mathcal{H} : E(M, \mathbb{R}^n)/\mathbb{R}^n \to C^\infty(M, \mathbb{R}^n), \tag{7.1}$$

the constitutive map of the deformable medium. We will first exhibit its influence on the constitutive entities of the material forming the boundary of the body. According to Theorem 5.2, this map yields force densities

$$\begin{aligned} \Phi &: E(M, \mathbb{R}^n)/\mathbb{R}^n \to C^\infty(M, \mathbb{R}^n) \\ \text{and} \quad \varphi &: E(M, \mathbb{R}^n)/\mathbb{R}^n \to C^\infty(\partial M, \mathbb{R}^n). \end{aligned} \tag{7.2}$$

The latter, the force density acting on ∂M, is defined by

$$\varphi(dJ) = d\mathcal{H}(dJ)(N) \quad \forall dJ \in E(M, \mathbb{R}^n)/\mathbb{R}^n. \tag{7.3}$$

Let us split this force density φ into

$$\varphi(dJ) = \varphi_{\mathbb{R}^n}(dJ) + \psi(dJ) \quad \forall dJ \in E(M, \mathbb{R}^n)/\mathbb{R}^n, \tag{7.4}$$

where $\varphi_{\mathbb{R}^n}(dJ)$ is characterized for each $dJ \in E(M, \mathbb{R}^n)/\mathbb{R}^n$ by the equation

$$\int_{\partial M} \langle \varphi_{\mathbb{R}^n}(dJ), z \rangle i_N \mu(J) = 0 \quad \forall z \in \mathbb{R}^n, \tag{7.5}$$

and $\psi : E(M, \mathbb{R}^n) \to \mathbb{R}^n$ a smooth map, which satisfies Eq. (7.4). Let us remark that even if $dJ_1 | \partial M = dJ_2 | \partial M$ for some $J_1, J_2 \in E(M, \mathbb{R}^n)$, we may not necessarily have $\varphi_{\mathbb{R}^n}(dJ_1) = \varphi_{\mathbb{R}^n}(dJ_2)$.

The condition given by Eq. (7.5) allows us to choose some map

$$h_{\mathbb{R}^n} : E(M, \mathbb{R}^n)/\mathbb{R}^n \to C^\infty(\partial M, \mathbb{R}^n), \tag{7.6}$$

such that for all $dJ \in E(M, \mathbb{R}^n)/\mathbb{R}^n$ the equation

$$\varphi_{\mathbb{R}^n}(dJ) = \Delta(J\partial M) h_{\mathbb{R}^n}(dJ) \tag{7.7}$$

holds. We may choose $h_{\mathbb{R}^n}$ such that

$$h_{\mathbb{R}^n}(dJ) \in C^\infty(\partial M, \mathbb{R}^n) \quad \text{with} \quad j \equiv J|\partial M \tag{7.8}$$

for all $J \in E(M, \mathbb{R}^n)$. This map depends smoothly on its variable J, as shown in the proof of Theorem 6.1.

Thus the constitutive law F is determined by a map

$$\mathcal{H} : E(M, \mathbb{R}^n)/\mathbb{R}^n \to C^\infty(M, \mathbb{R}^n), \tag{7.9}$$

yielding a force density Φ with boundary condition,

$$\begin{aligned} \Phi(J) &= \Delta(J)\mathcal{H}(dJ) \quad \forall J \in E(M, \mathbb{R}^n) \\ \text{with} \quad d\mathcal{H}(dJ)(N) &= \Delta(J\partial M) h_{\mathbb{R}^n}(dJ) + \psi(dJ) \end{aligned} \tag{7.10}$$

for some smooth maps

$$h_{\mathbb{R}^n} : E(M, \mathbb{R}^n)/\mathbb{R}^n \to C^\infty(\partial M, \mathbb{R}^n) \quad \text{and} \quad \psi : E(M, \mathbb{R}^n) \to \mathbb{R}^n. \tag{7.11}$$

This boundary condition in Eq. (7.10) obviously describes how the constitutive properties of the material forming the boundary of the body are encoded in $\mathcal{H}$.

On the other hand we have F_∂, which according to Theorem 6.1 is determined by a smooth map

$$h_\partial : O_\partial/\mathbb{R}^n \to C^\infty(\partial M, \mathbb{R}^n). \tag{7.12}$$

The force density defined on $O_\partial/\mathbb{R}^n$ and associated with h_∂ will be denoted in the sequel by φ_∂. We choose an extension

$$\mathcal{H}_\partial : E(M,\mathbb{R}^n)/\mathbb{R}^n \to C^\infty(M,\mathbb{R}^n) \tag{7.13}$$

of h_∂ by posing the following *Višik problem*, which according to [10] has a solution unique up to constants:

$$\phi_\partial(dJ) = 0 = \Delta(J)\mathcal{H}_\partial(dJ) \tag{7.14}$$

$$\text{and} \quad d\mathcal{H}_\partial(dJ)(N) = \Delta(J|\partial M)h_\partial(d(J|\partial M)) = \varphi_\partial(dJ), \tag{7.15}$$

$$\text{together with} \quad \mathcal{H}_\partial(dJ)|\partial M = h_\partial(d(J|\partial M)), \tag{7.16}$$

all holding for any $J \in E(M,\mathbb{R}^n)$. Again, $\mathcal{H}_\partial$ depends smoothly on its variable. This is due to the fact that the constitutive law, determined by $\mathcal{H}_\partial$, only depends on its integral over the boundary ∂M and therefore is a reformulation of F_∂. Prop. 5.1 yields the smoothness of $\mathcal{H}_\partial$. F^0 shall denote the constitutive law on $E(M,\mathbb{R}^n)$ determined by $\mathcal{H}_\partial$.

$h_{\mathbb{R}^n} - R^* h_\partial$ and ψ show how the material forming the boundary of the body is affected by the fact that the boundary material is implemented into the body. Without loss of generality we may think of $R^* h_\partial$ being an additive part of $h_{\mathbb{R}^n}$. This motivates us to write only h instead of $h_{\mathbb{R}^n}$ in the sequel.

What we have done in this section may be formulated in:

7.1 Theorem:

Any smoothly deformable medium is characterized by a constitutive map

$$\mathcal{H} : E(M,\mathbb{R}^n) \to C^\infty(M,\mathbb{R}^n), \tag{7.17}$$

which itself determines two smooth maps

$$h : E(M,\mathbb{R}^n) \to C^\infty(\partial M,\mathbb{R}^n) \quad \text{and} \quad \psi : E(M,\mathbb{R}^n) \to \mathbb{R}^n, \tag{7.18}$$

which are linked to $\mathcal{H}$ by the boundary condition

$$d\mathcal{H}(dJ)(N) = \Delta(J|\partial M)h(dJ) + \psi(dJ). \tag{7.19}$$

h is unique up to $\mathbb{R}^n$-valued smooth maps of $E(M,\mathbb{R}^n)$, and ψ is unique. Moreover, $\mathcal{H}$ satisfies the integrability conditions

$$0 = \int_M \langle \Delta(J)\mathcal{H}(dJ), z\rangle \mu(J) + \int_{\partial M} \langle \partial\mathcal{H}(dJ)(N), z\rangle i_N \mu(J) \tag{7.20}$$

for all $J \in E(M,\mathbb{R}^n)$ and $z \in \mathbb{R}^n$.

Eq. (7.20) may be equivalently formulated as

$$0=\int_M \Delta(J)\mathcal{H}(dJ)\mu(J)+\int_{\partial M}\psi(dJ)i_N\mu(J), \tag{7.21}$$

a boundary condition holding for both $\mathcal{H}$ and ψ. The constitutive law on $E(M,\mathbb{R}^n)$, describing the constitutive properties of the materials forming the body and its boundary, is thus given via the formula

$$F(J)(L)=\int_M \langle\Delta(J)\mathcal{H}(dJ),L\rangle\mu(J)+\int_{\partial M}\langle d\mathcal{H}(dJ)N,L\rangle i_N\mu(J), \tag{7.22}$$

or equivalently by

$$\begin{aligned}F(J)(L)=&\int_M \langle\Delta(J)\mathcal{H}(J),L\rangle\mu(J)\\&+\int_{\partial M}\langle\Delta(J|\partial M)h(J|\partial M)+\psi(dJ)),L\rangle i_N\mu(J)\\&\forall J\in E(M,\mathbb{R}^n) \text{ and } L\in C^\infty(M,\mathbb{R}^n).\end{aligned} \tag{7.23}$$

The work of any distortion $l\in C^\infty(\partial M,\mathbb{R}^n)$ of the deformable material forming the boundary attached to the body is given by

$$F_{\partial M}(dJ)(l)=\int_{\partial M}\langle\Delta(J|\partial M)h(dJ),l\rangle i_N\mu(J) \tag{7.24}$$

for any $J\in E(M,\mathbb{R}^n)$.

Any constitutive properties describing the constitutive properties of the deformable medium of the boundary detached from the body, which is given by a smooth map $h_\partial\in C^\infty(O_\partial/\mathbb{R}^n, C^\infty(\partial M,\mathbb{R}^n))$, is additively incorporated into h via the map $R^*h_\partial\in C^\infty(E(M,\mathbb{R}^n)/\mathbb{R}^n, C^\infty(\partial M,\mathbb{R}^n))$. Hence $h-R^*h_\partial$ and ψ describe how the constitutive properties of the material forming the boundary of the body is affected by the fact that this material is incorporated into the material forming the whole body.

Simple Examples:

Given $L\in C^\infty(M,\mathbb{R}^n)$ and $J\in E(M,\mathbb{R}^n)$, according to Eq. (4.50) we may set $\hat{L}$ for the component in $C^\infty(M,\mathbb{R}^n)$ of any $L\in C^\infty(M,\mathbb{R}^n)$.

Example 1: Let $\mathcal{H}(dJ)=\hat{J}$ for all $J\in E(M,\mathbb{R}^n)$, then $d\hat{J}=dJ$ and

$$\begin{aligned}\int_M dJ\cdot dL\,\mu(J)&=\int_M\langle\Delta(J)J,L\rangle\mu(J)+\int_{\partial M}\langle dJ(N),l\rangle i_n\mu(J)\\&=\int_M \operatorname{tr}A(dL,dJ)\mu(J)=\int_M \operatorname{tr}\nabla(J)X(L,J)\mu(J)\\&=\int_M \operatorname{div}_J X(L,J)\mu(J)=\int_{\partial M}\langle N(j),l\rangle\mu(j)\\&=D\left(\int_M\mu(J)\right)(L).\end{aligned} \tag{7.25}$$

Here $l := L|\partial M$ and $j := J|\partial M$. The above calculation shows

$$\Phi(dJ) = \Delta(J)J = 0 \quad \text{and} \quad \varphi(dJ) = N(j) \quad \forall J \in E(M, \mathbb{R}^n); \tag{7.26}$$

$l = z$ with $z \in \mathbb{R}^n$ evidently implies

$$\int_M \langle N(j), z\rangle \mu(j) = 0 \quad \forall z \in \mathbb{R}^n. \tag{7.27}$$

This shows that in this example we have $\varphi = \varphi_{\mathbb{R}^n}$. The map $h_{\mathbb{R}^n}$ in this case is thus given by

$$N(j) = \Delta(j) h_{\mathbb{R}^n}(dJ) \quad \forall J \in E(M, \mathbb{R}^n) \text{ and } j := J|\partial M. \tag{7.28}$$

Since $h_{\mathbb{R}^n}$ here depends only on dj, we have the situation that $F = F^0$. Now let us turn our attention to h_∂ on O_∂, given by $h_\partial(dj) = \hat{j}\ \forall j \in O_\partial$. Here $\hat{j} \in C^\infty(\partial M, \mathbb{R}^n)$ is the projection of j along $\mathbb{R}^n$. One easily verifies the following calculation:

$$\begin{aligned}
\int_{\partial M} \langle \Delta(j)j, l\rangle \mu(j) &= \int_{\partial M} dj \cdot dl\, \mu(j) \\
&= \int_{\partial M} (\operatorname{tr}(\nabla(j) X(l,j) + \theta(l,j) \cdot W(j)) \\
&= \int_{\partial M} (\operatorname{div}_j X(l,j) + \theta(l,j) \cdot H(j)) \mu(j) \\
&= \int_{\partial M} \theta(l,j) \cdot H(j) \mu(j) \\
&= \int_{\partial M} \langle \theta(l,j) \cdot H(j) \cdot N(j), l\rangle \mu(j) \\
&= D\left(\int_{\partial M} \mu(j)\right)(l).
\end{aligned} \tag{7.29}$$

Hence $h_{\mathbb{R}^n}$ is given by

$$0 = \Delta(J) \mathcal{H}_\partial(dJ) \quad \forall J \in E(M, \mathbb{R}^n), \tag{7.30}$$

together with

$$d\mathcal{H}_\partial(dJ)(N) = \Delta(j)j = H(j) \cdot N(j) \tag{7.31}$$

for all $J \in E(M, \mathbb{R}^n)$ and $J := J|\partial M$. Here again $F = F^0$ (cf. Sect. 6).

Example 2: Next let us consider quite another influence of the boundary by looking at $h_\partial : O_{\partial/\mathbb{R}^n} \to C^\infty(M, \mathbb{R}^n)$, given by $h_\partial(dj) = N(j)\ \forall j \in O_\partial$. Then the formula

$$\begin{aligned}
\Delta(j)N(j) &= d^* dN(j) = d^* dj\, W(j) \\
&= -dj \operatorname{grad}_j H(j) + (\operatorname{tr} W(j))^2 \cdot N(j)
\end{aligned} \tag{7.32}$$

holds for any $j \in O_\partial$.

In this case $\mathcal{H}_\partial$ is given by the system

$$0 = \Delta(J)\mathcal{H}\partial(dJ) \quad \forall J \in E(M, \mathbb{R}^n), \tag{7.33}$$

$$d\mathcal{H}_\partial(N) = \Delta(j) \cdot N(j) \quad \forall J \in E(M, \mathbb{R}^n) \text{ and } j := J|\partial M. \tag{7.34}$$

Let us point out that $\Delta(j)N(j) \neq 0$ even if $j(\partial M) \subset \mathbb{R}^n$ is minimal, i.e. even if $H(j) = \text{const}$. In the special case of $\dim \partial M = 2$ a topological constant, the Euler characteristic $\chi(\partial M)$, enters the constitutive law F determined by $N(j)$ for each $j \in O_\partial$. It is hidden in

$$F(j)(N(j)) = \int_{\partial M} \langle \Delta(j)N(j), N(j)\rangle \mu(j) = \int_{\partial M} \operatorname{tr} W(j)^2 \mu(j), \tag{7.35}$$

which may be seen as follows:

By the Cayley-Hamilton Theorem (cf.[4]) and the Gauss-Bonnet Theorem (cf.[13]), $F(j)(N(j))$ in Eq. (7.35) can be expressed by

$$F^\partial(j)(N(j)) = -4\pi\chi(\partial M) + \int_{\partial M} H(j)^2 \mu(j). \tag{7.36}$$

Here we also have $F = F^0$.

8. A General Decomposition of Constitutive Laws

In this section we will exhibit a decomposition of the constitutive map $\mathcal{H}$. This decomposition is based on two specific one-forms $E(M, \mathbb{R}^n)$ and $E(\partial M, \mathbb{R}^n)$, the derivatives of the *volume function* $\mathcal{V} : E(M, \mathbb{R}^n) \to \mathbb{R}$, assigning to any $J \in E(M, \mathbb{R}^n)$ the volume

$$\mathcal{V}(J) = \int_M \mu(J), \tag{8.1}$$

and of the *area function* $\mathcal{A} : E(\partial M, \mathbb{R}^n) \to \mathbb{R}$, sending any $j \in E(\partial M, \mathbb{R}^n)$ into

$$\mathcal{A}(j) = \int_{\partial M} \mu(j). \tag{8.2}$$

As we know from the previous examples, these derivatives are

$$\begin{aligned} D\mathcal{V}(J)(L) &= \int_{\partial M} \langle N(J|\partial M), L\rangle i_N \, \mu(J) \\ &= \int_M dJ \cdot dL \, \mu(J) = \int_{\partial M} \langle dJ(N), L\rangle i_N \, \mu(J) \end{aligned} \tag{8.3}$$

and

$$D\mathcal{A}(j)(l) = \int_{\partial M} H(j)\langle N(j), l\rangle \mu(j) = \int_{\partial M} dj \cdot dl \, \mu(j), \tag{8.4}$$

holding for all $J \in E(M, \mathbb{R}^n)$, all $L \in C^\infty(M, \mathbb{R}^n)$, all $j \in E(\partial M, \mathbb{R}^n)$ and all $l \in C^\infty(\partial M, \mathbb{R}^n)$. We will show in this section that $D\mathcal{V}$ and $R^*D\mathcal{A}$, multiplied with appropriate $\mathbb{R}$-valued maps, are all part of any constitutive law F defined on $E(M, \mathbb{R}^n)$. Let us first concentrate on $D\mathcal{V}$ and see how it is encoded in any constitutive map $\mathcal{H}$. To this end let F be determined by some

$$\mathcal{H} \in C^\infty(E(M, \mathbb{R}^n)/\mathbb{R}^n, C^\infty(M, \mathbb{R}^n)). \tag{8.5}$$

As we know from the previous section, it determines two maps

$$\begin{aligned} h \in C^\infty(E(\partial M, \mathbb{R}^n)/\mathbb{R}^n, C^\infty(\partial M, \mathbb{R}^n)) \\ \text{and} \quad \psi \in C^\infty(E(M, \mathbb{R}^n)/\mathbb{R}^n), \end{aligned} \tag{8.6}$$

such that both are linked to $\mathcal{H}$ by the equation

$$d\mathcal{H}(dJ)(N) = \Delta(J|\partial M)h(dJ) + \psi(dJ), \tag{8.7}$$

which holds for any $J \in E(M, \mathbb{R}^n)$.

Let us consider the real Hilbert-space H_J consisting of all $L, K : M \to \mathbb{R}^n$, for which

$$\langle\langle L, K \rangle\rangle := \int_M \langle L, K \rangle \mu(J) \quad \text{exists.} \tag{8.8}$$

Recalling that $C^\infty_J(M, \mathbb{R}^n)$ is the collection of all $L \in C^\infty(M, \mathbb{R}^n)$ such that

$$\int_M \langle L, z \rangle \mu(J) = 0 \quad \forall z \in \mathbb{R}, \tag{8.9}$$

we restate the splitting

$$C^\infty(M, \mathbb{R}^n) = \mathbb{R}^n \times C^\infty_J(M, \mathbb{R}^n). \tag{8.10}$$

This is a splitting as Fréchet spaces, since the functional which assigns to every $J \in C^\infty(M, \mathbb{R}^n)$ a real $\int_M \langle\ , z \rangle \mu(J)$ is continuous on $C^\infty(M, \mathbb{R}^n)$ for any $z \in \mathbb{R}$. Moreover, Eq. (8.10) is orthogonal with respect to $\langle\langle\ ,\ \rangle\rangle$, defined in Eq. (8.8).

The projection $\hat{J}$ along $\mathbb{R}^n$ of $J \in E(M, \mathbb{R}^n)$ in $C^\infty(M, \mathbb{R}^n)$ satisfies

$$d\hat{J} = dJ \quad \text{and} \quad \Delta(J)\hat{J} = \Delta(J)J = 0. \tag{8.11}$$

Since both $\mathcal{H}(dJ)$ and $\bar{J}$ belong to H_J, we may take the component of $\mathcal{H}(dJ)$ in H_J along $\hat{J}$. This component has the form $\pi^1(dJ) \cdot \hat{J}$ for some real number $\pi^1(dJ)$. Thus there is some $\mathcal{H}^1 \in C^\infty(E(M, \mathbb{R}^n)/\mathbb{R}^n, C^\infty(M, \mathbb{R}^n))$ such that

$$\mathcal{H}(dJ) = \pi^1(dJ) \cdot \hat{J} + \mathcal{H}^1(dJ) \tag{8.12}$$

is an orthogonal decomposition in H_J. We leave it to the reader to show that $\pi^1(dJ)$ and $\mathcal{H}^1(dJ)$ vary smoothly with dJ.

Clearly, due to Eqs. (8.7) and (8.11) we have

$$\Delta(J)\mathcal{H}(dJ) = \Delta(J)\mathcal{H}^1(dJ) \tag{8.13}$$

and

$$d\mathcal{H}(dJ)(N) = \pi^1(dJ)N(J|\partial M) + d\mathcal{H}^1(dJ)(N), \tag{8.14}$$

both holding for all $J \in E(M, \mathbb{R}^n)$.

Let us denote by F^1 the constitutive law on $E(M, \mathbb{R}^n)$ determined by the map π^1. Eq. (8.13) then yields

$$F = \pi^1 \cdot D\mathcal{V} + F^1. \tag{8.15}$$

The map

$$\pi^1 \cdot N : E(M, \mathbb{R}^n)/\mathbb{R}^n \to C^\infty(\partial M, \mathbb{R}^n), \tag{8.16}$$

assigning to any dJ the map $\pi(dJ) \cdot N(j)$ with $j := J|\partial M$, yields a part of h. This part, called h_N, is produced by regarding $\pi(dJ) \cdot N(j)$ as a force density along ∂M, which according to Eq. (7.28) has to satisfy the equation

$$\Delta(j)h_N(dJ) = \pi^1(dJ)N(j) \quad \forall j := J|\partial M \tag{8.17}$$

for all $J \in E(M, \mathbb{R}^n)$. The map $h_N(dJ) \in C^\infty(\partial M, \mathbb{R}^n)$ varies smoothly with $J \in E(M, \mathbb{R}^n)$. Hence we have the splitting

$$h(dJ) = \pi^1(dJ)h_N(dJ)) + h^1(dJ) \quad \forall dJ \in E(M, \mathbb{R}^n)/\mathbb{R}^n \tag{8.18}$$

for some smooth $h^1(dJ) \in C^\infty(\partial M, \mathbb{R}^n)$. This example shows what we had in mind when we claimed that $D\mathcal{V}$ is part of F.

To get the full decomposition we broaden our scope a little and introduce first of all the Hilbert space A_i consisting of all maps $\gamma_1, \gamma_2 : TM \to \mathbb{R}$ linear on the fibres of TM, for which the right hand side of

$$G_{\mathbb{R}^n}(dj)(\gamma_1, \gamma_2) := \int_{\partial M} \gamma_1 \cdot \gamma_2 \, \mu(j) \tag{8.19}$$

exists. Clearly, $dh_N(dJ)$, dj and $dN(j)$ all belong to A_i and are generically linearly independent. The set O_3 of all $j \in E(\partial M, \mathbb{R}^n)$ for which these three differentials are linearly independent form a dense open set. If, however, $j(\partial M)$ is an $(n-1)$-sphere in $\mathbb{R}^n$, $N(j)$ is a real multiple, say r, of j and $h_N(J)$ is hence $\frac{r}{n-1} \cdot j$. This can be confirmed by looking at Eqs. (7.28) and (7.31). In the case of linear independence the three differentials mentioned above are, however, in general (with respect to $G_{\mathbb{R}^n}(dj)$) not orthogonal to each other. We therefore orthogonalize them by using the method of Schmidt.

Next we extend all maps $h_N(dJ)$, $\hat{\jmath}$ and $N(\jmath)$ to all of M in the following way: Given $f \in C^\infty(\partial M, \mathbb{R}^n)$, we solve the Višik problem

$$\Delta(J)f = 0, \qquad df_M(N) - \Delta(\jmath)f = 0, \tag{8.20}$$

with $f_M \subset C^\infty(M, \mathbb{R}^n)$ and $\jmath := J|\partial M$, where $J \in E(M, \mathbb{R}^n)$. Clearly,

$$(h_N(dJ))_M = \hat{J}. \tag{8.21}$$

All the splittings and extensions executed to construct $\hat{\jmath}_M$ and $N(\jmath)_M$ depend smoothly on $\jmath \in E(\partial M, \mathbb{R}^n)$. The decomposition of $\mathcal{H}$ mentioned above is then described by

8.1 Theorem:

Let F be a constitutive law on $E(M, \mathbb{R}^n)$, determined by

$$\mathcal{H} \in C^\infty(E(M, \mathbb{R}^n)/\mathbb{R}^n, C^\infty(M, \mathbb{R}^n)). \tag{8.22}$$

Then $\mathcal{H}$ uniquely determines three smooth maps

$$a_1, a_2, a_3 : O_3 \subset E(M, \mathbb{R}^n)/\mathbb{R}^n \to \mathbb{R} \tag{8.23}$$

and also uniquely determines two smooth maps

$$h, h^2 : O_3 \subset E(M, \mathbb{R}^n)/\mathbb{R}^n \to C^\infty(\partial M, \mathbb{R}^n), \tag{8.24}$$

such that the following splitting holds for any $dJ \in O_3 \subset E(M, \mathbb{R}^n)$:

$$h(dJ) = a_1(dJ) \cdot h_N(dJ) + a_2(dJ)\hat{\jmath} + a_3(dJ) \cdot N(\jmath) + h^2(dJ) \tag{8.25}$$

with $\jmath := \jmath/\partial M$. The differential $dh^2(dJ)$ is orthogonal with respect to $G(d\jmath)_{\partial M}$ to the space spanned by $dh_N(dJ)$, $d\jmath$ and $dN(\jmath)$. The map $\mathcal{H}(dJ)$ decomposes into

$$\mathcal{H}(dJ) = a_1(dJ) \cdot \hat{J} + a_2(dJ) \cdot \hat{\jmath}_M + a_3(dJ) \cdot N_M(dJ) + \mathcal{H}^2(dJ), \tag{8.26}$$

where $\jmath := J|\partial M$.

The constitutive law F splits accordingly into

$$F(J)(L) = a_1(dJ) \cdot D\mathcal{V}(J)(L) + a_2(dJ) \cdot D\mathcal{A}(\jmath)(l) + F_N(dJ)(dL) + F^2(J)(L), \tag{8.27}$$

with $J := J|\partial M$, $l := L|\partial M$ and

$$F_N(J)(L) = a_3(dJ) \cdot \int_M dN_M(dJ) \cdot dL\, \mu(J). \tag{8.28}$$

References

1. E. Binz, J. Sniatycki, H. Fischer: ***Geometry of Classical Fields***, Mathematics Studies **154** (North-Holland, Amsterdam 1988)
2. A. Frölicher, A. Kriegl: ***Linear Spaces and Differentiation Theory*** (John Wiley, Chichester, England 1988)
3. M.W. Hirsch: ***Differential Topology*** (Springer,Berlin 1976)
4. W. Greub: ***Lineare Algebra I***, Graduate Texts in Mathematics **23** (Springer, Berlin, Heidelberg, New York 1981)
5. E. Hellinger: "Die allgemeinen Ansätze der Mechanik der Kontinua", Enzykl. Math. Wiss. **4/4**, (1914)
6. M. Epstein, R. Segev: "Differentiable Manifolds and the Principle of Virtual Work in Continuum Mechanics", Journal of Mathematical Physics **5.21**, (1980)
7. E. Binz: "On the Notion of the Stress Tensor Associated with $\mathbb{R}^n$-invariant Constitutive Laws Admitting Integral Representations", Reports on Mathematical Physics **27**, (1989)
8. E. Binz, G. Schwarz, D. Socolescu: "On a Global Differential Geometric Description of the Motions of Deformable Media", to appear in ***Infinite Dimensional Manifolds, Groups, and Algebras, Vol. II, Ed. H.D. Doebner, ed. J. Hennig*** (World Scientific, Singapur 1990)
9. R. Abraham, J.E. Marsden, T. Ratiu: ***Manifolds, Tensor Analysis and Applications*** (Addison Wesley, Reading Massachusetts 1983)
10. L. Hörmander: ***Linear Partial Differential Operations***, Grundlehren der mathematischen Wissenschaften **116** (Springer, Berlin, Heidelberg, New York 1976)
11. F. John: ***Partial Differential Equations***, Applied Mathematical Science **1**, (1978)
12. Y. Matsushima: ***Vector Bundle Valued Canonical Forms***, Osaka Journal of Mathematics **8**, (1971)
13. W. Greub, S. Halperin, J. Vanstone: ***Connections, Curvature and Cohomology I,II*** (Academic Press, New York 1972-73)

Two Lectures on Fermions and Gravity

Friedrich W. Hehl [1], *Jürgen Lemke* [2], *and Eckehard W. Mielke* [3]

Institute for Theoretical Physics, University of Cologne,
D(W)-5000 Köln 41, Federal Republic of Germany

Preface

In these two lectures we want to provide information on the behavior of elementary particles in *special relativity* (SR), such as electrons, protons, or neutrons, in order to get ideas of the underlying principles of the *gravitational* interaction of fermions. For tangible matter and for the electromagnetic field, Einstein's gravitational theory, *general relativity* (GR), describes all phenomena very well and has been verified experimentally with ever increasing accuracy. In contrast therefrom, not too much is known experimentally for fermions and their gravitational interaction, apart from the celebrated Colella-Overhauser-Werner (or COW) experiment [21] using a neutron interferometer in a gravitational field.

When Einstein had finalized special relativity and looked, starting from SR, for a proper way to formulate a gravitational theory, he took the *equivalence principle* as his guiding principle. According to this principle, gravitational forces can be *simulated* within SR by going over from an inertial to a non-inertial reference frame. In other words, the structure of inertial forces is closely related to that of gravitational forces. The formulation of the principle of equivalence was Einstein's answer to the experimentally well-established, but theoretically in those days ill-understood proportionality of the inertial and the gravitational mass of matter.

Turning now to fermions, we will describe them, in the one-particle approximation, by means of a semi-classical matter field $\Psi(x^i)$, where the $\{x^i\}$ are the four coordinates of the underlying flat four-dimensional Minkowski spacetime M_4 of SR and Ψ is a spinorial representation of the Poincaré (inhomogeneous Lorentz) group. Let us stress again: Our hard core results are special-relativistic. However, after applying the equivalence principle to the fermionic *matter fields*, these results are instrumental in uncovering the structure of the gravitational interaction of fermions.

---

[1] This work was supported by the German-Israeli Foundation for Scientific Research and Development (GIF), Jerusalem and Munich

[2] supported by a graduate scholarship of the State of Nordrhein-Westfalen

[3] supported by the Deutsche Forschungsgemeinschaft (DFG), Bonn

Whereas most of the material is traditional, see, for instance, the textbooks of Misner, Thorne, and Wheeler [94], Rindler [111], Sexl and Urbantke [124, 125], Straumann [130], or Thirring [132], there are some new aspects stressed and worked out here: In Lecture I it is mainly the accelerated and rotating laboratory frame of reference as expressed in laboratory coordinates (see Sect. 9 and Eq. (9.8)) and, in Sects. 17 and 18, the Dirac equation formulated in that frame [52]. In particular, we recover the spin-rotation effect, predicted by Mashhoon [80] for a COW type experiment with polarized neutrons.

In Lecture II the relocalization of energy-momentum and angular momentum is treated in Sects. 10 and 11 such that the Belinfante-Rosenfeld symmetrization and other improved expressions come out as special cases. The decomposition of the energy-momentum and the spin current of a matter field into a convective and a polarization part was developed in Cologne. In Sect. 15 we display this procedure explicitly for the Dirac field in terms of exterior differential forms. In this special-relativistic context, we derive *gravitational moment* 2-forms of the Dirac field. Moreover, the gyro-gravitational ratio for a Dirac particle is addressed in Sect. 16, an issue rarely discussed in the literature.

We talked in Bad Honnef also on a continuum theoretical approach to spacetime physics, according to which spacetime may be regarded as a generalized continuum carrying stress and hyperstress [87, 51]. We pointed out the relation of these stresses to the currents of energy-momentum, spin momentum, and dilation. The workout of these considerations is, however, left to the future. Our mathematical formalism is mostly based on concepts described by Schouten [117] and by Choquet-Bruhat et al. [18].

For the description of the Dirac field we used the conventions of the standard text of Bjorken-Drell [8] as far as possible (metric $(+,-,-,-)$). We tried hard to get all signs correct and to put in the c's (velocity of light) whereever necessary. But these lecture notes were worked out under time pressure. Hence there may have crept in some slips. If this is the case, we apologize in advance and we ask to let us know our mistakes. Address: Institut für Theoretische Physik der Universität, D-5000 Köln 41, Germany, electronically hehl@thp.uni-koeln.de.

We also tried to find all literature relevant to our subject. Any omission is unintentional. We would like to ask "forgotten authors" to send us copies of their articles. In a possible update, we may then refer to them.

Lecture I: Inertial Properties of a Massive Fermion

Abstract

In this lecture we will recount how Einstein arrived at GR by applying the equivalence principle to classical point particles and to the electromagnetic field. We will contrast this with the corresponding approach as applied to *matter fields*, in particular, to the Dirac field. The results of this discussion are collected in Table 1. It turns out that the notion in SR of a local reference frame of four orthonormal vectors occupies a central position. Even more so, for the discussion of the COW experiment and some of its related effects, a so-called *laboratory frame* in SR, which represents a subclass of the local reference frames, is still more instrumental. We will define it in Sects. 8 and 9.

Contents

I.1 Special Relativity in an Inertial Frame

As long as it is possible, in a certain experiment, to neglect the gravitational forces, the theory of space and time is represented by special relativity theory (SR). Its Minkowskian spacetime M_4 is a four-dimensional differentiable manifold X_4 with a flat pseudo-Riemannian metric.

In the M_4, we can introduce a global Cartesian coordinate system given by $\{x^i\} = \{ct, \boldsymbol{x}\} = \{ct, x, y, z\} = \{x^0, x^1, x^2, x^3\}$. In these coordinates, the line element ds is given by the quadratic form

$$ds^2 \stackrel{*}{=} c^2 dt^2 - dx^2 - dy^2 - dz^2 \stackrel{*}{=} o_{ij}\, dx^i dx^j, \tag{1.1}$$

with $o_{ij} := \text{diag}\,(1, -1, -1, -1)$. Summation over repeated indices is understood. The star indicates that this formula is only valid in a special coordinate system.

The line element is invariant under rigid Lorentz rotations $\Lambda^i{}_{i'}$ and under rigid spacetime translations b^i, i.e., $x^{i'} \stackrel{*}{=} \Lambda^{i'}{}_i\, x^i + b^{i'}$. In particular, $o_{i'j'} \stackrel{*}{=} \Lambda^i{}_{i'} \Lambda^j{}_{j'}\, o_{ij}$. Taken together, these transformations build up the ten-parameter Poincaré (or inhomogeneous Lorentz) group, the group of pseudo-Euclidean motions in the M_4. Incidentally, we use the kernel index method of Schouten [117].

In SR, according to Einstein, a Cartesian coordinate system represents a global inertial frame of reference. By means of a Poincaré transformation, we can go over to a new Cartesian coordinate system $\{x^{i'}\}$, that is, to a new inertial frame, that moves with constant three-velocity relative to the old frame.

Operationally, we can determine whether we are in an inertial frame by the following method [76]: Let three force-free mass points be ejected from a common origin into three different non-coplanar directions. If they move on *straight lines* relative to our frame, we are in an inertial frame of reference.

Consider a (point-like) observer in spacetime. He, she, or it ("Schrödinger's cat"?) traces a future pointing timelike curve[1] C in the M_4. We can parametrize C by means of the proper time τ of the observer, that is, by the time read in the respective instantaneous rest frame of the observer: $C = C(\tau)$. In Cartesian coordinates x^i, C is described by $x^i = x^i(\tau)$. The *proper time* interval $d\tau$ is given by the Poincaré-invariant expression

$$d\tau := \frac{ds}{c} = \sqrt{dt^2 - \frac{dx^2 + dy^2 + dz^2}{c^2}} = dt\sqrt{1 - \frac{v^2}{c^2}} = \frac{dt}{\gamma(v)}, \tag{1.2}$$

where $\boldsymbol{v}$ denotes the three-velocity and $\gamma(v) := (1 - v^2/c^2)^{-1/2}$ the Lorentz factor. The (four-)velocity of the observer is

$$u^i := \frac{dx^i}{d\tau} \quad \Rightarrow \quad u^i = \{\gamma c, \gamma \boldsymbol{v}\}. \tag{1.3}$$

[1] In Sachs and Wu [113], the curve C itself is called the "observer".

Since $o_{ij}u^i u^j = u_i u^i = c^2$, we recognize by differentiation that the acceleration of the observer

$$a^i :\stackrel{*}{=} \frac{du^i}{d\tau} \equiv \overset{\bullet}{u}{}^i \quad \Rightarrow \quad a^i = \gamma\{c\frac{d\gamma}{dt}, \frac{d\gamma}{dt}\boldsymbol{v} + \gamma\boldsymbol{a}\} \tag{1.4}$$

is always perpendicular to the velocity:

$$o_{ij}a^i u^j = a_i u^i = 0\,. \tag{1.5}$$

The equation of motion of a force-free point mass m in Cartesian coordinates reads

$$m\frac{d^2x^i}{d\tau^2} \stackrel{*}{=} 0\,. \tag{1.6}$$

Hence the particle, as seen from an inertial frame, moves on a straight worldline (Fig. 1).

Originally Einstein used "reference frame" and "coordinate system" synonymously. Nowadays, it is common convention in special and general relativity that a *simple reference frame* is given by a (timelike) vector field, each of whose integral curves is the world line of an observer, see [113]. We may visualize these concepts by means of a fluid. The world lines of the fluid mass points represent the observers and their *velocity field* u^i constitute the simple reference frame. We call such a reference frame "simple", in distinction to the more general notion of a *local* reference frame (see Sect. 4), which will involve four vector fields.

It is convenient to introduce coordinates which are *adapted* to the observers. Since an observer moves on a timelike curve C, we may choose the adapted coordinate system $\{x^{\bar{i}}\}$ such that the C's are the coordinate lines obtained by putting the spatial coordinates $x^{\overline{A}} = const.$, where $\overline{A}, \overline{B} \cdots = 1, 2, 3$. We will stay within adapted coordinates, if we reparametrize C and rename the spatial coordinates according to (see [54])

$$x^{\bar{\bar{0}}} = x^{\bar{\bar{0}}}(x^{\bar{i}}), \quad x^{\overline{\overline{K}}} = x^{\overline{\overline{K}}}(x^{\overline{K}}), \quad (\bar{i} = 0, 1, 2, 3;\, \overline{K} = 1, 2, 3) \tag{1.7}$$

with four arbitrary functions.

If a simple reference frame is represented in an M_4 by an autoparallel (timelike) vector field, any observer is free of acceleration and the simple reference frame is an inertial frame. Then, all the vectors point along the x^0-axes (Fig. 2).

I.2 Standard Measurement Hypothesis

Let us introduce, besides the Cartesian coordinates $\{x^i\}$, arbitrary curvilinear coordinates $\{x^{i'}\}$ according to

$$x^i \to x^{i'} = x^{i'}(x^i), \quad p^i{}_{i'} := \frac{\partial x^i}{\partial x^{i'}}, \quad p^i{}_{i'j'} := \frac{\partial^2 x^i}{\partial x^{i'}\partial x^{j'}}, \quad \det p^i{}_{i'} \neq 0\,, \tag{2.1}$$

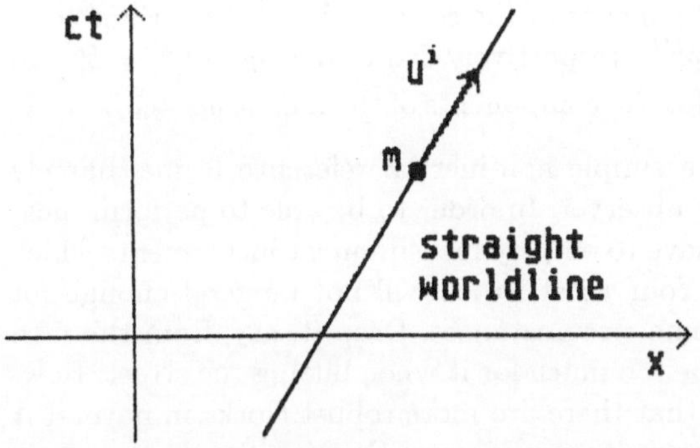

Fig. 1. Minkowski diagram of a force-free point particle.

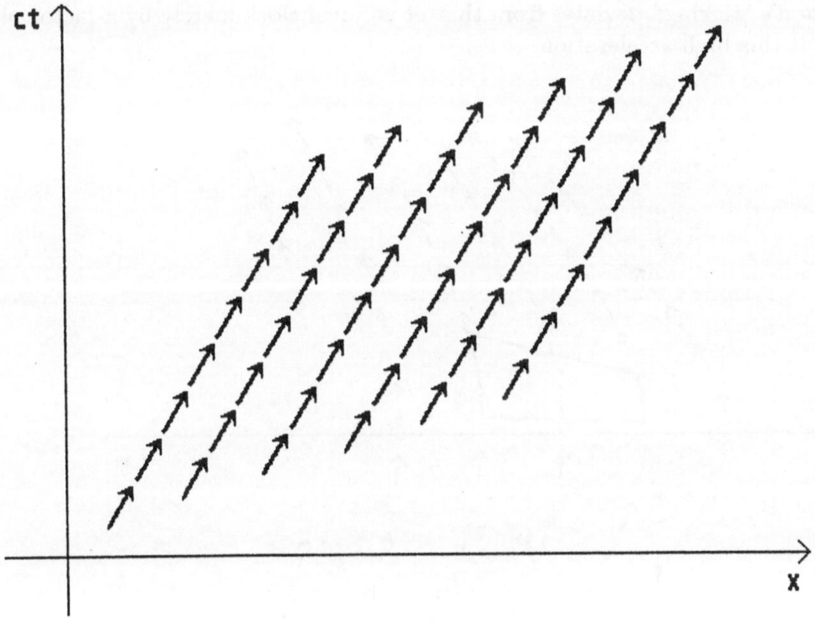

Fig. 2. This simple reference frame is inertial.

where the functions $x^{i'}(x^i)$ are at least twice differentiable and the $p^i{}_{i'}$ are non-singular. Then the coordinates loose their direct significance for distance measurements and the new coordinate system in general represents, using Einstein's language, a non-inertial reference frame. The line element reads

$$ds^2 = g_{i'j'}\,dx^{i'}\,dx^{j'} \quad \text{with} \quad g_{i'j'} = \frac{\partial x^i}{\partial x^{i'}}\frac{\partial x^j}{\partial x^{j'}}\,o_{ij}\,. \tag{2.2}$$

We can raise and lower indices by means of the co- and the contravariant components of the metric $g_{i'j'}$ and $g^{k'l'}$, respectively. Note that $g_{i'j'}\, g^{j'k'} = \delta^{k'}_{i'}$. In Cartesian coordinates, we have for the components of the metric $g_{ij} \stackrel{*}{=} o_{ij} \stackrel{*}{=} o^{ij}$.

In Fig. 3 we have depicted a simple non-inertial reference frame, namely that of a uniformly accelerating observer. In order to be able to perform measurements in such a frame, we have to employ measurement instruments which are insensitive to acceleration. Your wrist watch will not be good enough for time measurements under extreme circumstances. Drop it, say, from the fifth floor, and the acceleration will be too much for it when hitting the street. However, we know from experience that there are more robust clocks in nature: A muon, for instance, doesn't change its decay time, that is, its inherent clock mechanism is not affected, even if it is subject to an acceleration as high as $10^{18} g$, where $g \approx 981\, cm/s^2$ ([4a], see also [111]). Eisele [31] estimated, using the Fermi limit of the Weinberg-Salam model of electroweak interaction, that the muon's "ticking" deviates from that of an ideal clock merely by a factor of 10^{-25} at this high acceleration.

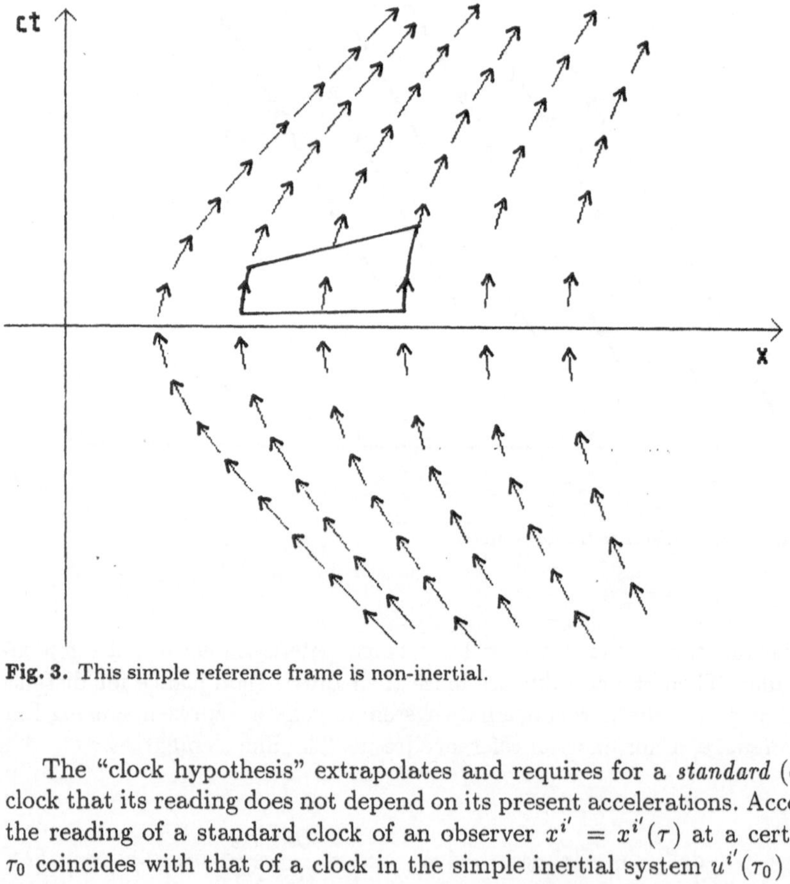

Fig. 3. This simple reference frame is non-inertial.

The "clock hypothesis" extrapolates and requires for a *standard* (or ideal) clock that its reading does not depend on its present accelerations. Accordingly, the reading of a standard clock of an observer $x^{i'} = x^{i'}(\tau)$ at a certain time τ_0 coincides with that of a clock in the simple inertial system $u^{i'}(\tau_0) = \text{const}$.

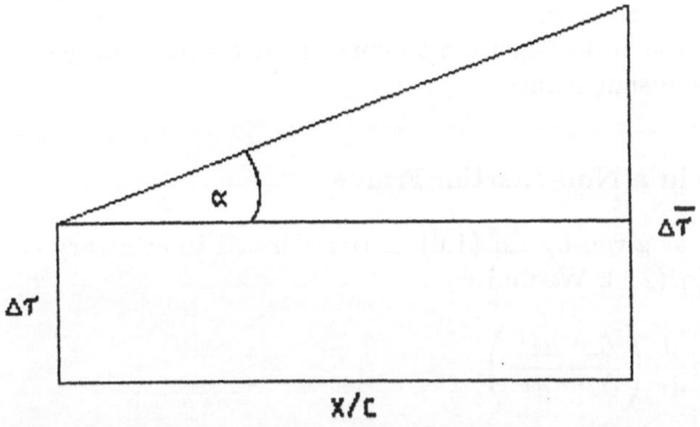

Fig. 4. Magnified view of the trapezoid in Fig. 3: By an elementary consideration we can derive the time-dilation along neighboring worldlines: $\tan\alpha = a\Delta\tau/c = c(\Delta\bar{\tau} - \Delta\tau)/x \Rightarrow \Delta\bar{\tau}/\Delta\tau = 1 + ax/c^2$.

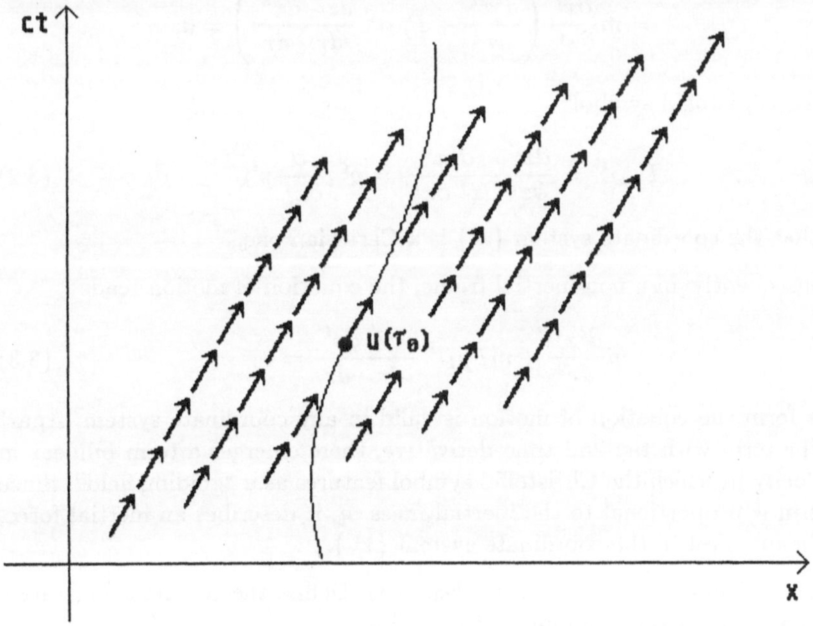

Fig. 5. The velocity of the observer at the time τ_0 determines a simple inertial frame. According to the clock hypothesis, at the observer's time τ_0 the rates of the inertial and the observer's clock coincide.

(Fig. 5). Incidentally, the success of the Hafele-Keating experiment of transporting and comparing atomic clocks in westward and eastward direction around the world in tourist jet planes attests to the validity of the clock hypothesis for atomic clocks under moderate accelerations.

The *standard measurement hypothesis* will be assumed to be valid henceforth not only for time measurements, but for all other measurements as well.

Then non-inertial, that is accelerated observers can perform measurements like in an inerital frame and the transformation to curvilinear coordinates gets a straightforward operational significance.

I.3 Special Relativity in a Non-Inertial Frame

The equation of motion, as given by Eq. (1.6), is transformed to arbitrary coordinates according to Eq. (2.1). We find

$$\begin{aligned} m\frac{d^2x^i}{d\tau^2} &= m\frac{d}{d\tau}\left(\frac{\partial x^i}{\partial x^{i'}}\frac{dx^{i'}}{d\tau}\right) \\ &= m\frac{\partial x^i}{\partial x^{i'}}\left(\frac{d^2x^{i'}}{d\tau^2}+\frac{\partial x^{i'}}{\partial x^l}\frac{\partial^2 x^l}{\partial x^{j'}\partial x^{k'}}\frac{dx^{j'}}{d\tau}\frac{dx^{k'}}{d\tau}\right) \\ &= m\frac{\partial x^i}{\partial x^{i'}}\left(\frac{d^2x^{i'}}{d\tau^2}+\Gamma_{j'k'}{}^{i'}\frac{dx^{j'}}{d\tau}\frac{dx^{k'}}{d\tau}\right) = 0, \end{aligned} \tag{3.1}$$

with the Christoffel symbol[2]

$$\Gamma_{j'k'}{}^{i'} \stackrel{*}{=} \frac{\partial x^{i'}}{\partial x^i}\frac{\partial^2 x^i}{\partial x^{j'}\partial x^{k'}} = p^{i'}{}_i\frac{\partial}{\partial x^{j'}}p^i{}_{k'}\,. \tag{3.2}$$

Note that the coordinate system $\{x^i\}$ is a Cartesian one.

Consequently, in a non-inertial frame, the equation of motion reads

$$m\frac{d^2x^{i'}}{d\tau^2}+m\Gamma_{j'k'}{}^{i'}\frac{dx^{j'}}{d\tau}\frac{dx^{k'}}{d\tau}=0\,. \tag{3.3}$$

In this form the equation of motion is valid in any coordinate system. Apart from the term with the 2nd time derivative, there emerges a term bilinear in the velocity in which the Christoffel symbol features as a "guiding field". Since this term is proportional to the inertial mass m, it describes an inertial force, which is manifest in the coordinate system $\{x^{i'}\}$.

There is another way of writing Eq. (3.3). Define the covariant derivative ";" of the components of a four-vector $w^{i'}$ by

$$w^{i'}{}_{;j'} := w^{i'}{}_{,j'} + \Gamma_{j'k'}{}^{i'}\,w^{k'} = \nabla_{j'}\,w^{i'}\,, \tag{3.4}$$

and the absolute derivative of $w^{i'}$ with respect to the velocity field $u^{j'}$ of the observer by

$$\frac{D_{(u)}w^{i'}}{d\tau} := u^{j'}\,w^{i'}{}_{;j'} = \frac{dx^{j'}}{d\tau}w^{i'}{}_{;j'} = \frac{dw^{i'}}{d\tau} + u^{j'}\,\Gamma_{j'k'}{}^{i'}\,w^{k'}\,. \tag{3.5}$$

2 The Christoffel symbol is the analogue of a potential in a gauge theory. The relation in Eq. (3.2) corresponds to the "pure gauge" case, in other words, it represents the case of a vanishing gauge field strength, see Eq. (3.13).

Then the equation of motion can alternatively be written as

$$m\,\frac{D_{(u)}u^{i'}}{d\tau} = 0\,. \tag{3.6}$$

Hence, in order to get the equation of motion in an arbitrary coordinate system, take the equation of motion in Cartesian coordinates and replace $d/d\tau$ by $D/d\tau$, i.e. the ordinary by the absolute derivative. This "comma goes to semicolon rule" (see [94]) is valid quite generally. Later, we will use it in order to derive the Dirac equation for a non-inertial observer.

It is not very practical to use the relation in Eq. (3.2) of the Christoffel symbol for explicit computations. In order to obtain a more general expression, we differentiate the expression in Eq. (2.2) and find

$$g_{i'j',k'} = (p^{i}{}_{i'k'}p^{j}{}_{j'} + p^{i}{}_{i'}p^{j}{}_{j'k'})o_{ij}\,. \tag{3.7}$$

We want to resolve this equation with respect to one of its terms on the right hand side. Therefore we form the linear combination

$$\frac{1}{2}(g_{i'j',k'} + g_{j'k',i'} - g_{k'i',j'}) = p^{j}{}_{j'}p^{i}{}_{k'i'}o_{ij} = g_{j'l'}p^{l'}{}_{i}p^{i}{}_{k'i'}\,. \tag{3.8}$$

After raising the index j' and, remembering Eq. (3.2), we find for the Christoffel symbol

$$\Gamma_{j'k'}{}^{i'} \equiv \left\{ \begin{matrix} i' \\ j'k' \end{matrix} \right\} := \frac{1}{2}g^{i'l'}\left(\frac{\partial g_{j'l'}}{\partial x^{k'}} + \frac{\partial g_{k'l'}}{\partial x^{j'}} - \frac{\partial g_{j'k'}}{\partial x^{l'}}\right) = \Gamma_{k'j'}{}^{i'}\,. \tag{3.9}$$

Consequently, the Christoffel symbol is symmetric in its lower indices and has a dimension of $length^{-1}$. Evidently in this formula, in contrast to Eq. (3.2), the Cartesian coordinates do not occur any more.

If we introduce still another curvilinear coordinate system $\{x^{i''}\}$, the Christoffel symbol transforms inhomogeneously according to

$$\Gamma_{j''k''}{}^{i''} = p^{i''}{}_{i'}\,p^{j'}{}_{j''}\,p^{k'}{}_{k''}\,\Gamma_{j'k'}{}^{i'} + p^{i''}{}_{i'}\,\partial_{j''}\,p^{i'}{}_{k''}\,, \tag{3.10}$$

as can be verified by using Eq. (3.9) or, more conveniently, Eq. (3.4). Compare in this context the book of Schouten [117], in which the Ricci calculus is displayed in detail and the advantage of the kernel index method explained. Note that Eq. (3.2) is a special case of this formula, provided one identifies the $\{x^{i'}\}$ coordinates with Cartesian ones. Such a transformation formula is familiar from the behavior of a connection in gauge theory under gauge transformations.

Up to now we have studied the equation of motion which results from a transformation from Cartesian to arbitrary coordinates. This procedure should be invertible. Given the equation of motion in arbitrary coordinates, it should be possible to transform back to Cartesian coordinates, in which the Christoffel symbol vanishes over the whole spacetime.

Let $\{x^{i'}\}$ denote an arbitrary coordinate system. What are the conditions on the coordinate transformation $x^{i''} = x^{i''}(x^{i'})$ which lead to a vanishing Christoffel symbol everywhere? From the transformation law in Eq. (3.10) we find

$$\Gamma_{i''j''}{}^{k''} \stackrel{*}{=} 0 \quad \Longleftrightarrow \quad \Gamma_{i'j'}{}^{k'} \stackrel{*}{=} \frac{\partial^2 x^{k''}}{\partial x^{i'}\partial x^{j'}}\frac{\partial x^{k'}}{\partial x^{k''}} . \tag{3.11}$$

Hence $x^{i''}$ is a Cartesian coordinate system and we can drop the double primes. Accordingly, one has to solve the 2nd order partial differential equation

$$\frac{\partial^2 x^k}{\partial x^{i'}\partial x^{j'}} = \Gamma_{i'j'}{}^{k'}\frac{\partial x^k}{\partial x^{k'}} . \tag{3.12}$$

The integrability condition of Eq. (3.12) reads

$$\begin{aligned} R_{i'j'k'}{}^{l'} &:= \partial_{i'}\Gamma_{j'k'}{}^{l'} - \partial_{j'}\Gamma_{i'k'}{}^{l'} + \Gamma_{i'm'}{}^{l'}\Gamma_{j'k'}{}^{m'} - \Gamma_{j'm'}{}^{l'}\Gamma_{i'k'}{}^{m'} \\ &= 0 . \end{aligned} \tag{3.13}$$

It is identically fulfilled. The $R_{ijk}{}^{l}$'s, which have the dimension of $length^{-2}$, are the components of the Riemannian *curvature* tensor. They obey a tensorial transformation law with respect to the coordinate transformations in Eq. (2.1). Since we are still in the *flat* Minkowskian spacetime of SR, the curvature is identically zero.

	Einstein's approach	**gauge approach**
elementary object in SR	mass point m	Dirac spinor $\Psi(x)$
inertial frame	Cart.coord. x^i $ds^2 \stackrel{*}{=} o_{ij}\, dx^i dx^j$	orthon. hol. tetrads $e_\alpha = \delta^i_\alpha \partial_i, \quad e_\alpha \cdot e_\beta = o_{\alpha\beta}$
force-free motion in IF	$\dot{u}^i \stackrel{*}{=} 0$	$(i\gamma^i \partial_i - m)\Psi \stackrel{*}{=} 0$
non-inertial frame	arb. curvilinear coord. $x^{i'}$	orthon. anhol. tetrads $e_\alpha = e^i{}_\alpha \partial_i$ coframe $\vartheta^\alpha = e_i{}^\alpha dx^i$
force-free motion in NIF	$\dot{u}^i + \left\{ {i \atop jk} \right\} u^j u^k = 0$	$[i\gamma^\alpha e^i{}_\alpha(\partial_i + \Gamma_i) - m]\Psi = 0$ $\Gamma_i := \frac{1}{2}\Gamma_i^{\beta\gamma} \rho_{\beta\gamma}$
non-inertial objects	$\left\{ {i \atop jk} \right\}$ 40	$\vartheta^\alpha, \quad \Gamma^{\alpha\beta} = -\Gamma^{\beta\alpha}$ 16 + 24
constraints in SR	$R(\partial\{\}, \{\}) = 0$ 20	$T(\partial e, e, \Gamma) = 0, \; R(\partial\Gamma, \Gamma) = 0$ 24 + 36
global IF	$g_{ij} \stackrel{*}{=} o_{ij}\,, \left\{ {i \atop jk} \right\} \stackrel{*}{=} 0$	$(e_i{}^\alpha, \Gamma_i^{\alpha\beta}) \stackrel{*}{=} (\delta_i^\alpha, 0)$
switch on gravity	$R \neq 0$ *Riemann*	$T \neq 0, \quad R \neq 0$ *Riemann – Cartan*
local IF	$g_{ij}\vert_P = o_{ij}\,, \left\{ {i \atop jk} \right\}\vert_P = 0$	$(e_i{}^\alpha, \Gamma_i^{\alpha\beta})\vert_P = (\delta_i^\alpha, 0)$
field equations	$Ric - \frac{1}{2}tr(Ric) \sim mass$	$Ric - \frac{1}{2}tr(Ric) \sim mass$ $Tor + 2\; tr(Tor) \sim spin$

Table 1: Einstein's approach to GR as compared to the gauge approach using a Dirac matter field referred to a local frame. IF means inertial frame, NIF non-inertial frame. So far, we have discussed Einstein's approach only up to the "global" IF. Gravity will be switched on in Sect. 10. The gauge approach will be touched upon in Sect. 17.

I.4 Local Reference Frame of an Observer

In a laboratory on earth, we are in a frame which is subject to the gravitational acceleration $\boldsymbol{g}$ caused by the earth and which rotates relative to the global Copernican inertial frame of the planetary system. According to Einstein's equivalence principle, which we will discuss further down, the gravitational acceleration can be locally simulated by an equal, but opposite acceleration $\boldsymbol{a} = -\boldsymbol{g}$ of the laboratory with respect to the Copernican frame.

In order to understand physical effects in an earthbound laboratory, it is then appropriate to study, in a *flat* Minkowski spacetime M_4, an *accelerated* and *rotating*, that is, a non-inertial frame of reference. It will be characterized by its 3-vectors of acceleration $\boldsymbol{a}$ and angular velocity $\boldsymbol{\omega}$ with respect to the Copernican frame. Let us start to develop a transparent mathematical formalism for such an accelerated and rotating observer.

The global Copernican frame can be covered by Cartesian coordinates x^i. Arbitrary curvilinear coordinates will again be denoted by $x^{i'}$. Then the line element reads

$$ds^2 = g_{i'j'}\, dx^{i'} dx^{j'} \stackrel{*}{=} o_{ij}\, dx^i dx^j\,. \tag{4.1}$$

The accelerated observer's worldline C in the M_4 with $x^{i'} = x^{i'}(\tau)$ yields its velocity and its acceleration, respectively:

$$u^{i'} := \frac{dx^{i'}}{d\tau} \quad \text{and} \quad a^{i'} := \frac{D_{(u)} u^{i'}}{d\tau} = \frac{D^2_{(u)} x^{i'}}{d\tau^2}\,. \tag{4.2}$$

The velocity vector $u^{i'}$ is the *tangent* vector of the world line, whereas the acceleration $a^{i'}$ points into the *normal* direction. The simple reference frame of Sect. 1 has been formulated with the help of the tangent vector field $u^{i'}$.

Following basically E.Cartan [17], we introduce a *local frame* which will be represented by a set of four linearly independent vectors, one timelike and three spacelike vectors, a so-called tetrad or vierbein $e_\alpha = \{e_{\hat{0}}, e_{\hat{1}}, e_{\hat{2}}, e_{\hat{3}}\}$. This tetrad can be decomposed with respect to the tangent vectors $\partial_{i'}$ along the coordinate lines, the *natural* basis, according to

$$e_\alpha = e^{i'}{}_\alpha\, \partial_{i'} = e^i{}_\alpha \partial_i\,, \tag{4.3}$$

where $e^{i'}{}_\alpha$ and $e^i{}_\alpha$ are the tetrad components with respect to curvilinear or Cartesian coordinates, respectively. It is convenient and appropriate in an M_4 to require that the tetrad field e_α be *orthonormal* anywhere:

$$e_\alpha \cdot e_\beta := g_{i'j'}\, e^{i'}{}_\alpha e^{j'}{}_\beta = o_{ij}\, e^i{}_\alpha e^j{}_\beta = o_{\alpha\beta} = \mathrm{diag}\,(+,-,-,-). \tag{4.4}$$

We will equip the observer O' of C with such a tetrad. From now on, e_α *will always be chosen orthonormal.* Such a reference frame e_α is usually introduced in general relativity (GR), if one discusses experiments in an "Einstein elevator", whether the elevator is falling freely or not. But in GR, the vierbein

e_α does not aquire a fundamental meaning from the point of view of the underlying principles of the theory. In contrast, if one wants to describe fermionic fields in accelerated frames in the flat M_4 of SR, such a local frame is already an absolute necessity: Spinors, describing fermionic fields, transform according to a two-valued representation of the Lorentz group and as such are defined with respect to an orthonormal, oriented frame e_α. Consequently, since we are interested in fermionic fields, we cannot manage without the local orthonormal frame field e_α, the "simple" reference frame u^i of Sect. 1 being insufficient.

If a frame e_α, a vector basis, is given, we can always introduce a *coframe* ϑ^β, a 1-form basis, according to

$$e_\alpha \rfloor \vartheta^\beta = \delta_\alpha^\beta\,, \tag{4.5}$$

where $\rfloor$ denotes the *interior product* (or contraction) of a vector with a 1-form. If we expand the coframe

$$\vartheta^\beta = e_{i'}{}^\beta dx^{i'}\,, \tag{4.6}$$

the relation in Eq. (4.5) can be rewritten in terms of the respective components as follows: $e^{i'}{}_\alpha\, e_{i'}{}^\beta = \delta_\alpha^\beta$. Accordingly, frame e_α and coframe ϑ^β are dual to each other. Since we chose e_α to be orthonormal, see Eq. (4.4), the same is true, via Eq. (4.5), for ϑ^β.

In performing measurements in our local frame e_α, we have to refer all quantities at O' to the frame e_α. The frame components of a velocity vector $u^{i'}$ would read

$$u^\alpha := e_{i'}{}^\alpha\, u^{i'}\,, \text{ etc.} \tag{4.7}$$

This prescription extends analogously to all other tensor fields.

In treating connections, however, that is the Christoffel symbol $\Gamma_{i'j'}{}^{k'}$ in our M_4, one has to be careful. The frame components of the Christoffel read (see Schouten [117])

$$\Gamma_{\alpha\beta}{}^\gamma = e^{i'}{}_\alpha\, e^{j'}{}_\beta\, e_{k'}{}^\gamma\, \Gamma_{i'j'}{}^{k'} - e^{i'}{}_\alpha\, e^{j'}{}_\beta\, \partial_{i'} e_{j'}{}^\gamma\,, \tag{4.8}$$

compare the related formula in Eq. (3.10). At times, this relation is erroneously regarded as a new requirement for the covariant constancy of $e_{j'}{}^\gamma$ with respect to the combined Christoffel connection $\Gamma_{i'j'}{}^{k'}$ and spin connection $\Gamma_{\alpha\beta}{}^\gamma$, i.e.

$$\partial_{i'} e_{j'}{}^\gamma - \Gamma_{i'j'}{}^{k'}\, e_{k'}{}^\gamma + \Gamma_{\alpha\beta}{}^\gamma\, e_{i'}{}^\alpha\, e_{j'}{}^\beta = 0\,, \tag{4.9}$$

a relation which we find from Eq. (4.8). Let us substitute the explicit form of the Christoffel symbol in Eq. (3.9) into Eq. (4.8). Then, by some extensive reordering of terms and remembering the orthonormality $g_{\alpha\beta} \overset{*}{=} o_{\alpha\beta}$, see Eq. (4.4), we can express the frame components of the connection,

$$\Gamma_{\alpha\beta\gamma} := g_{\gamma\delta}\Gamma_{\alpha\beta}{}^\delta \overset{*}{=} \frac{1}{2}(-C_{\alpha\beta\gamma} + C_{\beta\gamma\alpha} - C_{\gamma\alpha\beta})\,, \tag{4.10}$$

in terms of the object of anholonomicity

$$C_{\alpha\beta}{}^{\delta} := 2e^{i'}{}_{\alpha}\, e^{j'}{}_{\beta}\, \partial_{[i'} e_{j']}{}^{\delta} = -C_{\beta\alpha}{}^{\delta}\,, \quad C_{\alpha\beta\gamma} := g_{\gamma\delta} C_{\alpha\beta}{}^{\delta}\,. \tag{4.11}$$

Note that

$$\Gamma_{\alpha(\beta\gamma)} \stackrel{*}{=} 0 \quad \text{or} \quad \Gamma_{\alpha\beta\gamma} \stackrel{*}{=} -\Gamma_{\alpha\gamma\beta}\,, \tag{4.12}$$

i.e., the "anholonomic" or tetrad connection in the M_4 has only 24 independent components. The inversion of Eq. (4.10) reads

$$C_{\alpha\beta\gamma} \stackrel{*}{=} -2\Gamma_{[\alpha\beta]\gamma}\,. \tag{4.13}$$

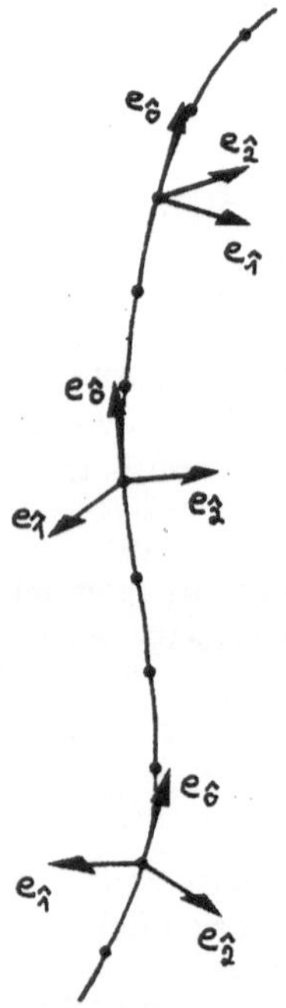

Fig. 6. The local orthonormal reference frame $e_\alpha(\tau)$ of an observer O', cf. [94]. Its clock is represented by the zeroths leg $e_{\hat{0}}$ of the tetrad, which is tangential to its world line, whereas the e_{Ξ}'s are the (mutually orthogonal) unit measuring rods.

The observer O' at time τ has the velocity $u^{i'}(\tau) = \{c, 0, 0, 0\}$. Therefore its *standard clock* can be represented by the zeroth leg $e_{\hat{0}}(\tau)$ of the tetrad, provided $e_{\hat{0}}(\tau)$ is identified with the observer's velocity $u^{i'}(\tau)$. We will always apply this identification in future. Then the orthogonal spacelike triad of the vectors e_{Ξ} at O', with $\Xi = 1, 2, 3$, span a spacelike hyperplane which is normal to $e_{\hat{0}}(\tau)$, see Fig. 6. Accordingly,

$$e_\alpha(\tau) = \left\{ \frac{u^{i'}(\tau)}{c} \partial_{i'}, e_{\hat{1}}(\tau), e_{\hat{2}}(\tau), e_{\hat{3}}(\tau) \right\}\,. \tag{4.14}$$

I.5 Generalized Fermi Derivative of a Vector

Assume that the local frame e_α in Eq. (4.14) evolves along a timelike curve C parametrized by proper time τ. We want to introduce a transport of that frame along C, which fulfills certain requirements. Since a linear connection underlies the M_4, it is near at hand to define a transport by means of the connection. This is done, if one defines the absolute derivative of a vector by

$$\frac{D_{(u)}w^i}{d\tau} := \frac{d_{(u)}w^i}{d\tau} + u^k\, \Gamma_{kj}{}^i\, w^j \,. \tag{5.1}$$

Parallel transport along C requires the vanishing of the absolute derivative. This transport, however, is not sufficient for the frame in Eq. (4.14), since it does *not* carry the timelike leg $e_{\hat{0}}(\tau)$ into $e_{\hat{0}}(\tau + d\tau)$. Rather, the absolute derivative of its components $e^i{}_{\hat{0}} = u^i$ yields, by definition, the acceleration of the observer

$$\frac{D_{(u)}e^i{}_{\hat{0}}}{d\tau} = \frac{D_{(u)}u^i}{d\tau} =: a^i \,. \tag{5.2}$$

Accordingly, the connection has to be supplemented by an additional tensorial piece. We introduce the generalized Fermi derivative of a vector by

$$\frac{\tilde{F}_{(u)}w^i}{d\tau} := \frac{D_{(u)}w^i}{d\tau} + \Omega_j{}^i(u)\, w^j = \frac{d_{(u)}w^i}{d\tau} + \Big(u^k\, \Gamma_{kj}{}^i + \Omega_j{}^i(u)\Big) w^j \,, \tag{5.3}$$

where the additional tensorial piece $\Omega_j{}^i$, which has dimension $time^{-1}$, in analogy to the connection piece in the parenthesis, is assumed to depend *linearly* on the velocity u^i.

Similarly as for the absolute derivative, we postulate the Leibniz rule to hold for $\tilde{F}_{(u)}/d\tau$. Furthermore, for a scalar field Φ along C, we postulate

$$\frac{\tilde{F}_{(u)}\Phi}{d\tau} := \frac{d_{(u)}\Phi}{d\tau} \,, \tag{5.4}$$

that is, the generalized Fermi derivative reduces to the ordinary derivative in this special case. An arbitrary covariant vector v_i, if transvected with w^i, yields a scalar $\Phi = w^i v_i$. Accordingly,

$$\frac{\tilde{F}_{(u)}(w^i v_i)}{d\tau} = \frac{\tilde{F}_{(u)}w^i}{d\tau}\, v_i + w^i \frac{\tilde{F}_{(u)}v_i}{d\tau} = \frac{d_{(u)}(w^i v_i)}{d\tau} = \frac{D_{(u)}(w^i v_i)}{d\tau} \,. \tag{5.5}$$

If we substitute this into Eq. (5.3), we find

$$\frac{\tilde{F}_{(u)}v_i}{d\tau} = \frac{D_{(u)}v_i}{d\tau} - \Omega_i{}^j(u)\, v_j \,. \tag{5.6}$$

Like for the absolute derivative, the laws in Eqs. (5.3) and (5.6) can be generalized for arbitrary tensors. For a 2nd rank covariant tensor t_{ij} we have, for instance,

$$\frac{\tilde{F}_{(u)}t_{ij}}{d\tau} = \frac{D_{(u)}t_{ij}}{d\tau} - \Omega_i{}^k\, t_{kj} - \Omega_j{}^k\, t_{ik}\,. \tag{5.7}$$

So far, for $\tilde{F}_{(u)}/d\tau$ we have only required the properties a derivative should have. More specifically, however, we want a transport in accordance with Eq. (4.14), which, like a Lie derivative, carries $u^i(\tau)$ into $u^i(\tau + d\tau)$:

$$\frac{\tilde{F}_{(u)}u^i}{d\tau} \overset{!}{=} 0 = \frac{D_{(u)}u^i}{d\tau} + \Omega_j{}^i\, u^j = a^i + \Omega_j{}^i\, u^j\,, \tag{5.8}$$

$$\text{or} \quad \Omega_j{}^i\, u^j = -a^i\,. \tag{5.9}$$

Obviously, the 16 components of $\Omega_j{}^i$ are somewhat constrained by these four relations.

Although we restrict our considerations to the M_4, we only have to require the metric-compatibility of the spactime connection[3]:

$$\frac{D_{(u)}g_{ij}}{d\tau} = 0\,. \tag{5.10}$$

Since $\tilde{F}_{(u)}/d\tau$ should not disturb the metric-compatibility, we postulate

$$\frac{\tilde{F}_{(u)}g_{ij}}{d\tau} \overset{!}{=} 0\,. \tag{5.11}$$

Using Eqs. (5.7) and (5.10), this yields

$$\Omega^{(ij)} = 0\,, \tag{5.12}$$

i.e., Ω^{ij} has at most six independent components.

Keeping in mind that $\Omega^{ij} = -\Omega^{ji}$ should be linear in u^i, the most general ansatz,

$$\Omega^{ij} = \frac{2}{c^2} b^{[i}u^{j]} + \eta^{ijkl}\, \frac{u_k}{c}\omega_l\,, \tag{5.13}$$

contains two vectors b^i and ω_l. The constraint in Eq. (5.9) identifies b^i as acceleration:

$$\Omega^{ij} = \overset{FW}{\Omega}{}^{ij} + \overset{R}{\Omega}{}^{ij} \quad \text{with} \quad \overset{FW}{\Omega}{}^{ij} := \frac{2}{c^2} a^{[i}u^{j]}\,, \quad \overset{R}{\Omega}{}^{ij} := \eta^{ijkl} \frac{u_k}{c}\omega_l\,. \tag{5.14}$$

Because of the constraint

$$a^i u_i = 0\,, \tag{5.15}$$

the Fermi-Walker part $\overset{FW}{\Omega}{}^{ij}$ depends on three independent components of a^i and describe Lorentz *boosts*, as we will see later. As will be discussed in the

[3] Hence our considerations, leading finally to the generalized Fermi derivative in Eq. (5.17), remain valid in the Riemannian spacetime of GR as well as in the Riemann-Cartan spacetime of the Poincaré gauge theory.

next section, for the *spatial rotation* part $\overset{R}{\Omega}{}^{ij}$ three independent components are left over. Hence it is possibe to require similarly[4]

$$\omega^i u_i = 0 \,. \tag{5.16}$$

Summing up, we have found the *generalized Fermi derivative*

$$\frac{\tilde{F}_{(u)} w^i}{d\tau} := \frac{D_{(u)} w^i}{d\tau} + \Omega_j{}^i\, w^j = \frac{D_{(u)} w^i}{d\tau} + (\overset{FW}{\Omega}{}_j{}^i + \overset{R}{\Omega}{}_j{}^i) w^j \,. \tag{5.17}$$

It is straightforward to generalize Eq. (5.17) to anholonomic coordinates. The frame or anholonomic components of a vector read $w^\alpha := e_i{}^\alpha\, w^i$, furthermore $w^i = e^i{}_\alpha\, w^\alpha$. The generalized Fermi derivative $\tilde{F}_{(u)} w^\alpha / d\tau$, similarly as $\tilde{F}_{(u)} w^i / d\tau$, contains an absolute derivative, whose definition is standard,

$$\frac{D_{(u)} w^\alpha}{d\tau} := u^\beta \nabla_\beta w^\alpha = u^\beta\, e^i{}_\beta\, \nabla_i\, w^\alpha \,, \tag{5.18}$$

and an additional tensor piece which translates directly into anholonomic form. Therefore we get for the Fermi derivative of the frame components of a vector

$$\frac{\tilde{F}_{(u)} w^\alpha}{d\tau} = \frac{D_{(u)} w^\alpha}{d\tau} + \Omega_\beta{}^\alpha\, w^\beta \,. \tag{5.19}$$

I.6 Fermi-Walker Transport as a Standard of non-Rotation

If we put $\overset{R}{\Omega}{}_j{}^i = 0$ in Eq. (5.17), we recover the conventional Fermi derivative, denoted without a tilde,

$$\frac{F_{(u)} w^i}{d\tau} := \frac{D_{(u)} w^i}{d\tau} + \overset{FW}{\Omega}{}_j{}^i\, w^j \,. \tag{6.1}$$

We call a vector, and, more generally, a tensor, *Fermi-Walker transported* provided its Fermi derivative vanishes[5]: $F_{(u)} w^i / d\tau = 0$. However, if the additional rotational degree of freedom inherent in $\overset{R}{\Omega}{}_j{}^i$ is admitted, then we arrive at the *generalized* Fermi derivative $\tilde{F}_{(u)}/d\tau$. If this derivative vanishes, $\tilde{F}_{(u)}/d\tau = 0$, a tensor is "generalized Fermi transported".

In order to find the exact meaning of ω_i, we are going to study some properties of a vector z^i which is subject to a generalized Fermi transport[6]:

---

4 Multiply the expression in Eq. (5.13) by u^m and η_{rijm}. Then we find: $\omega_l = u_l(u^i\omega_i) + \frac{1}{2}\eta_{lijk}\Omega^{ij}u^k$. The first piece on the right hand side does not contribute to Eq. (5.13). Therefore, without restricting the generality of Eq. (5.13), we can put $u^i\omega_i = 0$.

5 According to [25], there is a relation between Fermi-Walker transport and the Berry phase in quantum mechanics. For experiments on the Aharonov-Casher effect see [19].

6 We follow here a presentation which was suggested to us by McCrea [85].

$$\frac{\tilde{F}_{(u)} z^i}{d\tau} = 0 \qquad \text{or} \qquad \frac{D_{(u)} z^i}{d\tau} = (\overset{FW}{\Omega}{}^i{}_j + \overset{R}{\Omega}{}^i{}_j) z^j \,. \tag{6.2}$$

Then

(a) if $z^i = u^i$, Eq. (6.2) is an identity;
(b) if z^i and w^i both satisfy Eq. (6.2), then $z^i w_i =$ constant along $\mathcal{C}$, therefore
  (i) if z^i is initially orthogonal to u^i it remains so, and
  (ii) if a tetrad is initially orthonormal and propagated in accordance with Eq. (6.2), it remains orthonormal;
(c) if z^i is orthogonal along $\mathcal{C}$, i.e. $z^i u_i = 0$, and $e^i{}_\alpha$ (with $e^i{}_{\hat{0}} = u^i/c$) is Fermi-Walker propagated along $\mathcal{C}$, then, at every event on $\mathcal{C}$, the three-vector $\boldsymbol{z}$ is rotating with *angular velocity* $\boldsymbol{\omega}$ relative to the orthonormal triad $\boldsymbol{e}_\Xi = \{e^i{}_\Xi\}$, where $\boldsymbol{\omega} = \omega^\Xi \boldsymbol{e}_\Xi$ and $\omega^\Xi = e_i{}^\Xi \omega^i$.

The items (a) and (b) are consequences of Eq. (5.8) and (5.11), respectively. The proof of (c) is a bit more involved: The components of $\boldsymbol{z}$ with respect to the Fermi-Walker propagated tetrad are

$$z^\alpha = e_i{}^\alpha z^i, \qquad \text{with} \qquad z^{\hat{0}} = 0 \,. \tag{6.3}$$

Hence

$$\begin{aligned} \frac{D_{(u)} z^\alpha}{d\tau} &= e_i{}^\alpha \frac{D_{(u)} z^i}{d\tau} + z^i \frac{D_{(u)} e_i{}^\alpha}{d\tau} \\ &= \overset{FW}{\Omega}{}^{ij} (e_i{}^\alpha z_j + e_j{}^\alpha z_i) + e_i{}^\alpha z_j \overset{R}{\Omega}{}^{ij} \\ &= e_i{}^\alpha z_j \overset{R}{\Omega}{}^{ij} = \overset{R}{\Omega}{}^{\alpha\beta} z_\beta \,. \end{aligned} \tag{6.4}$$

For $\alpha = \hat{0}$ we get $Dz^{\hat{0}}/d\tau = 0$, which we know already, and for $\alpha = \Xi$, using $\eta^{\hat{0}\Xi\Upsilon\Lambda} = \epsilon^{\Xi\Upsilon\Lambda}$ we get

$$\frac{D_{(u)} z^\Xi}{d\tau} = e_i{}^\Xi z_j \eta^{ijkl} e_k{}^{\hat{0}} \omega_l = -\epsilon^{\Xi\Upsilon\Lambda} \omega_\Upsilon z_\Lambda \,. \tag{6.5}$$

$$\text{Thus} \quad \frac{D_{(u)} \boldsymbol{z}}{d\tau} = \boldsymbol{\omega} \times \boldsymbol{z} \,, \tag{6.6}$$

which proves (c).

If $\omega^i = 0$, then $\boldsymbol{z}$ points in a constant direction with respect to the triad $\boldsymbol{e}_\Xi$. Such a vector can be realized by means of the spin of a *gyroscope*, see [130]. Furthermore, if $\mathcal{C}$ becomes a geodesic, then $\boldsymbol{z}$ does not rotate relative to a Cartesian coordinate system. Therefore the Fermi-Walker transport defines a standard of non-rotation for an observer who is accelerated. Accordingly, the matrix $\overset{R}{\Omega}{}^{\alpha\beta}$ generates a spatial rotation, as measured by means of the angular velocity $\boldsymbol{\omega}$, and $\overset{FW}{\Omega}{}^{\alpha\beta}$ a time-space "rotation", that is a boost, as measured by means of the acceleration $\boldsymbol{a}$. Consequently we can now understand the physical meaning of Eq. (6.2).

Each of the four vectors of a tetrad can be generalized Fermi transported. Then the analogue of Eq. (6.2) reads

$$\frac{D_{(u)} e^i{}_\alpha}{d\tau} = (\overset{FW}{\Omega}{}^i{}_j + \overset{R}{\Omega}{}^i{}_j) e^j{}_\alpha \,. \tag{6.7}$$

If we use Eq. (4.9), we can evaluate the left hand side and find

$$\overset{FW}{\Omega}{}_\alpha{}^\beta + \overset{R}{\Omega}{}_\alpha{}^\beta = -u^k \, \Gamma_{k\alpha}{}^\beta \,. \tag{6.8}$$

Therefore some components of the connection are determined by Eq. (5.14). Along $\mathcal{C}$, we have

$$\Gamma_{0\Xi\Lambda} = -\Gamma_{0\Lambda\Xi} = \epsilon_{\Xi\Lambda\Upsilon} \frac{\omega^\Upsilon}{c} \,, \quad \Gamma_{0\hat{0}\Xi} = -\Gamma_{0\Xi\hat{0}} = \frac{a_\Xi}{c^2} \,. \tag{6.9}$$

Here, again, the fact shows up in a most direct way that $\boldsymbol{a}$ and $\boldsymbol{\omega}$ generate Lorentz transformations of the local frame.

I.7 Hyperbolic Motion

As an explicit example, we will study the motion of a non-rotating observer who is uniformly accelerated in the x^1-direction only (Fig. 7). In the instantaneous rest frame (denoted by a zero) of the observer, the velocity and the acceleration are

$$u^{i'} \overset{\circ}{=} \{c, 0, 0, 0\} \quad \text{and} \quad a^{i'} \overset{\circ}{=} \{0, \boldsymbol{a}\} \,, \tag{7.1}$$

respectively, where $\boldsymbol{a}$ is the three-acceleration with norm $a := |\boldsymbol{a}|$, compare Eqs. (1.3) and (1.4). Hence we have

$$a_i a^i = -\boldsymbol{a} \cdot \boldsymbol{a} \,. \tag{7.2}$$

The norm a is called the *proper acceleration* of the observer (cf. Rindler [111]). Specifically, we have $a_{i'} \overset{\circ}{=} \{0, a, 0, 0\}$ with $a =$ constant.

Let us now determine the velocity of the observer. The acceleration is perpendicular to the velocity, hence we make the ansatz

$$a^i \sim \{u^1, u^0, 0, 0\} \,. \tag{7.3}$$

Using $a_i a^i = -a^2$ and $u_i u^i = c^2$, we get

$$a^i \overset{*}{=} \frac{du^i}{d\tau} = \frac{a}{c} \{u^1, u^0, 0, 0\} \,. \tag{7.4}$$

Alternatively, this equation could have been obtained from Eq. (6.2) for $z^i = u^i$ with $\boldsymbol{\omega} = 0$. The solution of Eq. (7.4) with initial condition $u^i|_{\tau=0} = \{c, 0, 0, 0\}$ is

$$u^i = c \left\{ \cosh \frac{a\tau}{c}, \sinh \frac{a\tau}{c}, 0, 0 \right\} . \tag{7.5}$$

Then

$$a^i = a\left\{\sinh\frac{a\tau}{c}, \cosh\frac{a\tau}{c}, 0, 0\right\}. \tag{7.6}$$

The worldline of the accelerated observer follows from the differential equation

$$\frac{dx^i}{d\tau} = u^i = c\left\{\cosh\frac{a\tau}{c}, \sinh\frac{a\tau}{c}, 0, 0\right\} \tag{7.7}$$

by integration. For the initial conditions $x^i|_{\tau=0} = \{0, c^2/a, 0, 0\}$, we obtain

$$x^i = \{ct, x, 0, 0\} = \frac{c^2}{a}\left\{\sinh\frac{a\tau}{c}, \cosh\frac{a\tau}{c}, 0, 0\right\}. \tag{7.8}$$

Thus, in a Minkowski diagram, the worldline of the accelerated observer is a hyperbola in the tx-plane (Fig. 8). One can infer from Fig. 8 that the accelerated observer has no causal relationship to the "walled up" part of spacetime. There are other parts from which the observer can only receive signals, but not send to, or vice versa.

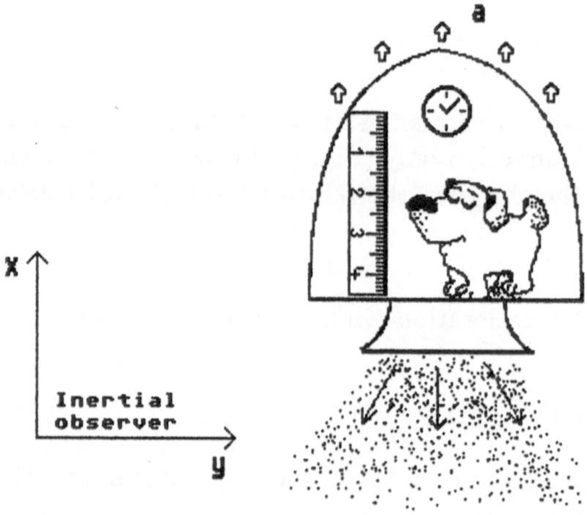

Fig. 7. Accelerated observer with standard clocks and rods.

The local frame in Eq. (4.14) of the observer is spanned by $e^i{}_{\hat 0} = u^i/c$, $e^i{}_{\hat 1} = a^i/|a^k|$ and by two trivial legs:

$$\begin{aligned} e_{\hat 0} &= e^i{}_{\hat 0}\,\partial_i = \cosh\frac{a\tau}{c}\,\partial_0 + \sinh\frac{a\tau}{c}\,\partial_1\,, \\ e_{\hat 1} &= e^i{}_{\hat 1}\,\partial_i = \sinh\frac{a\tau}{c}\,\partial_0 + \cosh\frac{a\tau}{c}\,\partial_1\,, \\ e_{\hat 2} &= e^i{}_{\hat 2}\,\partial_i = \partial_2\,, \\ e_{\hat 3} &= e^i{}_{\hat 3}\,\partial_i = \partial_3\,. \end{aligned} \tag{7.9}$$

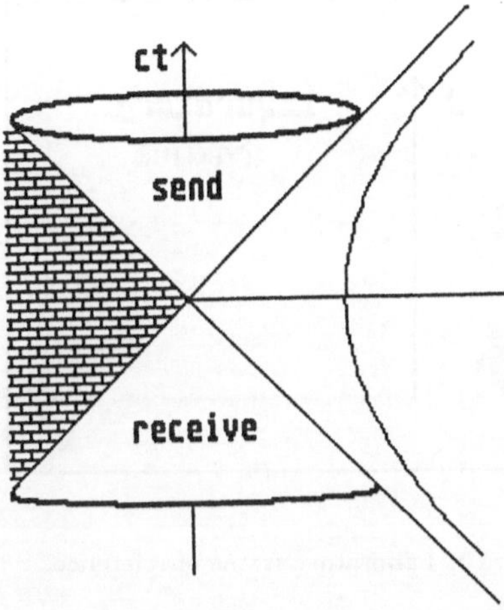

Fig. 8. Worldline of the uniformly accelerated observer.

By inversion we find the local coframe

$$\begin{aligned}
\vartheta^{\hat{0}} &= e_i{}^{\hat{0}}\, dx^i = \quad \cosh\frac{a\tau}{c}\, dx^0 - \sinh\frac{a\tau}{c}\, dx^1\,, \\
\vartheta^{\hat{1}} &= e_i{}^{\hat{1}}\, dx^i = -\sinh\frac{a\tau}{c}\, dx^0 + \cosh\frac{a\tau}{c}\, dx^1\,, \\
\vartheta^{\hat{2}} &= e_i{}^{\hat{2}}\, dx^i = \quad dx^2\,, \\
\vartheta^{\hat{3}} &= e_i{}^{\hat{3}}\, dx^i = \quad dx^3\,.
\end{aligned} \tag{7.10}$$

The time-space rotation generated by a is made manifest in Eqs. (7.9) and (7.10).

I.8 Laboratory Frame of Reference, Acceleration Length

Up to now, we considered a "pointlike" observer moving along a worldline C. We introduced Fermi-Walker transported frame as a standard of non-rotation. If we think, for example, of a laboratory on earth (Fig. 9), we certainly want to leave C and measure spatial separations and angles in our laboratory. Taking the floor and two neighboring walls as coordinate planes, we will find it convenient to introduce spatial Cartesian coordinates $x^{\underline{A}}$ in our non-inertial laboratory. Furthermore, the time will be measured by an observer located in the origin of the coordinate system (Fig. 10) and moving along C. A local frame in which such a coordinate system is used will be called a *laboratory frame of reference* (*"lab frame"*). In this lab frame, the three-vectors of acceleration $\boldsymbol{a}$ and angular velocity $\boldsymbol{\omega}$ can be measured by means of an accelerometer and a gyroscope, respectively.

Accordingly, we are going to construct coordinates based on our accelerated and rotating laboratory. Consider the worldline C given by $X^i = X^i(\tau)$, where

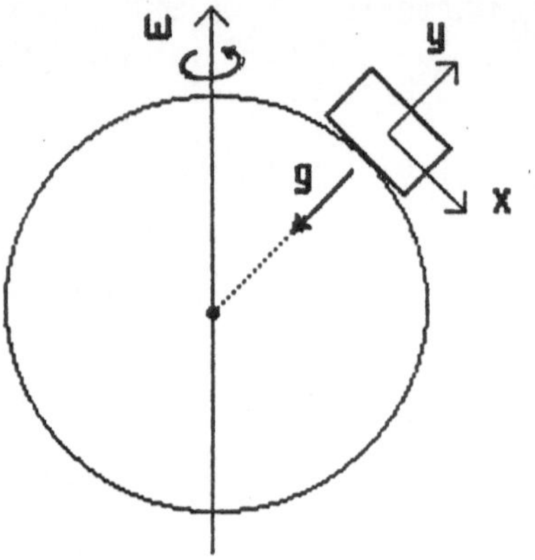

Fig. 9. Accelerated and rotating laboratory.

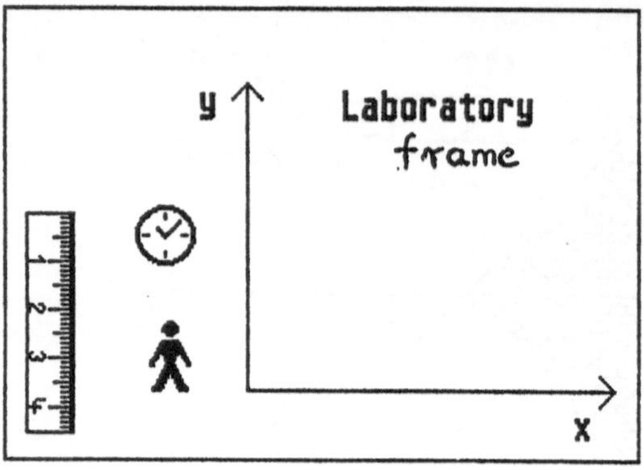

Fig. 10. Laboratory frame of reference.

τ is the proper time along C and X^i are Cartesian coordinates, and a local frame $e^i{}_\alpha$ on C (with $e^i{}_{\hat{0}} = u^i/c$, see Eq. (4.14)) propagating in accordance with Eq. (6.7):

$$\frac{D_{(u)}e^i{}_\alpha}{d\tau} = (\overset{FW}{\Omega}{}^i{}_j + \overset{R}{\Omega}{}^i{}_j)e^j{}_\alpha\,. \tag{8.1}$$

Let P be an event with Cartesian coordinates x^i. Assign another set of coordinates $\{x^{\bar{i}}\}$ to P as follows:

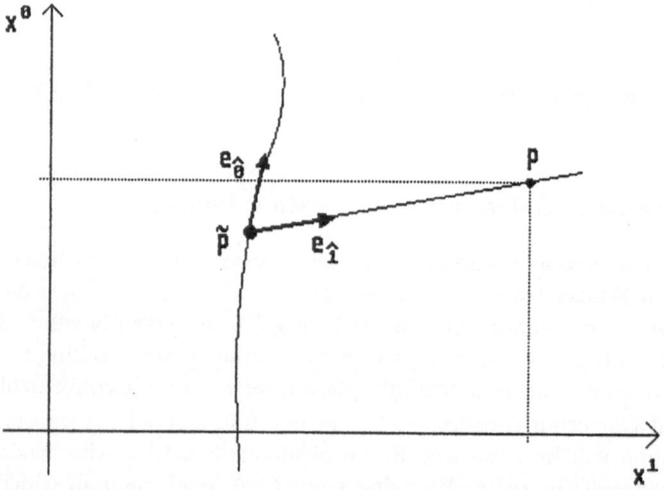

Fig. 11. Construction of the lab coordinate system for an accelerated and rotating observer.

Draw the line from P to C which cuts C orthogonally at $\tilde{P}$ (say). Let $\tilde{\tau}$ be the value of τ at $\tilde{P}$. Define the new *lab coordinates* $x^{\bar{i}}$ of P as[7]

[7] In this and the next section, spatial anholonomic components will be denoted by capital Latin letters $A, \ldots = \hat{1}, \hat{2}, \hat{3}$.

$$x^{\overline{0}} = c\tau, \qquad x^{\overline{A}} = [x^i - X^i(\tilde{\tau})]\, e_i{}^{\overline{A}}(\tilde{\tau}), \tag{8.2}$$

that is, the $x^{\overline{A}}$ are the components of the vector connecting $\tilde{P}$ and P as referred to the local triad of covectors $e^{\overline{A}}$. This construction is only possible in a Minkowski spacetime. In a Riemannian spacetime one has to take the spacelike geodesics emanating perpendicularly from $\tilde{P}$ for the construction of the lab coordinates thereby finding an expression which, for P sufficiently near to $\tilde{P}$, contains additionally higher order deviations caused by the curvature of spacetime. The transformation inverse to the one given in Eq. (8.2) is

$$\boxed{x^i = X^i(x^{\overline{0}}) + e^i{}_{\overline{A}}(x^{\overline{0}})\, x^{\overline{A}}\,.} \tag{8.3}$$

From Eqs. (8.3) and (8.1), together with $u^{\overline{A}} = 0$, we get

$$\begin{aligned} dx^i &= \frac{u^i}{c}\, dx^{\overline{0}} + \left(\frac{de^i{}_{\overline{A}}}{d\tau}\right) x^{\overline{A}}\, \frac{dx^{\overline{0}}}{c} + e^i{}_{\overline{A}}\, dx^{\overline{A}} \\ &= \{u^i + \frac{1}{c^2}(a^i u^j - a^j u^i) e_{j\overline{A}} x^{\overline{A}} + \eta^{ijkl} \frac{u_k}{c} \omega_l\, e_{j\overline{A}}\, x^{\overline{A}}\} \frac{dx^{\overline{0}}}{c} + e^i{}_{\overline{A}} dx^{\overline{A}} \\ &= \{u^i (1 + \frac{\boldsymbol{a}\cdot\overline{\boldsymbol{x}}}{c^2}) + \eta^{ijkl} \frac{u_k}{c} \omega_l e_{j\overline{A}} x^{\overline{A}}\} \frac{dx^{\overline{0}}}{c} + e^i{}_{\overline{A}} dx^{\overline{A}}\,. \end{aligned} \tag{8.4}$$

Hence

$$\begin{aligned} ds^2 = o_{ij}\, dx^i\, dx^j = \{(1 + \frac{\boldsymbol{a}\cdot\overline{\boldsymbol{x}}}{c^2})^2 - (\frac{\boldsymbol{\omega}}{c} \times \overline{\boldsymbol{x}})^2\}\, (dx^{\overline{0}})^2 \\ - 2 (\frac{\boldsymbol{\omega}}{c} \times \overline{\boldsymbol{x}})_{\overline{A}} dx^{\overline{0}} dx^{\overline{A}} + o_{\overline{AB}} dx^{\overline{A}} dx^{\overline{B}}\,, \end{aligned} \tag{8.5}$$

where $(\boldsymbol{\omega} \times \overline{\boldsymbol{x}})_{\overline{A}} = \epsilon_{\overline{ABC}}\, \omega^{\overline{B}}\, x^{\overline{C}}$ and $\boldsymbol{a} = a^{\overline{A}}\, e_{\overline{A}}$, $a^{\overline{A}} = e_i{}^{\overline{A}}\, a^i$.

By the framed formula in Eq. (8.3), we expressed the Cartesian coordinates of an event in terms of the lab coordinates. In Fig. 12 this is done for two different times at the worldine C. Which coordinates should we assign to the event Q? There are two possibilities because two $x^{\overline{1}}$-coordinate lines intersect. A similar situation arises in Fig. 8 where the lab coordinates cannot be extended into the walled up part.

This problem is inherent to any accelerated observer, and it can only be remedied by making the laboratory sufficiently small. The lab coordinate system $x^{\overline{i}}$ of Eq. (8.2) is only useful in the immediate vicinity of the laboratory observer. To get a more precise idea of the words "small" and "immediate vicinity", let us construct, as *simple example*, the coordinate system of the non-rotating observer of Sect. 7 who is uniformly accelerated merely in the 1-direction (Fig. 8). We insert Eq. (7.9) into Eq. (8.3). This yields for the hyperbolic motion

$$x^i = \left\{ (\frac{c^2}{a} + x^{\overline{1}}) \sinh \frac{a x^{\overline{0}}}{c},\, (\frac{c^2}{a} + x^{\overline{1}}) \cosh \frac{a x^{\overline{0}}}{c},\, x^{\overline{2}},\, x^{\overline{3}} \right\}. \tag{8.6}$$

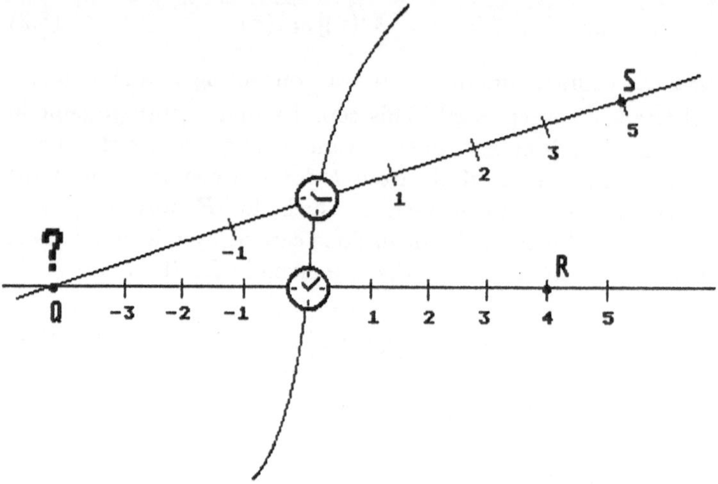

Fig. 12. Problems with the coordinate extension of the lab frame.

By differentiation we have

$$
\begin{aligned}
dx^0 &= (1+\frac{ax^{\overline{1}}}{c^2})\cosh\frac{ax^{\overline{0}}}{c^2}\,dx^{\overline{0}} + \sinh\frac{ax^{\overline{0}}}{c^2}\,dx^{\overline{1}}\,,\\
dx^1 &= (1+\frac{ax^{\overline{1}}}{c^2})\sinh\frac{ax^{\overline{0}}}{c^2}\,dx^{\overline{0}} + \cosh\frac{ax^{\overline{0}}}{c^2}\,dx^{1'}\,,\\
dx^2 &= dx^{\overline{2}}\,,\\
dx^3 &= dx^{\overline{3}}\,.
\end{aligned}
\tag{8.7}
$$

Substitute this into Eq. (7.10). Then we find the coframe in terms of lab coordinates as

$$
\begin{aligned}
\vartheta^{\hat{0}} &= (1+\frac{ax^{\overline{1}}}{c^2})\,dx^{\overline{0}}\,,\\
\vartheta^{A} &= dx^{\overline{A}}\,.
\end{aligned}
\tag{8.8}
$$

By inversion one gets the frame:

$$
\begin{aligned}
e_{\hat{0}} &= \frac{1}{1+ax^{\overline{1}}/c^2}\,\partial_{\overline{0}}\,,\\
e_A &= \partial_{\overline{A}}\,.
\end{aligned}
\tag{8.9}
$$

The metric in lab coordinates reads

$$
ds^2 = (1+\frac{ax^{\overline{1}}}{c^2})^2\,(dx^{\overline{0}})^2 - (dx^{\overline{1}})^2 - (dx^{\overline{2}})^2 - (dx^{\overline{3}})^2\,. \tag{8.10}
$$

We recognize that $x^{\overline{0}}$ coincides with the proper time only in those regions of spacetime in which the inequality

$$\frac{|x^{\overline{1}}|}{\ell} \ll 1, \qquad \text{with} \qquad \ell := \frac{c^2}{a}, \tag{8.11}$$

is fulfilled. We call ℓ the acceleration length. The problems with the intersection of the coordinate lines, as in Fig. 12, do not arise if Eq. (8.11) is fulfilled. Then the metric and the frame do not become singular. The geometrical meaning of the $g_{\overline{00}}$-component of the metric becomes evident in Figs. 3 and 4. It turns out to be the streching factor of the $x^{\overline{0}}$-coordinate line. Therefore any acceleration is linked with a length.

For the gravitational acceleration on earth and the "acceleration" of a neutron within the nucleus (because of the uncertainty principle) we find, respectively [94]:

$$\begin{aligned} g &= 981\ cm/s^2 &&\Rightarrow\ \ell \ll 1 \text{ lightyear}, \\ a_n &= 10^{28}\ g &&\Rightarrow\ \ell \ll 10^{-10}\ cm. \end{aligned} \tag{8.12}$$

Hence a lab on the earth is a safe place. Degeneracies of the type depicted in Fig. 12 occur only at distances far beyond reach. This may not so for the neutron in the nucleus. Caianiello [15] argues that there should be for a particle of mass m a maximal acceleration in nature, which is reached when the acceleration length ℓ is of the same order as the Compton wavelengh of the particle $\hbar/mc$, see also Mashhoon [81].

The degenerate behavior of the lab coordinates at too big distances shows also up in the Christoffel symbol belonging to the metric in Eq. (8.10). By straightforward computation we get

$$\left\{ \begin{matrix} \overline{1} \\ \overline{00} \end{matrix} \right\} = \frac{a}{c^2}\left(1 + \frac{a x^{\overline{1}}}{c^2}\right), \quad \left\{ \begin{matrix} \overline{0} \\ \overline{01} \end{matrix} \right\} = \frac{a/c^2}{1 + a x^{\overline{1}}/c^2}. \tag{8.13}$$

The equation of motion for the uniform acceleration in 1-direction takes the form

$$\frac{du^{\overline{i}}}{dT} + \Gamma_{\overline{kj}}{}^{\overline{i}} u^{\overline{k}} u^{\overline{j}} = 0, \tag{8.14}$$

where T denotes the proper time of the particle and $u^{\overline{i}} = (\gamma c, \gamma v^{\overline{A}})$. The zero component reads

$$\frac{d\gamma}{d\tau} = -2\, \frac{a/c^2}{1 + a x^{\overline{1}}/c^2} \gamma^2 v^{\overline{1}}. \tag{8.15}$$

Inserting this into the spatial components, we finally get

$$\frac{dv^{\overline{A}}}{d(x^{\overline{0}}/c)} = -a^{\overline{A}}\left(\frac{1 + a x^{\overline{1}}}{c^2}\right) + 2\, \frac{a/c^2}{1 + a x^{\overline{1}}/c^2} v^{\overline{1}} v^{\overline{A}}. \tag{8.16}$$

The leading term is well known from Newtonian mechanics.

I.9 The Accelerated and Rotating Lab Frame

If we first turn to a frame *uniformly rotating* around its 3-axis, then, in analogy to Eq. (7.9), one has

$$\begin{aligned} e_{\hat{0}} &= \partial_0 \,, \\ e_{\hat{1}} &= \cos(\omega\tau)\,\partial_1 + \sin(\omega\tau)\,\partial_2 \,, \\ e_{\hat{2}} &= -\sin(\omega\tau)\,\partial_1 + \cos(\omega\tau)\,\partial_2 \,, \\ e_{\hat{3}} &= \partial_3 \,. \end{aligned} \tag{9.1}$$

Introducing (rotating) lab coordinates, one arrives at

$$\begin{aligned} e_{\hat{0}} &= \partial_{\overline{0}} + \frac{\omega}{c}\left(x^{\overline{2}}\,\partial_{\overline{1}} - x^{\overline{1}}\,\partial_{\overline{2}}\right), \\ e_A &= \partial_{\overline{A}} \,. \end{aligned} \tag{9.2}$$

Hence also this special case can be explicitly solved.

Let us right away treat the general case displayed in Eq. (6.7) of an arbitarily accelerated and rotating lab frame and express this frame in terms of the lab coordinates. We follow again a deduction of McCrea [85]. Starting with the vector frame, we have

$$\begin{aligned} e_{\hat{0}} &= \frac{u^i}{c}\frac{\partial}{\partial x^i} = \frac{u^i}{c}\frac{\partial x^{\overline{k}}}{\partial x^i}\frac{\partial}{\partial x^{\overline{k}}} \\ &= e^{\overline{k}}{}_{\hat{0}}\,\frac{\partial}{\partial x^{\overline{k}}}\,, \quad \text{where } e^{\overline{k}}{}_{\hat{0}} = \frac{u^i}{c}\frac{\partial x^{\overline{k}}}{\partial x^i}\,, \end{aligned} \tag{9.3}$$

$$\begin{aligned} \text{and} \quad e_A &= e^i{}_A\,\frac{\partial}{\partial x^i} = e^i{}_A\,\frac{\partial x^{\overline{k}}}{\partial x^i}\frac{\partial}{\partial x^{\overline{k}}} \\ &= e^{\overline{k}}{}_A\,\frac{\partial}{\partial x^{\overline{k}}}\,, \quad \text{where } e^{\overline{k}}{}_A = e^i{}_A\,\frac{\partial x^{\overline{k}}}{\partial x^i}\,. \end{aligned} \tag{9.4}$$

One way to get $(\partial x^{\overline{k}}/\partial x^i)$ is to go back to Eq. (8.2) and use the result

$$\frac{\partial\tau}{\partial x^i} = (1 + \frac{\boldsymbol{a}\cdot\overline{\boldsymbol{x}}}{c^2})^{-1}\,u_i\,. \tag{9.5}$$

This yields

$$\frac{\partial x^{\overline{0}}}{\partial x^j} = (1 + \frac{\boldsymbol{a}\cdot\overline{\boldsymbol{x}}}{c^2})^{-1}\,\frac{u_j}{c^2}\,, \tag{9.6}$$

$$\begin{aligned} \frac{\partial x^{\overline{A}}}{\partial x^j} &= e_j{}^A + (1 + \frac{\boldsymbol{a}\cdot\overline{\boldsymbol{x}}}{c^2})^{-1}\eta_{ikmn}\,[x^i - X^i(\tau)]\,e^{kA}\,\frac{u^m}{c}\,\frac{\omega^n}{c}\,u_j \\ &= e_j{}^A + (1 + \frac{\boldsymbol{a}\cdot\overline{\boldsymbol{x}}}{c^2})^{-1}\,(\frac{\boldsymbol{\omega}}{c}\times\overline{\boldsymbol{x}})^A\,\frac{u_j}{c}\,. \end{aligned} \tag{9.7}$$

Substituting Eqs. (9.6) and (9.7) into Eqs. (9.3) and (9.4), we get the local reference frame, which we will apply to the Dirac equation [52] later on:

$$\boxed{\begin{aligned} e_{\hat{0}} &= \frac{1}{1+\boldsymbol{a}\cdot\overline{\boldsymbol{x}}/c^2}\left[\partial_{\overline{0}} - \left(\frac{\boldsymbol{\omega}}{c}\times\overline{\boldsymbol{x}}\right)^{\overline{B}}\partial_{\overline{B}}\right], \\ e_A &= \partial_{\overline{A}}. \end{aligned}} \tag{9.8}$$

For completeness we also display the coframe, that is, the 1-form basis, which one finds by inverting Eq. (9.8):

$$\begin{aligned} \vartheta^{\hat{0}} &= (1+\frac{\boldsymbol{a}\cdot\overline{\boldsymbol{x}}}{c^2})\, dx^{\overline{0}} = N dx^{\overline{0}}, \\ \vartheta^A &= dx^{\overline{A}} + \left(\frac{\boldsymbol{\omega}}{c}\times\overline{\boldsymbol{x}}\right)^{\overline{A}} dx^{\overline{0}} = dx^{\overline{A}} + N^{\overline{A}} dx^{\overline{0}}. \end{aligned} \tag{9.9}$$

In the (3 + 1)-decomposition of spacetime, N and $N^{\overline{A}}$ are known as *lapse function* and *shift vector*, respectively. The metric we gave already in Eq. (8.5). In *curved* spacetime, coupled inertial and gravitational effects occur in the 3rd order approximation of the metric; for related work, see [26,77,78,98,99, 100,119].

Starting with the coframe, we can read off the connection coefficients (for vanishing torsion) by using Cartan's 1st structure equation $d\vartheta^\alpha = -\Gamma_\beta{}^\alpha \wedge \vartheta^\beta$ with $\Gamma_\beta{}^\alpha = \Gamma_{\bar{i}\beta}{}^\alpha\, dx^{\bar{i}}$. However, by construction, the connection projected in spacelike directions vanishes. We have spatial Cartesian lab coordinates, after all. Conseqently, the connection coefficients displayed in Eq. (6.9) are the only non-vanishing ones:

$$\begin{aligned} \Gamma_{\overline{0}\hat{0}A} &= -\Gamma_{\overline{0}A\hat{0}} = \frac{a_A}{c^2}, \\ \Gamma_{\overline{0}AB} &= -\Gamma_{\overline{0}BA} = \epsilon_{ABC}\frac{\omega^C}{c}. \end{aligned} \tag{9.10}$$

It is important to note that the first index is holonomic, whereas the 2nd and the 3rd indices are anholonomic. If we transform the first index, by means of the frame coefficients $e^{\bar{i}}{}_\alpha$, into an anholonomic one, then we find the totally anholonomic connection coefficients as follows:

$$\begin{aligned} \Gamma_{\hat{0}\hat{0}A} &= -\Gamma_{\hat{0}A\hat{0}} = \frac{a_A/c^2}{1+\boldsymbol{a}\cdot\overline{\boldsymbol{x}}/c^2}, \\ \Gamma_{\hat{0}AB} &= -\Gamma_{\hat{0}BA} = \frac{\epsilon_{ABC}\,\omega^C/c}{1+\boldsymbol{a}\cdot\overline{\boldsymbol{x}}/c^2}. \end{aligned} \tag{9.11}$$

These connection coefficients will enter the Dirac equation referred to an arbitrary local frame in Sect. 17.

We used the EXCALC package on exterior differential forms [120] of the computer algebra system REDUCE, compare also the lectures of McCrea [84] and [128], for checking the correctness of the connection. Clearly, according to Eq. (4.13), the anholonomicity coefficients are trivially related to the connection. Hence we will not display them explicitly.

I.10 Equivalence Principle for Macroscopic Matter

Up to now we have considered a free classical point particle. Its only inherent property of importance seemed to be its inertial mass $m_{\text{inert}} (\equiv m)$. But already in Newton's mechanics, there enter two distinct types of mass (see [111]): Besides the inertial mass, which occurs in Newton's second law and measures the particle's resistance to acceleration, the *gravitational mass* occurs in Newton's law of attraction and can be regarded as the *gravitational charge* of a particle.

According to the attraction law, the gravitational field, produced by a fixed particle, is given by

$$\boldsymbol{g} = -\nabla\varphi = -G\frac{M_{\text{grav}}}{|\boldsymbol{R}|^3}\boldsymbol{R}\,. \tag{10.1}$$

Its gravitational potential $\varphi = -GM_{\text{grav}}/|\boldsymbol{R}|$ is singular at the position of the particle $\boldsymbol{R} = 0$. The gravitational force which this particle exerts on another particle with inertial mass m_{inert} and gravitational mass m_{grav}, is given by Newton's second law

$$\boldsymbol{F} = m_{\text{grav}}\,\boldsymbol{g}\,. \tag{10.2}$$

This law cannot be compatible with the foundations of special relativity because the principle of action at a distance is invoked in Eqs. (10.1) and (10.2). The equations of motion are

$$m_{\text{inert}}\frac{d^2\boldsymbol{x}}{dt^2} = m_{\text{grav}}\,\boldsymbol{g}\,. \tag{10.3}$$

In principle, the path of different particle in a gravitational field could be dependent on their composition (even on the nuclear scale). The *weak equivalence principle* asserts, however, that the path does only depend on the starting point and the velocity at that point, i.e. that all particles with the same initial conditions take the same path. This principle of the "universality of free fall" (cf. [134]), however, has to be tested experimentally:

Pendulum experiments by Newton and Bessel as well as the more precise torsion balance experiments by Eötvös, Dicke, Braginski and coworkers (c.f. [94]) have proved the universality of the free fall towards the earth and the sun with a precision up to $1 : 10^{12}$. A recent reanalysis of Eötvös results by Fischbach [33] seems to indicate that a very weak dependence of gravity on the baryon number of the test bodies may be present and has led to speculations on a "fifth force"[8]. However, a replication of the experiment alleged to Galilei with the aid of laser interferometry [103] has reconfirmed that the universality of free fall is better than 5×10^{-10}. Also Adelberger's group [131], by placing 1.3 ton lead (Pb) source close to a beryllium-aluminium (Be/Al) torsion balance, has seen no sign of a composition-dependent gravitational force in the intermediate range of 0.3 to 10 m. With the use of a polarized dysprosium-iron compound, Ni et al. [59, 101, 102] could give upper bounds for a possible spin dependence of the free fall. Within the limits provided by these experiments, the ratio of m_{grav} and m_{inert} is the same for all bodies, i.e. $m^1_{\text{grav}}/m^1_{\text{inert}} = m^2_{\text{grav}}/m^2_{\text{inert}} = \ldots$.

[8] "At this writing honesty compels me to say that there exist no battle-tested evidence whatsoever for the existence of any such fifth force, any departure from identical free fall for all things." [143].

Consequently, by choosing suitable units, we are allowed to set

$$m_{\text{grav}} = m_{\text{inert}} \tag{10.4}$$

for *all* bodies. This result has far-reaching consequences. In the equation of motion the mass drops out. It can now be written as

$$\frac{d^2\boldsymbol{x}}{dt^2} = \boldsymbol{g}\,, \tag{10.5}$$

which looks like the equation of motion for an accelerated observer – and we can get rid of acceleration by going over to a suitable reference frame. Consequently, gravity can locally be transformed to zero: This is realized in the famous freely falling, non-rotating "Einstein elevator". Incidentally, such an elevator will soon find technological application in the 144 *m* high *free fall tower* at Bremen [13].

Accordingly, the weak equivalence principle allows us to *eliminate* the gravitational force on a particle by free fall; on the other hand, we can locally *simulate* gravity by acceleration.

These two key "observations" were deduced in the context of the motion of a particle. In his *strong equivalence principle*, Einstein extended this concept to the rest of physics: All local, freely falling, non-rotating laboratories are fully equivalent for the performance of all physical experiments (see [111]).

In this section, we have discussed gravity only in the context of Galileian spacetime. This is due to the incompatibility of Newton's attraction law with the foundations of special relativity. The equivalence principle may serve as a starting point for a relativistic theory of gravitation. As a consequence, gravitation should act on particles, similarly as in the case of non-inertial force, via the term (see Eq. (3.3))

$$m\left\{ {k \atop ij} \right\} u^i u^j\,. \tag{10.6}$$

This suggests to relate the Christoffel symbol to the gravitational field $\boldsymbol{g}$ and the metric to the gravitational potential φ. Of course this cannot be the full story, since in SR we have seen that it is possible to transform the Christoffel symbol to zero over the whole spacetime. Then we would be left without gravitation which should not be possible for inhomogenous fields. It appears as if we are missing a crucial criterion which distinguishes gravity from ficticious inertial forces.

I.11 Gravity as Curvature of Spacetime

Let us consider the gravitational field of a macroscopic object, as for example that of the earth. Let us envision some "elevators" falling freely from different directions towards the center of the earth (Fig. 13).

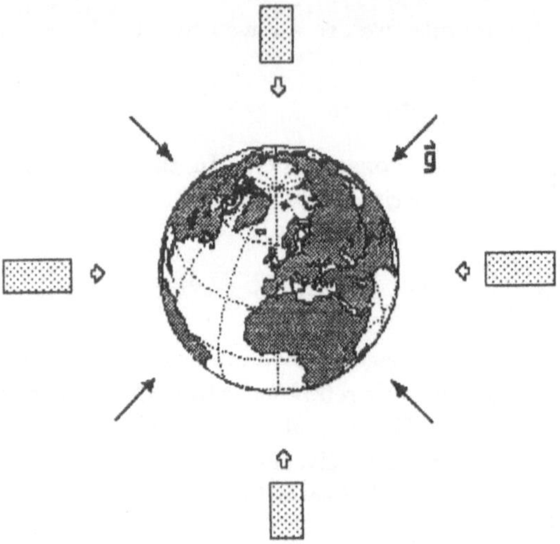

Fig. 13. "Einstein elevators" falling towards the center of the earth.

According to the principle of equivalence, the observers in the elevators, within restricted time[9] and space intervals, do not sense a gravitational field. Therefore, within these limits, the elevator represent *local inertial frames* (LIFs). In contradiction to SR, where inertial frames are supposed to move with constant velocity relative to each other, these LIFs are accelerated relative to each other. Hence it is not possible to constitute a global inertial frame, the local inertial systems do *not* match together. The only remedy is to start with the Minkowski spacetime M_4 of SR and cut it into little grains. In each grain special relativity is still valid, but, due to the presence of gravity, the grains cannot be configurated such as to yield a global M_4.

The next step is to give these findings a geometrical representation. To this end, remember the integrability condition in Eq. (3.13) of vanishing curvature for the transformation of an accelerated to a global inertial observer. Since such observers do not exist any more in general, it is therefore natural to drop the condition in Eq. (3.13), that is, we represent gravity by a non-vanishing *curvature of spacetime*. Of course, we have to make sure that it is still possible to transform gravity to zero locally. In order to make the degree of locality more precise (see [125]), we expand the metric components in a Taylor series at the point P:

$$g_{ij} = g_{ij}|_P + \frac{1}{1!} g_{ij,k}|_P \, x^k + \frac{1}{2!} g_{ij,kl}|_P \, x^k x^l + \dots \, . \tag{11.1}$$

[9] Before they crash onto the earth.

The Christoffel symbol obeys the inhomogenous transformation law of Eq. (3.10). Therefore, by applying a suitable coordinate transformation, it can always be transformed to zero at a spacetime event P, i.e.:

$$\left\{ {k \atop ij} \right\}|_P \stackrel{*}{=} 0 . \tag{11.2}$$

A coordinate system for which this holds is called *geodesic* at P. The Christoffel symbol is constructed such that the metric is covariant constant, i.e.

$$g_{ij;k} = g_{ij,k} - \left\{ {l \atop ki} \right\} g_{lj} - \left\{ {l \atop kj} \right\} g_{il} = 0 . \tag{11.3}$$

This implies that

$$\left\{ {k \atop ij} \right\}|_P = 0 \quad \Leftrightarrow \quad g_{ij,k}|_P = 0 . \tag{11.4}$$

Moreover, within the geodesic coordinate systems, it is still possible to pick a coordinate transformation such as to make the metric at P Minkowskian (*Riemannian normal coordinates*)

$$g_{ij}|_P \stackrel{*}{=} o_{ij} . \tag{11.5}$$

Now the Taylor series of the metric reduces to

$$g_{ij} \stackrel{*}{=} o_{ij} + \frac{1}{2} g_{ij,kl}|_P \, x^k x^l + \dots . \tag{11.6}$$

Since in geodesic coordinates, dropping now the P,

$$\left\{ {k \atop ij} \right\}_{,l} + \left\{ {k \atop jl} \right\}_{,i} + \left\{ {k \atop li} \right\}_{,j} \stackrel{*}{=} 0 , \tag{11.7}$$

$$g_{ij,kl} \stackrel{*}{=} - g_{nj} \left\{ {n \atop kl} \right\}_{,i} , \quad \text{and} \quad R_{l(ij)k} \stackrel{*}{=} \frac{3}{2} g_{nk} \left\{ {n \atop ij} \right\}_{,l} , \tag{11.8}$$

we finally obtain

$$g_{ij} \stackrel{*}{=} o_{ij} - \frac{1}{3} R_{iklj} \, x^k x^l + \dots . \tag{11.9}$$

Hence, in the immediate vicinity of the point P, where

$$\| R_{iklj} \, x^k x^l \| \ll 1 , \tag{11.10}$$

we can replace the Riemannian spacetime by a "Minkowskian grain". Then at P, in Riemannian normal coordinates, the gravitational field vanishes in accordance with the equivalence principle. At the surface of the earth, for instance, we have $\|R_{ijkl}\| \approx (2GM/c^2R^3)$, where $R = 6.4 \times 10^8 \, cm$ is the radius of the earth and $2GM/c^2 = 0.9 \, cm$ its Schwarzschild radius. For an object of $\ell \approx 1 \, cm$, we have $\|R_{iklj} \, x^k x^l\| \approx (2GM/3R^3c^2) \times 1 \times 1 \approx 10^{-27}$. Hence the deviation from o_{ij} will be negligible.

The reader should always bear in mind that the equivalence principle is merely a locally valid principle. For a larger portion of spacetime, it is *always* possible to distinguish between gravity and acceleration. This is due to the measurability of the curvature tensor (geodesic deviation). It has a homogeneous transformation property and therefore *cannot* be made zero by a coordinate transformation.

I.12 Field Equation of Gravity

For a heuristic derivation of Einstein's field equation we study the motion of the earth in the gravitational fied of the sun. The exterior gravitational field of the sun is very weak, hence we can linearize the metric:

$$g_{ij}(t,\boldsymbol{x}) = o_{ij} + h_{ij}(t,x,y,z) \qquad \text{with } h_{ij} \ll 1, \quad O(h^2)=0\,. \tag{12.1}$$

The velocity of the earth relative to the sun is only about 1% of the velocity of light. Consequently, we are allowed to set $dt \approx d\tau$ and $u^k \approx (c, \boldsymbol{v})$ for the velocity vector of the earth. Furthermore, we know that the gravitational field of the sun is quasistatic, i.e. $\partial_0 h_{ij} \ll \partial_A h_{ij}$. Then, from Eq. (3.3) we obtain, in lowest order, via a straightforward calculation, the equation of motion,

$$\frac{d^2 x_A}{dt^2} = \frac{c^2}{2}\frac{\partial h_{00}}{\partial x^A}\,, \qquad A = 1,2,3\,. \tag{12.2}$$

In order to relate this result to Newton's non-relativistic equation of motion, we merely have to set $h_{00} = 2\varphi/c^2$, where φ is the Newtonian gravitational potential. It satisfies the 2nd order Poisson equation

$$\Delta\varphi = 4\pi G\rho(\boldsymbol{r},t)\,, \qquad \Delta := \frac{\partial^2}{\partial x^2} + \frac{\partial^2}{\partial y^2} + \frac{\partial^2}{\partial z^2}\,, \tag{12.3}$$

where the mass density ρ is the source of gravity. This supports our earlier interpretation of the metric tensor as the gravitational potential. In this approxiation, the metric of the sun becomes

$$ds^2 = (1 + \frac{2\varphi}{c^2})\, c^2 dt^2 - dx^2 - dy^2 - dz^2\,. \tag{12.4}$$

In general, we have to find a field equation for all components of the metric. Generalizing Eq. (12.3), it should have the form of a wave equation for the metric,

$$''(\frac{\partial^2}{c^2\partial t^2} - \Delta)''\, g_{ij} + \text{nonlin. terms} = -\kappa T_{ij}\,, \qquad \kappa := \frac{8\pi G}{c^4}\,. \tag{12.5}$$

The energy-momentum tensor T_{ij} of matter, which comprises the mass density, acts as the source. The nonlinear terms indicate that gravity couples, via self-interaction, to its own gravitational energy-momentum current.

From the definition given in Eq. (3.13) we see that the curvature tensor $R_{ijk}{}^{l}$ has the correct differential order, i.e., $R = R(\partial\partial g_{ij}, \partial g_{ij}, g_{ij})$. However, it would not make sense to take ${R_{ijk}}^l = 0$ as field equation because this would lead us back to flat Minkowski spacetime. The correct equation is expected to involve ten conditions which control the ten components g_{ij} of the metric. Furthermore, the energy-momentum tensor should be covariantly conserved, i.e. ${T_{ij;}}^j = 0$. Up to a possible cosmological term, this requirement in four dimensions leads uniquely (Lovelock's theorem [79]) to the Einstein equation

$$\boxed{R_{ij} - \tfrac{1}{2} g_{ij} {R_k}^k = -\kappa T_{ij} \,, \qquad \kappa := 8\pi G/c^4 \,,} \tag{12.6}$$

where $R_{ij} := {R_{lij}}^l$ denotes the Ricci tensor of dimension $length^{-2}$.

The theory of ***general relativity*** (GR), briefly described here, was finally formulated by Einstein 1915/16 and is nowadays experimentally well-established in the macroscopic scale. The following tests lead to an agreement with GR within a precision of typically 1%:

(1) Redshift (Pound-Rebka-Snider),
gravitational time dilation (Hafele-Keating).
(2) Lightbending (→ graviational lensing, Huchra's lens [127]).
(3) Perihelion advance (Mercury, but also in highly relativistic binary pulsars).
(4) Delay of radar echos (Shapiro effect).
(5) Emission of gravitational waves
(from speed-up of the Hulse-Taylor binary pulsar).

The precession of a spinning top in the graviational field will be measured in the not too distant future (see [143] p. 232). For the moon, however, the de Sitter precession has already been verified [126]. Further details can be found in the text books [94, 125, 130].

I.13 Is the Equivalence Principle Valid for Elementary Particles?

In the standard model of elementary particle physics, only the leptons e^-, μ^-, τ^-, the corresponding neutrinos ν_e, ν_μ, ν_τ, and the (colored) quarks q^c and their respective antiparticles are regarded as fundamental point-like objects. They all have spin $S = \hbar/2$ (helicity in the case of massless fields). Proton, antiproton, neutron, and all the other (excited) hadrons are composed of quarks, which are glued together by the virtual exchange of intermediate Yang-Mills type gauge bosons. (The scalar Higgs boson predicted by the Weinberg-Salam model still has to be seen.)

What are the consequences of these new developments for gravity, which Einstein could not have forseen? Is the equivalence principle still valid in some form? Is it still possible to mimick gravity locally by acceleration? There is only one way to find out: To perform experiments with elementary particles in an accelerated reference frame or in a gravitational field. The following experiments have tested the classical trajectories of elementary particles:

(1) Thermal *neutrons* in vacuum fall like mass points (5 cm on 100 m horizontal travel); polarization of the neutrons does not affect the parabolic trajectory [24, 67].
(2) By applying nuclear resonance spectroscopy to ^7Li, which has nuclear spin $S = 3\hbar/2$, a possible *anisotropy of the inertial mass* of the proton is estimated to be less than 5×10^{-23} [29]. This lower bound has been further improved by comparing nuclear spin-precession frequencies of ^{201}Hg ($S = 3\hbar/2$) and ^{199}Hg ($S = \hbar/2$) isotopes [74].
(3) *Electrons* also seem to fall like mass points, as Witteborn and Fairbanks [144] have shown by using a path of about 80 cm. However, the electrons have to be shielded from external electric fields by a Faraday cage – and the redistribution, due to the gravitational field, of the charge carriers within the metallic walls of the cage seemingly has been not properly taken into account (cf. [115]). The inertial and the gravitational mass of a proton, which is kept in a trap, coincide within 4% accuracy [105].
(4) The weight of a *spin polarized test body*, namely the antiferromagnetic $Dy_6\,Fe_{26}$ alloy with a net spin of $S \sim$ Avog. $\times\,(\hbar/2) \approx 10^{-4} erg\, s$ was compared with the weight of (unpolarized) brass, and this in spin-up and spin-down position, respectively. It behaves, up to the present accuracy [59, 101], like a spinless test body. A corresponding Dicke type experiment will be set up soon [102].
(5) Experiments to test the universaliy of free fall for *antimatter* are under construction [38]: At CERN, antiprotons p^- from the antiproton storage ring LEAR are cooled down to $10^{-3} eV$ kinetic energy. Then their motion will be observed in a vertical drift chamber and compared with those of hydrogen ions H^-.

There exist some theoretical estimates for the *gravitational interaction of antimatter*: If the positrons of the virtual pairs pulled out of the vacuum by the Coulomb field of the nucleus violated the weak equivalence principle, they would contribute an effect to Eötvös type experiments with different elements which is four orders of magnitude larger than the present accuracy, cf. [114]. A more stringent bound comes from decay rates of the $(1/2)(K^0 - \overline{K}^0)$ system. Since it is a coherent superposition of neutral kaons and antikaons, its decay channels function as a very sensitive "natural interferometer" for the investigation of the gravitational coupling to antimatter as compared to matter. Taking the gravitational potential of our own galaxy into account, Good [39] found that the gravitational mass of the K^0 and the $\overline{K}^0$ relative to the common inertial mass m_K differs at most by a few parts in 10^{-13}. This improved value has been given by Fitch [34], see also [95]. The equivalence principle for neutrinos ν and antineutrinos $\overline{\nu}$ has been verified on the basis of data from the supernova SN 1987A, see [104a]. Incidentally, the inertial mass of the proton and the antiproton [34a] coincide within an accuracy of 10^{-8}.

All these observations are consistent with the conventional concept of mass, and no specific effect due to spin or antimatter have been found. In that sense no rethinking of the equivalence principle is necessary at this stage.

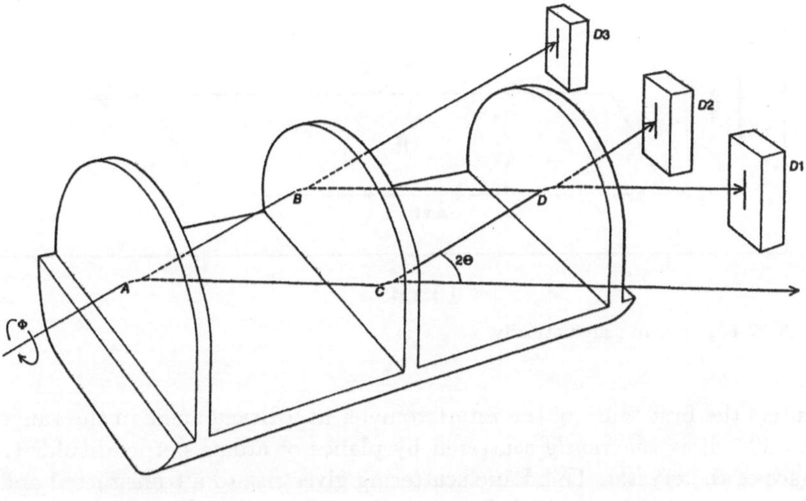

Fig. 14. The neutron interferometer of the COW-experiment, see [44].

I.14 Neutron Interferometry: Is the Equivalence Principle Valid for Matter Waves?

Up to now, we have considered mainly the *classical* trajectories of *structureless* point particles. However, it is known since de Broglie's introduction of *matter waves* with wave length $\lambda = 2\pi\hbar/p$, where p is the momentum of the corresponding particle, that matter can also produce *interference patterns*. This was experimentally demonstrated for electrons, Helium atoms and other particles since the late twenties. In Schrödinger's non-relativistic quantum mechanics (QM), the state of an elementary particle is described by a complex wave function $\psi(t, \boldsymbol{x})$. In the Born interpretation [12], the norm $|\psi(t, \boldsymbol{x})| := \sqrt{\psi^*(t, \boldsymbol{x})\psi(t, \boldsymbol{x})}$ (* stands for complex conjugation) multiplied by $d^3\boldsymbol{x}$ represents the probability density to find the particle at the time t in the volume element $[\boldsymbol{x}, \boldsymbol{x} + d\boldsymbol{x}]$.

The availability of nearly perfect single silicon crystals of $\sim 10\,cm$ length has provided a new tool for X-ray and neutron interferometry. This had first been demonstrated by Bonse and Hart in 1965 for X-rays. After Bonse (1974) and Rauch, Treimer, and Bonse (1974), see [43], had shown that this device also works for neutrons, Colella, Overhauser, and Werner (in the following abbreviated by COW) "...used a neutron interferometer to observe the quantum-mechanical phase shift of neutrons caused by their interaction with the Earth's gravitational field" [21], see also Fig. 14. Their celebrated experiment is sketched schematically in Fig. 15.

They used neutrons cooled to room temperature such that their resulting mean velocity $v_n \simeq 10^{-5}c$ is non-relativistic. Their mass is $m_n = 1.67 \times 10^{-24}g$, and the de Broglie wave length $\lambda_n := 2\pi\hbar p \approx 2 \times 10^{-8}cm$. A beam of $1\,cm$

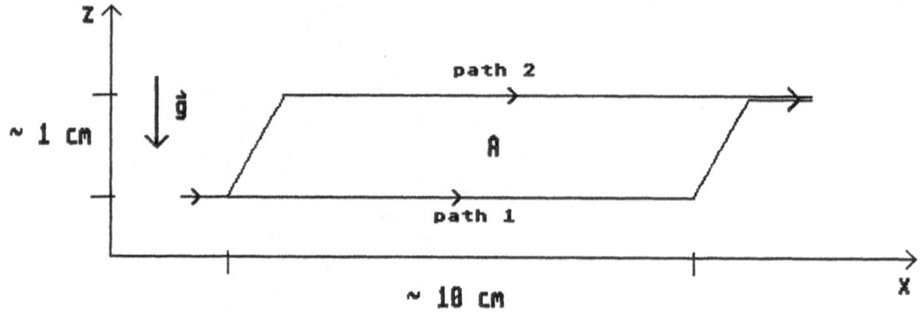

Fig. 15. COW experiment schematically.

width enters the first "ear" of the interferometer at a Bragg angle in the range of 20° to 30°. It is *coherently* scattered by planes of atoms perpendicular to the surface of the crystal. This Laue scattering gives rise to a transmitted and a diffracted beam, with opposite Bragg angles. Due to the Borrmann effect, the beam travels through the crystal at first along the planes and the splitting occurs actually only after it emerges from the ear again.

When the interferometer gets rotated in the gravitational field of the earth, the upper and lower beams travel $1cm$ apart and encounter a potential difference of $\Delta\varphi/E_{\rm kin} = (m_n g \Delta h/(1/2)m_n v^2) \approx 10^3/(3\times 10^5)^2 \approx 10^{-8}$, which is only a tiny fraction of the kinetic energy. Nevertheless, this leads to a measurable effect on the *phase* of the neutron's coherent wave which oscillates about $10\,cm/\lambda_n \approx 10^9$ times during the horizontal flight. Although the oscillation rate of the upper beam is "redshifted" merely by a factor of 10^{-7}, the upper beam manages it to make $\theta_{\rm grav} \sim 10^9/10^7 = 100$ more oscillations than the lower beam. This phase shift can be observed by the interference pattern of the recombined beams.

In the actual experiment, side effects have to be taken care of: Gravity produces distortions in the single crystal. Contributions from this can be eliminated by comparing X-ray and neutron interference patterns in the same interferometer. Moreover, the neutron beam is itself bent into a parabolic path with $4\times 10^{-7}cm$ loss in altitude. This yields, however, no significant influence on the phase.

In the COW experiment, the single-crystal interferometer is at rest with respect to the laboratory, whereas the neutrons are subject to the gravitational potential. In order to compare this with the effect of acceleration relative to the lab frame, Bonse and Wroblewski [11] let the interferometer oscillate horizontally by driving it via a pair of standard loudspeaker magnets. Thus these experiments test the effect of acceleration and local gravity on the wave aspect of matter and prove its equivalence up to an accuracy of $\sim 4\%$.

I.15 Bonse-Wroblewski Experiment Quantum-Mechanically

In order to understand this effect theoretically, we treat the neutron as a non-relativistic particle, with negligible spin and velocity $v_n \ll c$, which fulfills the Schrödinger equation

$$i\hbar \frac{\partial \psi}{\partial t} = H\psi \,, \tag{15.1}$$

where H is the Hamilton operator. If H is time-independent, we get

$$\psi(t, \boldsymbol{x}) = \phi(\boldsymbol{x}) \exp\left(-i\frac{Ht}{\hbar}\right) \quad \text{and} \quad H\phi(\boldsymbol{x}) = E\phi(\boldsymbol{x}) \,. \tag{15.2}$$

For a free particle, $H_0 = -(\hbar^2/2m)\Delta$, and the (unnormalized) solution of the eigenvalue equation is

$$\phi_0(\boldsymbol{x}) = e^{i\boldsymbol{k}_0 \cdot \boldsymbol{x}} \,, \tag{15.3}$$

where $\boldsymbol{k}_0$ is the wave vector with $\boldsymbol{k}_0 \cdot \boldsymbol{k}_0 = (2m/\hbar^2)E_0$. The Hamiltonian for the neutron in an accelerated and rotating frame is

$$H = H_0 + H_{\text{int}} = \frac{\boldsymbol{p}^2}{2m} + m\boldsymbol{a} \cdot \boldsymbol{x} - \boldsymbol{\omega} \cdot \boldsymbol{L} \,, \tag{15.4}$$

where $m(\equiv m_{\text{inert}})$ is the inertial mass of the neutron, $\boldsymbol{\omega}$ the angular velocity of the turn-table (including the earth's rotation), and $\boldsymbol{L} = \boldsymbol{x} \times \boldsymbol{p}$ the angular momentum of the neutron. We now have to solve

$$(H_0 + H_{\text{int}})\psi = E\psi \,. \tag{15.5}$$

In the semiclassical approximation, the trajectory and the energy of the neutron are unchanged. The interaction Hamiltonian H_{int} is just producing a phase shift. With the ansatz

$$\psi = \phi_0 e^{i\theta(\boldsymbol{x})} \,, \qquad \text{we obtain} \tag{15.6}$$

$$\left\{ \frac{-\hbar^2}{2m}(-\boldsymbol{k}_0 \cdot \boldsymbol{k}_0 + 2\boldsymbol{k}_0 \cdot \nabla\theta + \nabla\theta \cdot \nabla\theta + i\Delta\theta) + H_{\text{int}} \right\} \phi_0 = E_0 \phi_0 \,. \tag{15.7}$$

The inequalities $|\nabla\theta| \ll k_0$ and $|\Delta\theta| \ll \boldsymbol{k}_0 \cdot \boldsymbol{k}_0$ hold in the semiclassical approximation. Thus,

$$\frac{\hbar^2}{m} \boldsymbol{k}_0 \cdot \nabla\theta = H_{\text{int}} \,.$$

Integration along the particle's path yields

$$\theta = -\frac{m}{\hbar^2 k_0} \int_{\text{path}} H_{\text{int}} \, ds \,. \tag{15.8}$$

The total phase difference θ between the two different paths turns out to be

$$\begin{aligned} \theta &= -\frac{m}{\hbar^2 k_0} \int_{\text{path}_1} H_{\text{int}} \, ds + \frac{m}{\hbar^2 k_0} \int_{\text{path}_2} H_{\text{int}} \, ds \\ &= -\frac{m}{\hbar^2 k_0} \oint m\boldsymbol{a} \cdot \boldsymbol{x} \, ds + \frac{m}{\hbar^2 k_0} \oint \boldsymbol{\omega} \cdot \boldsymbol{L} \, ds \,. \end{aligned} \tag{15.9}$$

If we denote by $\boldsymbol{A}$ the vector normal to the effective interferometer area A, the integration yields

$$\theta = \theta_{\text{acc}} + \theta_{\text{Sagnac}} = \frac{m^2 a}{2\pi\hbar^2}\,\lambda A + \frac{2m}{\hbar}\boldsymbol{\omega}\cdot\boldsymbol{A}\,. \tag{15.10}$$

Bonse and Wroblewski found a result which was within $\sim 4\%$ of the theoretical prediction given by θ_{acc}.

I.16 The Colella-Overhauser-Werner (COW) Experiment

The COW experiment is different, inasmuch as the interferometer is at rest with respect to the lab frame. Let $\boldsymbol{R}$ be the radius vector of the lab with respect to the center of the earth. By rotating the apparatus in steps, the neutrons are subject to a changing gravitational attraction of the earth. Then, the interaction Hamiltonian reads

$$H_{\text{int}} = -G\,\frac{M_{\text{grav}} m_{\text{grav}}}{|\boldsymbol{R}+\boldsymbol{x}|} - \boldsymbol{\omega}\cdot\boldsymbol{L}\,, \tag{16.1}$$

where M_{grav} and m_{grav} are the gravitational mass of earth and neutron, respectively, and $\boldsymbol{x}$ the radius vector of the neutron in the lab. Since the dimensions of the experiment are small compared with the radius of the earth, $x \ll R$, we can perform an expansion in terms of x/R. By suppressing an unimportant additive constant, we find that we can approximate the interaction by

$$H_{\text{int}} = m_{\text{grav}}\,\boldsymbol{g}\cdot\boldsymbol{x} - \boldsymbol{\omega}\cdot\boldsymbol{L}\,, \tag{16.2}$$

where $g := |\boldsymbol{g}| = 981\,cm/s^2$ is the gravitational field on earth. For an interferometer which rotates stepwise with respect to the lab system, i.e. for $\boldsymbol{g} = g(\boldsymbol{A}/|\boldsymbol{A}|)\sin\alpha$, we find, similarly to the case of the accelerated frame,

$$\theta = \theta_{\text{grav}} + \theta_{\text{Sagnac}} = \frac{m\, m_{\text{grav}}\, g}{2\pi\hbar^2}\,\lambda A \sin\alpha + \frac{2m}{\hbar}\boldsymbol{\omega}\cdot\boldsymbol{A}\,. \tag{16.3}$$

Here α is the angle between the vector $\boldsymbol{A}$ normal to the interferometer area $|\boldsymbol{A}|$ and the vector $\boldsymbol{g}$. Since $g = GM_{\text{grav}}/R^2$, this is the first time that Planck's constant $\hbar$ and the gravitational constant G are jointly contained in a formula which can be tested experimentally. Observe that the θ_{grav} depends on the *gravitational mass* m_{grav}, in contradistinction to θ_{acc}, in which only the *inertial mass* m of the neutron occurs. This distinction allows us to check the equivalence principle on the level of quantum interference.

We obtain approximatively

$$\theta_{\text{grav}} \approx \frac{(2\times 10^{-24})^2 \times 10^3 \times 2 \times 10^{-8} \times 10}{6\times(10^{-27})^2} \sim 100\,, \tag{16.4}$$

which is roughly in compliance with our heuristic considerations above. The formula in Eq. (16.3) has been experimentally verified in the COW experiment within 1% precision.

For light, the phase shift induced by rotational motion of an interferometer was first observed 1913 by Sagnac (cf. [108]), and then, with the earth as "turntable", subsequently measured in the 1925 experiment of Michelson, Gale, and Pearson. For neutrons, the corresponding *"Michelson-Gale effect"* has been predicted by Page [104] and then found by Werner et al. [141] within 3% of the theoretical value. Atwood et al. [2] rotated their interferometer with an angular velocity of $|\boldsymbol{\omega}| \sim 30|\boldsymbol{\omega}|_{\text{earth}}$ and found the *"Sagnac effect"* for neutrons to be in agreement with theory in the 1% range. More recently, the "dragging of frames", observed for light in moving media by Fizeau in 1851, has been found for neutrons likewise [10].

Let us return to our basic question: Is the strong equivalence principle still valid? In order to be able to answer this question *affirmatively*[10], we have shown that the gravitational result in Eq. (16.3) can equally be obtained by transforming the Schrödinger equation to the frame of an accelerated and rotating observer and by imposing the "weak equivalence" between inertial and gravitational mass. In a conceptually more precise approach, we should start with the *special-relativistic* Dirac equation, transform it to a non-inertial frame (see [138]), and then go over to the non-relativistic approximation. Are there further physical effects coming up?

I.17 Dirac Particle in an Accelerated and Rotating Frame

According to the standard gauge model of weak and strong interactions, a neutron is not a fundamental particle, but consists of one up and two down quarks which are kept together via the virtual exchange of gluons, the vector bosons of quantum chromodynamics, in a permanent "confinement phase". This beautifully predicts, for instance, that the anomalous magnetic moment of the neutron is $\mu_n = -(2/3)\mu_p$. For studying the properties of the neutron in a non-inertial frame and in low-energy gravity, we may disregard its extension of $\sim 0.7\,fm$, its form factors, etc. In fact, for that purposes it is sufficient to treat it as a Dirac particle of spin $\hbar/2$ which is *otherwise structureless.*

In the Minkowski spacetime of SR in Cartesian coordinates, the field equation for a massive fermion is represented by the Dirac equation

$$i\hbar\gamma^i\partial_i\psi \overset{*}{=} mc\psi\,, \tag{17.1}$$

where the Dirac matrices γ^i fulfill the relation

$$\gamma^i\gamma^j + \gamma^j\gamma^i = 2o^{ij}\,. \tag{17.2}$$

For the conventions and the representation of the γ's, we essentially follow Bjorken-Drell [8].

With the rules of Sects. 3, 8, and 9, we could straightforwardly transform this equation from an inertial to an accelerated and rotating frame. However,

10 See Treder [136a]; for a somewhat different point of view, see Dehnen [27].

by analogy with the equation of motion in an arbitrary frame as well as from gauge theory, we can infer the result of this tranformation: In the non-inertial frame, the partial derivative in the Dirac equation is simply replaced by the *covariant derivative*

$$\partial_i \quad \Rightarrow \quad D_\alpha := \partial_\alpha + \frac{i}{4}\sigma^{\beta\gamma}\,\Gamma_{\alpha\beta\gamma}\,, \qquad \partial_\alpha := e^i{}_\alpha \partial_i \equiv e_\alpha\,, \tag{17.3}$$

where $\Gamma_{\alpha\beta\gamma}$ are the anholonomic components of the connection and x^i are the Cartesian coordinates of the lab system (which we called $x^{\bar{i}}$ previously; we drop the bar for convenience). The anholonomic Dirac matrices are defined by

$$\gamma^\alpha := e_i{}^\alpha\,\gamma^i \quad \Rightarrow \quad \gamma^\alpha\gamma^\beta + \gamma^\beta\gamma^\alpha = 2o^{\alpha\beta}\,. \tag{17.4}$$

The six matrices $\sigma^{\beta\gamma}$ are the infinitesimal generators of the Lorentz group and fulfill the commutation relation[11]

$$[\gamma^\alpha, \sigma^{\beta\gamma}] = 2i(o^{\alpha\beta}\gamma^\gamma - o^{\alpha\gamma}\gamma^\beta)\,. \tag{17.5}$$

Then, the Dirac equation formulated in the orthonormal frame of the accelerated and rotating observer reads

$$i\hbar\gamma^\alpha D_\alpha\psi = mc\psi\,. \tag{17.6}$$

Although there appears now a "minimal coupling" to the connection, there is no new physical concept involved in this equation. As already stressed before, the coupling to the connection arises simply from the change of the frame. Since we are still in SR, the curvature and the torsion $T^\alpha := D\vartheta^\alpha$ of spacetime both remain zero. Thus Eq. (17.6) is just a reformulation of the "Cartesian" Dirac equation (17.1).

However, the rewriting in terms of the covariant derivative provides us with a rather elegant way of explicitly calculating the Dirac equation in the non-inertial frame (9.8) of an accelerated, rotating observer [119, 52]: Using the anholonomic connection components of Eq. (9.11) as well as $\boldsymbol{\alpha} = -i\{\sigma^{\hat{0}\Xi}\}$, we find for the covariant derivative ($\boldsymbol{a}$ is a three-acceleration):

$$\begin{aligned} D_{\hat{0}} &= \frac{1}{1 + \boldsymbol{a}\cdot\boldsymbol{x}/c^2}\left(\partial_0 + \frac{1}{2c^2}\boldsymbol{a}\cdot\boldsymbol{\alpha} - \frac{i}{c\hbar}\boldsymbol{\omega}\cdot\boldsymbol{J}\right)\,,\\ D_\Xi &= \partial_\Xi\,. \end{aligned} \tag{17.7}$$

The total three-angular momentum operator

$$\boldsymbol{J} := \boldsymbol{L} + \boldsymbol{S} = \boldsymbol{x} \times \frac{\hbar}{i}\frac{\partial}{\partial \boldsymbol{x}} + \frac{1}{2}\hbar\boldsymbol{\sigma} = \boldsymbol{x}\times\boldsymbol{p} + \frac{1}{2}\hbar\boldsymbol{\sigma} \tag{17.8}$$

is built, in the canonical manner, from the orbital operator $\boldsymbol{L}$ and the spin $\boldsymbol{S} := (\hbar/2)\boldsymbol{\sigma}$. Observe that $\boldsymbol{\sigma} = \{\sigma^\Xi\} := \{(1/2)\epsilon^{\Xi\Lambda\Upsilon}\sigma_{\Lambda\Upsilon}\}$ can be constructed

[11] For Dirac spinors, the Lorentz generators can be represented by $\sigma^{\beta\gamma} := (i/2)(\gamma^\beta\gamma^\gamma - \gamma^\gamma\gamma^\beta)$, but we do not make use of this explicit form in the following. In this representation, we find $\boldsymbol{\alpha} = \gamma^{\hat{0}}\boldsymbol{\gamma}$ with $\boldsymbol{\gamma} = \{\gamma^\Xi\}$.

for any spinor representation Ψ of the Lorentz group. Thus our results above would also hold for any Lorentz-covariant Dirac type *1st order wave equation* as exemplified by the Rarita-Schwinger equation (see [139]) for particles with spin $(3\hbar/2)$.

The physical effects in our lab frame can most easily understood by going over to the Hamiltonian. After multiplying the Dirac equation by $\beta := \gamma^0$ and $c(1 + \boldsymbol{a} \cdot \boldsymbol{x}/c^2)$, we get

$$i\hbar \frac{\partial \psi}{\partial t} = H\psi \quad \text{with } H = \beta mc^2 + \mathcal{O} + \mathcal{E} \,. \tag{17.9}$$

After substituting the covariant derivatives, the operators $\mathcal{O}$ and $\mathcal{E}$, which are odd and even with respect to β, read, respectively [52]:

$$\mathcal{O} := c\boldsymbol{\alpha} \cdot \boldsymbol{p} + \frac{1}{2c} \{(\boldsymbol{a} \cdot \boldsymbol{x})(\boldsymbol{p} \cdot \boldsymbol{\alpha}) + (\boldsymbol{p} \cdot \boldsymbol{\alpha})(\boldsymbol{a} \cdot \boldsymbol{x})\} \,, \tag{17.10}$$

$$\mathcal{E} := \beta m(\boldsymbol{a} \cdot \boldsymbol{x}) - \boldsymbol{\omega} \cdot (\boldsymbol{L} + \boldsymbol{S}) \,. \tag{17.11}$$

Up to now, these are *exact* results[12]. For later purposes we introduce $\mathcal{O}_1 = c\boldsymbol{\alpha} \cdot \boldsymbol{p}$ and $\mathcal{O}_2 = \mathcal{O} - \mathcal{O}_1$.

The total angular momentum couples to the angular velocity three-vector $\boldsymbol{\omega}$ such that the relative "gyro-rotational moment" between the orbital and the spin operator is one. In order to determine the *"gyro-gravitational moment"*, we would need to consider the non-relativistic Pauli equation in a spacetime with non-vanishing Riemann-Cartan curvature (cf. [106]), or to perform a Gordon decomposition of the current into convective and polarization currents, as will be done in Leture II.

I.18 Non-relativistic Approximation via Foldy-Wouthuysen Transformation

In order to obtain a non-relativistic approximation, a unitary transformation is required which will remove from the equation all operators, such as $\boldsymbol{\alpha}$, which couple the "large" spinor component ϕ to the "small" spinor component in

$$\psi = \begin{pmatrix} \phi \\ \frac{\boldsymbol{\sigma} \cdot (\boldsymbol{p} - i\hbar \boldsymbol{\Gamma})}{2mc} \phi \end{pmatrix} . \tag{18.1}$$

Similarly as in QED, this can be achieved by applying successive Foldy-Wouthuysen transformations. First the unitary transformation

$$U := -\frac{i\beta}{2mc^2} \mathcal{O}_1 \tag{18.2}$$

[12] Compare our results with quantum electrodynamics (QED). There, $\mathcal{O}_{\mathrm{QED}} = c\boldsymbol{\alpha} \cdot (\boldsymbol{p} - (e/c)\boldsymbol{A})$ and $\mathcal{E}_{\mathrm{QED}} = e\varphi$ (see [8], $\boldsymbol{A}$ is here the vector potential, of course.)

yields, up to the order of c^{-2} the new Hamiltonian

$$\begin{aligned} H' = \beta mc^2 + \frac{\beta}{2m}\boldsymbol{p}^2 - \frac{\beta}{8m^3c^2}\boldsymbol{p}^4 + \frac{\beta}{2mc}\{\boldsymbol{\alpha}\cdot\boldsymbol{p}, \mathcal{O}_2\} \\ - \frac{1}{8m^2c^2}[\boldsymbol{\alpha}\cdot\boldsymbol{p}, [\boldsymbol{\alpha}\cdot\boldsymbol{p}, \mathcal{E}]] + \mathcal{O}', \end{aligned} \tag{18.3}$$

$$\text{where}\quad \mathcal{O}' := \mathcal{O}_2 + \frac{\beta}{2mc}[\boldsymbol{\alpha}\cdot\boldsymbol{p}, \mathcal{E}] - \frac{1}{3m^2c}(\boldsymbol{\alpha}\cdot\boldsymbol{p})\boldsymbol{p}^2 . \tag{18.4}$$

The second unitary transformation by means of

$$U' := -\frac{i\beta}{2mc^2}\mathcal{O}' \tag{18.5}$$

leads to

$$\begin{aligned} H' = \beta mc^2 + \frac{\beta}{2m}\boldsymbol{p}^2 - \frac{\beta}{8m^3c^2}\boldsymbol{p}^4 + \frac{\beta}{2mc}\{\boldsymbol{\alpha}\cdot\boldsymbol{p}, \mathcal{O}_2\} \\ - \frac{1}{8m^2c^2}[\boldsymbol{\alpha}\cdot\boldsymbol{p}, [\boldsymbol{\alpha}\cdot\boldsymbol{p}, \mathcal{E}]] . \end{aligned} \tag{18.6}$$

Evaluating the (anti-)commutators we eventualy find up to the order of c^{-2}

$$\begin{aligned} H' = \beta mc^2 + \frac{\beta}{2m}\boldsymbol{p}^2 - \frac{\beta}{8m^3c^2}\boldsymbol{p}^4 + \beta m(\boldsymbol{a}\cdot\boldsymbol{x}) - \boldsymbol{\omega}\cdot(\boldsymbol{L}+\boldsymbol{S}) \\ + \frac{\beta}{2m}\boldsymbol{p}\cdot\frac{\boldsymbol{a}\cdot\boldsymbol{x}}{c^2}\boldsymbol{p} + \frac{\beta\hbar}{4mc^2}\boldsymbol{\sigma}\cdot\boldsymbol{a}\times\boldsymbol{p} + O(\frac{1}{c^3}) . \end{aligned} \tag{18.7}$$

The different non-inertial effects of a fermion are displayed in the following table:

$\beta m(\boldsymbol{a}\cdot\boldsymbol{x})$	Redshift (Bonse-Wroblewski → COW)
$-\boldsymbol{\omega}\cdot\boldsymbol{L}$	Sagnac type effect (Page-Werner et al.)
$-\boldsymbol{\omega}\cdot\boldsymbol{S}$	Spin-rotation effect (Mashhoon)
$\beta\boldsymbol{p}\cdot(\boldsymbol{a}\cdot\boldsymbol{x})\,\boldsymbol{p}\,/\,(2mc^2)$	Redshift effect of kin. energy
$\beta\hbar\,\boldsymbol{\sigma}\cdot(\boldsymbol{a}\times\boldsymbol{p})\,/\,(4mc^2)$	New inertial spin-orbit coupling

Tab. 2 Inertial effects for a massive fermion of spin $\hbar/2$ in non-relativistic approximation.

Besides the rest mass and the usual kinetic term, we obtain terms which account for the ***redshift effect*** due to acceleration and the ***"Sagnac type" effect*** in the same manner as in the non-relativistic Schrödinger equation. Moreover, we find the spin-rotation effect which has been first proposed by Mashhoon [80] for the neutron interferometer. This term could have never been obtained in the simple Schrödinger picture.

For neutrons with $|\boldsymbol{S}| = \hbar/2$, this spin-rotation effect is much smaller than th Sagnac effect. The corresponding ratio of the phase shifts is given by

$$\theta_{\text{spin}}/\theta_{\text{Sagnac}} = \frac{\oint \boldsymbol{\omega} \cdot \boldsymbol{S}\, ds}{\oint \boldsymbol{\omega} \cdot \boldsymbol{x} \times \boldsymbol{p}\, ds} = \hbar/2pr_0 = \lambda_n/4\pi r_0 \approx 10^{-9}\,. \tag{18.8}$$

Here λ_n is the de Broglie wave length of the neutron and r_0 the "radius" of the effective area $A = \pi r_0^2$ of the interferometer. Notwithstanding this smallness, Werner [142] has already proposed a modified neutron interferometer which may see this tiny effect of the spin of the neutron. Nuclear resonance spectroscopy on a turn-table could possibly be another means to see the Mashhoon effect. Hamiltonians of the type given in Eq. (18.3), which in particular display the $\boldsymbol{\omega} \cdot (\boldsymbol{L} + \boldsymbol{S})$ coupling, can be found in earlier literature:

Schmutzer and Plebański [116] is the earliest reference we came across, a follow-up is Gorbatsievich [40]. Greenberger and Overhauser [42, 43] review some literature in an appendix; for further reading, see [1a,3,4,5,22a,60,61, 63,145,146] and the references given there.

Lecture II: Energy-Momentum and Spin Currents of Matter Fields

Abstract

In Lecture II we will concentrate on those currents of classical matter fields which are induced by the Poincaré symmetry group of the underlying Minkowski spacetime, namely the energy-momentum and the spin current 3-forms. The mathematical formalism used is that of exterior differential forms ("Cartan calculus") [14, 18, 37, 83, 121, 132, 136, 140] which may be a bit more demanding than the Ricci calculus of Lecture I, but which will hopefully increase the transparency of the structures displayed.

The Maxwell equations are formulated according to the metric-free procedure introduced by Kottler, E. Cartan, and van Dantzig. For the vacuum Maxwell field, which is massless and hence has only helicity, the spin current is zero, leaving us with a *symmetric* energy-momentum current as the sole external current for generating gravity. The Riemannian spacetime of GR behaves like a classical continuum carrying (Cauchy) *stress*, four-dimensional Cauchy stress being synonymous to the symmetric energy-momentum current mentioned.

The Dirac equation is put in exterior forms, then the Dirac-Maxwell theory is formulated as a $U(1)$-gauge theory. A short deduction of the Noether theorem is given, and the Noether currents of the Poincaré group are derived. The relocalization of the energy-momentum and the spin current is discussed, including a separate procedure if dilation invariance is prevailing.

The spin current of the Dirac field, besides the energy-momentum current, is an external current in its own right. We decompose both currents in their convective and polarization parts and derive, for the first time to our knowledge, the *gyro-gravitational* ratio of the Dirac electron. In continuum theoretical parlance, the momentum and the spin currents represent stress and *spin moment stress*, the latter of which cannot be accommodated in a classical continuum.

Contents

II.1 Maxwell Equations in Integral Form

We will be interested in electrodynamics within the framework of SR. The structure of Maxwell's theory will be displayed in a much more transparent way, however, if we follow Kottler, Cartan, and van Dantzig ([107, 109, 118, 129, 137] and references given, see also [88]) and put the basic framework of the theory into a general covariant form on a "bare" differentiable manifold. Only for the constitutive law relating the excitation 2-form H to the field strength 2-form F we do need a metric. Thereby we can recover straightforwardly, for example, the *conformal* invariance of the vacuum Maxwell theory.

Expressed by means of three-dimensional vector calculus, the integral form of the Maxwell equations reads

$$\oint_{\partial S} \boldsymbol{E}\cdot d\boldsymbol{r} = -\frac{d}{dt}\left(\int_S \boldsymbol{B}\cdot d\boldsymbol{f}\right), \qquad \int_{\partial V} \boldsymbol{B}\cdot d\boldsymbol{f} = 0, \tag{1.1}$$

$$\int_{\partial V} \boldsymbol{D}\cdot d\boldsymbol{f} = \int_V \rho\, dV, \qquad \oint_{\partial S} \boldsymbol{H}\cdot d\boldsymbol{r} = \int_S \boldsymbol{J}\cdot d\boldsymbol{f} + \frac{d}{dt}\left(\int_S \boldsymbol{D}\cdot d\boldsymbol{f}\right). \tag{1.2}$$

Here S denotes a two-dimensional spatial surface, V a three-dimensional spatial volume, and ∂S and ∂V the respective boundaries. Addititonally, there are constitutive laws connecting $\boldsymbol{D}$ and $\boldsymbol{H}$ with $\boldsymbol{E}$ and $\boldsymbol{B}$ and, in conducting media, $\boldsymbol{J}$ with $\boldsymbol{E}$ and $\boldsymbol{B}$. This is only a preliminary version of the Maxwell equations, since the three-dimensional Euclidian metric enters the scalar product, that we are denoting by a dot. A more appropriate formulation in terms of differentiable forms, which does *not* involve a metric, can be read off from these equations without difficulties.

The fields $\boldsymbol{E}$ and $\boldsymbol{H}$ are integrated along curves. Hence we will associate them with 1-forms. The fields $\boldsymbol{D}$, $\boldsymbol{B}$, and $\boldsymbol{J}$ are integrated over two-dimensional surfaces. This suggests that they are presented by 2-forms. Analogously, the charge density should be associated with a 3-form. Therefore we abandon Eqs. (1.1) and (1.2) and *postulate* the Maxwell equations in integral form according to

$$\oint_{\partial S} E = -\frac{d}{dt}\int_S B\,, \qquad \int_{\partial V} B = 0\,, \tag{1.3}$$

$$\int_{\partial V} \mathcal{D} = \int_V \hat{\rho}\,, \qquad \oint_{\partial S} \mathcal{H} = \int_S \mathcal{J} + \frac{d}{dt}\int_S \mathcal{D}\,. \tag{1.4}$$

This set of equations is invariant under 3-dimensional diffeomorphisms and *no* metric or connection are involved. The first set, Eqs. (1.3), interconnects the *intensive* quantities E, B (remember that they are related to forces), the second set, Eqs. (1.4), the *extensive* quantities (or densities) $\mathcal{D}$, $\mathcal{H}$, $\hat{\rho}$, $\mathcal{J}$. One mathematical way to pin down this difference is to connect the intensive quantities with *even* and the extensive quantities with *odd* (or twisted)[1] differential forms, compare [14, 88, 109]. It is implausible to introduce a magnetic charge and a magnetic current in Eqs. (1.3), because the magnetic charge would then have to be an intensive quantity, quite opposite to the intuitive picture one has in mind for a charge-like quantity[2]. In any case, if one introduced such intensive quantities, they would be quite dissimilar to the correspondig electric quantities. In three dimensions, intensive quantities will be denoted by Latin letters, extensive quantities by script letters or by a hat.

For the different fields in the Maxwell equations, in accordance with their operational definitions, we have then the expansions for the 1-forms

$$E = E_A dx^A = E_1 dx^1 + E_2 dx^2 + E_3 dx^3\,, \tag{1.5}$$

$$\mathcal{H} = \mathcal{H}_A dx^A = \mathcal{H}_1 dx^1 + \mathcal{H}_2 dx^2 + \mathcal{H}_3 dx^3\,, \tag{1.6}$$

for the 2-forms

$$\mathcal{D} = \frac{1}{2}\mathcal{D}_{AB} dx^A \wedge dx^B = \mathcal{D}_{23} dx^2 \wedge dx^3 + \mathcal{D}_{31} dx^3 \wedge dx^1 + \mathcal{D}_{12} dx^1 \wedge dx^2\,, \tag{1.7}$$

$$B = \frac{1}{2}B_{AB} dx^A \wedge dx^B = B_{23} dx^2 \wedge dx^3 + B_{31} dx^3 \wedge dx^1 + B_{12} dx^1 \wedge dx^2\,, \tag{1.8}$$

---

[1] Let us quote from Burke [14]: "We could make out a good case that the usual differential forms are actually the twisted ones, but the language is forced on us by history. Twisted differential forms are the natural representations for densities, and sometimes actually are called densities, which would be an ideal name were it not already in use in tensor analysis."

[2] It could be worthwhile to investigate whether a Dirac string (see the review [90]) behaves like an intensive quantity.

$$\mathcal{J} = \frac{1}{2}\mathcal{J}_{AB}dx^A \wedge dx^B = \mathcal{J}_{23}dx^2 \wedge dx^3 + \mathcal{J}_{31}dx^3 \wedge dx^1 + \mathcal{J}_{12}dx^1 \wedge dx^2 \,, \tag{1.9}$$

and for the charge 3-form[3]

$$\hat{\rho} = \frac{1}{3!}\,\hat{\rho}_{ABC}\; dx^A \wedge dx^B \wedge dx^C = \hat{\rho}_{123}\; dx^1 \wedge dx^2 \wedge dx^3 \,. \tag{1.10}$$

II.2 Maxwell Equations in Differential Form

We apply the Stokes theorem to Eqs. (1.3) and (1.4) and remember that the integration domains are arbitrary. Then we arrive at the Maxwell equations in differential form:

$$dE = -\frac{\partial B}{\partial t}\,, \qquad dB = 0\,, \tag{2.1}$$

$$d\mathcal{D} = \hat{\rho}\,, \qquad d\mathcal{H} = \mathcal{J} + \frac{\partial \mathcal{D}}{\partial t}\,. \tag{2.2}$$

It is possible to put these equations into a more compact four-dimensional form. We define the field strength, an intensive quantitiy, as (even) 2-form by

$$F := E \wedge dt + B\,, \tag{2.3}$$

and the extensive quantities, the excitation and the current as (twisted) 2- and 3-forms by

$$H := -\mathcal{H} \wedge dt + \mathcal{D} \quad \text{and} \quad j := -\mathcal{J} \wedge dt + \hat{\rho}\,, \tag{2.4}$$

respectively. Then the Maxwell equations in differential form finally read

$$\boxed{dF = 0\,, \qquad dH = j\,.} \tag{2.5}$$

The Poincaré lemma yields $ddH = 0$, that is, the electric current is conserved:

$$dj = 0\,. \tag{2.6}$$

Eqs. (2.3), (2.4) and (2.5) represent the framework of Maxwell's theory. They have to be supplemented by the constitutive law.

What insight did we gain by rewriting the Maxwell eqations in exterior calculus? Let us try to get some information from Eqs. (2.5) which is not obvious from the Maxwell equations as given in Eqs. (1.1) and (1.2). An important point is that the Maxwell equations in the form of Eqs. (2.5) are valid on every

[3] With the help of the Levi-Civita symbol, we can map these 2- and 3-forms into the more familiar vector or scalar ***densities***, respectively: $\mathcal{D}^A := \frac{1}{2}\epsilon^{ABC}\mathcal{D}_{BC}$, $B^A := \frac{1}{2}\epsilon^{ABC}B_{BC}$, $\mathcal{J}^A := \frac{1}{2}\epsilon^{ABC}\mathcal{J}_{BC}$, $\hat{\rho} := \frac{1}{3!}\epsilon^{ABC}\hat{\rho}_{ABC}$. This procedure, however, does not offer new insight to us. It rather brings us nearer to the quantities entering Eqs. (1.1) and (1.2), which we commonly learn in undergraduate courses: $\boldsymbol{D} = (\sqrt{^{(3)}g}\mathcal{D}^A)$ etc., where $^{(3)}g$ is the determinant of the three-dimensional Euclidian metric g_{CD}.

four-dimensional differentiable manifold. Eqs. (2.5) are invariant under arbitrary diffeomorphisms. We do neither need a connection nor a metric for the formulation of Eqs. (2.5). Therefore the Maxwell equations, in the form (2.5), are valid in a Weyl spacetime, a Riemann-Cartan spacetime or even in a metric-affine spacetime (see [48, 50]). In particular, Eqs. (2.5) are correct in GR and in SR in arbitrary non-inertial frames and coordinates.

In applications, one often needs a formulation in terms of the frame components of the field quantities:

$$F = \frac{1}{2}F_{\alpha\beta}\,\vartheta^\alpha \wedge \vartheta^\beta\,, \quad H = \frac{1}{2}H_{\alpha\beta}\,\vartheta^\alpha \wedge \vartheta^\beta\,, \quad j = \frac{1}{6}j_{\alpha\beta\gamma}\,\vartheta^\alpha \wedge \vartheta^\beta \wedge \vartheta^\gamma\,. \tag{2.7}$$

We introduce the anholonomicity 1-form $C^\alpha := d\vartheta^\alpha = \frac{1}{2}C_{\mu\nu}{}^\alpha\,\vartheta^\mu \wedge \vartheta^\nu$, see Eq. (I.4.11). Substitution of Eqs. (2.7) into Eqs. (2.5) yields the Maxwell equations in terms of frame components:

$$\partial_{[\alpha}F_{\beta\gamma]} - C_{[\alpha\beta}{}^\delta F_{\gamma]\delta} = 0\,, \quad \partial_{[\alpha}H_{\beta\gamma]} - C_{[\alpha\beta}{}^\delta H_{\gamma]\delta} = j_{\alpha\beta\gamma}\,. \tag{2.8}$$

This representation of the Maxwell theory can be used in GR or in SR, if one desires, for instance, to employ a lab frame derived in Lecture I; for interesting related work one should compare the article of Kretzschmar and Fugmann [71]. The object of anholonomicity in the lab frame is given by Eq. (I.4.13) together with Eq. (I.9.11). By substituting them into Eqs. (2.8), we find the Maxwell equations in terms of the components $F_{\alpha\beta}$ etc. of the electromagnetic field quantities with respect to the lab frame – and these are the quantities one observes in the laboratory. Therefore the $F_{\alpha\beta}$ etc. are often called *physical components* of F etc. As soon as one starts from Eqs. (2.5), the derivation of the set of Eqs. (2.8) is an elementary exercise. Many discussions on the Maxwell equations in SR in non-inertial frames would be appreciable shortened by using this formalism.

For the vacuum, the constitutive law of Maxwell's theory reads

$$H = {}^*F\,, \tag{2.9}$$

or, in terms of components,

$$H_{\alpha\beta} = \eta_{\alpha\beta}{}^{\mu\nu}\,F_{\mu\nu} \qquad \text{with} \qquad \eta_{\alpha\beta}{}^{\mu\nu} := \frac{\sqrt{|g|}}{2}\,\epsilon_{\alpha\beta\gamma\delta}\,g^{\gamma\mu}g^{\delta\nu}\,. \tag{2.10}$$

The Hodge $*$ depends on the orientation of the manifold. In Eq. (2.9) it maps an even into a twisted differential form. The constitutive tensor of the vacuum $\eta_{\alpha\beta}{}^{\mu\nu}$, because of its total antisymmetry, has only one independent component. The relation in Eq. (2.9), in contrast to what is often stated in textbooks, can and should be understood as a basic *microphysical* relation connecting the independent 2-forms H and F. If we go over to a post-classical extension of Maxwell's theory, to quantum electrodynamics in first approximation, a new effective constitutive relation of the Born-Infeld type arises, as was shown by Heisenberg and Euler [55].

Since the Maxwell equations are diffeomorphism invariant, the symmetry of the vacuum is determined by the invariance of the constitutive law in Eqs. (2.9) or (2.10). A *conformal change* of the metric,

$$g_{\alpha\beta} \longrightarrow \tilde{g}_{\alpha\beta} = e^{\lambda(x)} g_{\alpha\beta}\,, \tag{2.11}$$

yields $\tilde{g}^{\alpha\beta} = e^{-\lambda(x)} g^{\alpha\beta}$ and $\tilde{g} = e^{n\lambda(x)} g$ for the determinant $g := \det(g_{\alpha\beta})$. Therefore for spacetime dimension $n = 4$, the constitutive tensor of the vacuum is invariant under a conformal change due to Eq. (2.11):

$$\tilde{\eta}_{\alpha\beta}{}^{\mu\nu} = \eta_{\alpha\beta}{}^{\mu\nu}\,. \tag{2.12}$$

As has been discussed in [89], the conformal change due to Eq. (2.11) contains the 15-parameter group of conformal transformations of Cunningham and Bateman as a rather special case. Anyways, the statement in some books [37] that Eqs. (2.9) or (2.10) are at most Poincaré invariant, is clearly unsubstantiated.

Alternatively, we can recognize this conformal invariance from the behavior of the Hodge $*$ operator (cf. [28]). Consider the Hodge dual of a p-form $\alpha^{(p)}$ in n dimensions and employ a conformal change due to Eq. (2.11). We find:

$$\widetilde{{}^*\alpha^{(p)}} = e^{(n-2p)\lambda(x)/2}\, {}^*\tilde{\alpha}^{(p)}\,. \tag{2.13}$$

Because F is a 2-form which is invariant under a conformal change, $\tilde{F} = F$, its Hodge dual ${}^*F$, occuring in Eq. (2.9), is conformally invariant, too, and likewise the Maxwell vacuum equations $dF = 0$, $d^*F = 0$.

II.3 Maxwell Lagrangian

Let us try to find an appropriate special-relativistic Lagrangian. For that purpose we will introduce the electromagnetic potential. Since the field strength F is exact, $dF = 0$, we have at least locally

$$F = dA\,. \tag{3.1}$$

The potential is not determined uniquely. The field strength F is invariant under the *gauge transformation*

$$A \rightarrow A + d\Lambda \quad \Rightarrow \quad F \rightarrow dA + dd\Lambda = dA = F\,. \tag{3.2}$$

The total Lagrangian 4-form should consist of a free field part V of the Maxwell field and a matter part L_{mat} which describes the matter field Ψ and its coupling to A:

$$L = V + L_{\text{mat}} = V(A, dA) + L_{\text{mat}}(A, dA, \Psi, d\Psi)\,. \tag{3.3}$$

The field equations are at most of 2nd differential order, therefore the Lagrangian is assumed to be of 1st order in the fields.

Moreover, we require V to be *gauge-invariant*, that is

$$\delta V = V(A + \delta A, d[A + \delta A]) - V(A, dA) = 0\,, \tag{3.4}$$

where

$$\delta A = d\omega \quad \text{for} \quad \Lambda \approx 1 + \omega\,. \tag{3.5}$$

We obtain

$$\delta V = d\omega \wedge \frac{\partial V}{\partial A} = 0 \qquad \Leftrightarrow \qquad \frac{\partial V}{\partial A} = 0\,, \tag{3.6}$$

$$\text{or} \quad V = V(dA) = V(F)\,. \tag{3.7}$$

Hence the free field or *gauge* Lagrangian can depend on the gauge potential A only via the field strength $F = dA$. The matter Lagrangian should also be gauge-invariant. However, we postpone this discussion until Sect. 5. There we will investigate how Ψ transforms under gauge transformations.

The action reads

$$W = \int_M L\,. \tag{3.8}$$

The field equations for A are given by the stationary points of W under a variation δ of A which commutes with the exterior derivative, that is, $[\delta, d] := \delta d - d\delta = 0$, and vanishes at the boundary, i.e. $\delta A|_{\partial M} = 0$. Varying A yields:

$$\begin{aligned}\delta_A W = \int_M \delta_A L &= \int_M \delta A \wedge \frac{\partial L}{\partial A} + \delta dA \wedge \frac{\partial L}{\partial dA} \\ &= \int_M \delta A \wedge \left\{\frac{\partial L}{\partial A} - (-1)^1 d\frac{\partial L}{\partial dA}\right\} + d\left\{\delta A \wedge \frac{\partial L}{\partial dA}\right\} \\ &= \int_M \delta A \wedge \frac{\delta L}{\delta A} + \int_{\partial M} \delta A \wedge \frac{\partial L}{\partial dA}\,,\end{aligned} \tag{3.9}$$

whereby the variational derivative of the 1-form A is defined according to

$$\frac{\delta L}{\delta A} := \frac{\partial L}{\partial A} + d\frac{\partial L}{\partial dA}\,. \tag{3.10}$$

Stationarity of W leads to the *gauge field equation*

$$\frac{\delta L}{\delta A} = 0\,. \tag{3.11}$$

Keeping in mind the inhomogeneous Maxwell field equation in Eqs. (2.5), we define the field momentum conjugated to A and the matter current by

$$H := -\frac{\partial V}{\partial dA} = -\frac{\partial V}{\partial F} \qquad \text{and} \qquad j := \frac{\delta L_{\text{mat}}}{\delta A}\,, \tag{3.12}$$

respectively. Then we recover, indeed:

$$dH = j\,. \tag{3.13}$$

The homogeneous Maxwell equation is a consequence of working with the potential A, since $F = dA$ and $dF = ddA = 0$. We were also able to arrive at the inhomogeneus equation. The field momentum H and the current j are, however, only implicitly given. As we can see from Eqs. (3.12), only an explicit form of the Lagrangians V and $L_{\rm mat}$ promotes H and j to more than sheer placeholders. On the other hand, it is very satisfying to recover the structure of Maxwell's theory in such a neat way. Eqs. (3.12) represent the constitutive laws of Maxwell's theory.

For the vacuum, according to Eq. (2.9), we expect V to be quadratic in F. Let us assume more generally that the gauge Lagrangian V is homogeneous in F of degree k. Then, using Eqs. (3.12) and Euler's theorem for homogeneous functions, we have

$$F \wedge H = -F \wedge \frac{\partial V}{\partial F} = -kV\,. \tag{3.14}$$

For Maxwell's theory $k = 2$. Therefore the gauge Lagrangian, using also Eq. (2.9), reads[4]:

$$V = -\frac{1}{2} F \wedge H = -\frac{1}{2} F \wedge {}^*F\,. \tag{3.15}$$

We repeat an analogous variational procedure for the matter field Ψ, which is assumed to be a p-form. The variational derivative reads [23,36]

$$\frac{\delta L}{\delta \Psi} := \frac{\partial L}{\partial \Psi} - (-1)^p\, d \frac{\partial L}{\partial d\Psi}\,. \tag{3.16}$$

Since V does not depend on Ψ, we find for the *matter field equation*:

$$\frac{\delta L_{\rm mat}}{\delta \Psi} = 0\,. \tag{3.17}$$

All what is left to do now in this context, is to specify the matter Lagrangian explicitly.

II.4 Dirac Equation in Exterior Forms

We resume our considerations of Lecture I, Sect. 17 and will reformulate the Dirac equation by means of the calculus of exterior forms. It is well-known that the use of an orthonormal frame e_α or a coframe ϑ^α is indispensible in formulating the Dirac equation in a non-inertial frame or in a spacetime manifold which extends beyond the flat spacetime of SR.

Let us remind ourselves that the Dirac matrices γ_α obey the anticommutation relations

$$\gamma_\alpha \gamma_\beta + \gamma_\beta \gamma_\alpha = 2 o_{\alpha\beta} \mathbb{1}\,. \tag{4.1}$$

4 More generally, one could imagine the existence of a "duality rotated" Lagrangian of the type $\overset{\Theta}{V} = -(1/2)(\cos\Theta\, F \wedge {}^*F + \sin\Theta\, F \wedge F)$, where $F \wedge F = dC$, $C := A \wedge F$ (the Chern-Simons term), and Θ is the angle of the duality rotation.

The 16 elements $\{\mathbb{1}, \gamma_\alpha, \sigma_{\alpha\beta}, \gamma_5, \gamma_5\gamma_\alpha\}$ of 4×4 matrices form a basis of a *Clifford algebra.* The constant γ_α matrices can be converted into Clifford algebra-valued 1- or 3-forms, respectively:

$$\gamma := \gamma_\alpha \vartheta^\alpha , \qquad {}^*\gamma = \gamma^\alpha \eta_\alpha . \tag{4.2}$$

Here $\eta_\alpha := e_\alpha \rfloor \eta = {}^*\vartheta_\alpha$ is the coframe "density", and the volume 4-form is given by $\eta := (1/4!)\eta_{\alpha\beta\gamma\delta}\,\vartheta^\alpha \wedge \vartheta^\beta \wedge \vartheta^\gamma \wedge \vartheta^\delta$ (see Appendix). Another useful element of the Clifford algebra is the Lorentz generator

$$\sigma_{\alpha\beta} := \frac{i}{2}(\gamma_\alpha\gamma_\beta - \gamma_\beta\gamma_\alpha) , \qquad [\gamma_\alpha, \sigma_{\beta\gamma}] = 2i(o_{\alpha\beta}\,\gamma_\gamma - o_{\alpha\gamma}\,\gamma_\beta) , \tag{4.3}$$

with the associated 2-form given by

$$\sigma := \frac{1}{2}\sigma_{\alpha\beta}\,\vartheta^\alpha \wedge \vartheta^\beta = \frac{i}{2}\gamma \wedge \gamma . \tag{4.4}$$

For the metric we get the following equivalent presentations

$$g\,\mathbb{1} = o_{\alpha\beta}\,\vartheta^\alpha \otimes \vartheta^\beta\,\mathbb{1} = \gamma_{(\alpha}\gamma_{\beta)}\,\vartheta^\alpha \otimes \vartheta^\beta \quad \Leftrightarrow \quad g\,\mathbb{1} = \gamma \otimes \gamma + i\sigma , \tag{4.5}$$

if we keep in mind that the exterior product $\wedge$ is the antisymmetric part of the tensor product $\otimes$.

The Dirac wave function Ψ and its Dirac adjoint $\overline{\Psi} = \Psi^\dagger\beta$ are spinor valued 0-forms, where $\beta = \gamma_0$ in a Dirac basis and the dagger $\dagger$ denotes hermitian conjugation. The Dirac adjoint of Clifford algebra-valued forms is built according to the following definitions and rules:

$$\overline{\alpha}^{(p)} := \beta\alpha^{(p)\dagger}\beta \qquad \overline{\alpha^{(p)} \wedge \beta^{(q)}} = (-1)^{pq}\,\overline{\beta}^{(q)} \wedge \overline{\alpha}^{(p)} , \tag{4.6}$$

$$\overline{\gamma} = \beta\gamma^\dagger\beta = \gamma , \qquad \overline{\sigma} = \frac{i}{2}\overline{\gamma} \wedge \overline{\gamma} = \sigma . \tag{4.7}$$

With respect to the connection 1-form

$$\Gamma^{\alpha\beta} = \Gamma_\mu{}^{\alpha\beta}\,\vartheta^\mu = -\Gamma^{\beta\alpha} , \tag{4.8}$$

the $\overline{SO}(1,3) \cong SL(2,C)$-covariant exterior derivative D for spinors is introduced by

$$D\Psi = d\Psi + \frac{i}{4}\Gamma^{\alpha\beta} \wedge \sigma_{\alpha\beta}\Psi \quad \Rightarrow \quad \overline{D\Psi} = d\overline{\Psi} - \frac{i}{4}\Gamma^{\alpha\beta} \wedge \overline{\Psi}\sigma_{\alpha\beta} . \tag{4.9}$$

For an arbitrary frame, the Dirac Lagrangian is given by the *hermitian* $SO(1,3)$-gauge invariant 4-form

$$L_D = L(\vartheta^\alpha, \Psi, D\Psi) = \frac{i}{2}\left\{\overline{\Psi}\,{}^*\gamma \wedge D\Psi + \overline{D\Psi} \wedge {}^*\gamma\,\Psi\right\} + {}^*m\,\overline{\Psi}\Psi , \tag{4.10}$$

for which $L_D = \overline{L}_D = L_D^\dagger$, as required. The coframe ϑ^α necessarily occurs in the Dirac Lagrangian, even in SR. For the mass term, we use the short-hand notation ${}^*m = m\eta$. The hermiticity of the Lagrangian in Eq. (4.10) leads to a charge current which admits the usual probability interpretation.

The Dirac equation and its adjoint are obtained by varying L_D independently with respect to $\overline{\Psi}$ and Ψ:

$$\begin{aligned} i{}^*\gamma \wedge D\Psi + {}^*m\Psi - \frac{i}{2}(D{}^*\gamma)\Psi &= 0, \\ i\overline{D\Psi} \wedge {}^*\gamma + {}^*m\overline{\Psi} + \frac{i}{2}\overline{\Psi} D{}^*\gamma &= 0 . \end{aligned} \tag{4.11}$$

In order to identify the additional term $D{}^*\gamma$ in the Dirac equation, we note that

$$D\gamma^\alpha = 0\,, \qquad D\eta_\alpha = \eta_{\alpha\beta} \wedge T^\beta = {}^*T^\beta \wedge \vartheta_\alpha \wedge \vartheta_\beta\,, \tag{4.12}$$

$$\text{where} \quad T^\alpha := D\vartheta^\alpha = d\vartheta_\alpha + \Gamma_\beta{}^\alpha \wedge \vartheta^\beta \tag{4.13}$$

is the *torsion* 2-form of the Riemann-Cartan spacetime[5]. In Eqs. (4.12) we have used $\eta_{\alpha\beta} = {}^*(\vartheta_\alpha \wedge \vartheta_\beta)$ and the rule ${}^*\Phi^{(p)} \wedge \Psi^{(p)} = {}^*\Psi^{(p)} \wedge \Phi^{(p)}$ for ordinary forms of the same degree. Thus, the term

$$D{}^*\gamma = -{}^*T_\beta \wedge \vartheta^\beta \wedge \gamma = {}^*\gamma \wedge (e_\beta \rfloor T^\beta) \tag{4.14}$$

in the Dirac equation depends only on the trace or vector part of the torsion (cf.[135]). However, a further torsion piece is contained in the covariant exterior derivative. In order to separate out both contributions, we decompose the Riemann-Cartan connection $\Gamma_{\alpha\beta}$ (see Appendix) into the Riemannian (or Christoffel) connection $\Gamma^{\{\}}_{\alpha\beta}$ and the *contortion* 1-form $K_{\alpha\beta} = K_{\gamma\alpha\beta}\vartheta^\gamma = -K_{\beta\alpha}$. Thus

$$\Gamma_{\alpha\beta} = \Gamma^{\{\}}_{\alpha\beta} - K_{\alpha\beta}\,, \qquad T^\alpha = K^\alpha{}_\beta \wedge \vartheta^\beta . \tag{4.15}$$

By inserting this decomposition into Eq. (4.10), the Dirac Lagrangian splits into a Riemannian and a spin-contortion piece (cf. [53, 91]):

$$\begin{aligned} L_D &= L(\vartheta^\alpha, \Psi, D^{\{\}}\Psi) - \frac{1}{8}K^{\alpha\beta} \wedge \overline{\Psi}({}^*\gamma\,\sigma_{\alpha\beta} + \sigma_{\alpha\beta}\,{}^*\gamma)\Psi \\ &= L(\vartheta^\alpha, \Psi, D^{\{\}}\Psi) - \frac{i}{4}K^{\alpha\beta} \wedge \eta^\gamma\,\overline{\Psi}\gamma_{[\alpha}\,\gamma_\beta\,\gamma_{\gamma]}\Psi \\ &= L(\vartheta^\alpha, \Psi, D^{\{\}}\Psi) - \frac{i}{4}K^{[\alpha\beta\gamma]}\,\overline{\Psi}\gamma_\alpha\,\gamma_\beta\,\gamma_\gamma\Psi\,\eta . \end{aligned} \tag{4.16}$$

In these steps, we have employed the anticommutation relation

$$\sigma_{\alpha\beta}\,\gamma_\gamma + \gamma_\gamma\,\sigma_{\alpha\beta} = 2i\,\gamma_{[\alpha}\,\gamma_\beta\,\gamma_{\gamma]} \tag{4.17}$$

and the completeness relation $\vartheta^\alpha \wedge \eta_\beta = \delta^\alpha_\beta\,\eta$. Moreover, with

$$\eta = \frac{1}{6}\,{}^*\eta^{\alpha\beta\gamma} \wedge \eta_{\alpha\beta\gamma}\,, \qquad \sigma \wedge \gamma = \frac{i}{2}\,\gamma_{[\alpha}\,\gamma_\beta\,\gamma_{\gamma]}\,{}^*\eta^{\alpha\beta\gamma}\,, \tag{4.18}$$

$$\text{and} \quad K^{\alpha\beta\gamma}\eta_{\alpha\beta\gamma} = -e_\alpha \rfloor {}^*T^\alpha\,, \tag{4.19}$$

---

5 Although we are mainly interested in SR in these lectures, we here include torsion and curvature in our calculations. At the end they can be set to zero in order to obtain the special-relativistic results.

we eventually obtain

$$L = L(\vartheta^\alpha, \Psi, D^{\{\}}\Psi) - \frac{1}{12}(e_\alpha \rfloor {}^*T^\alpha) \wedge \overline{\Psi} \wedge \sigma \wedge \gamma \wedge \Psi \,. \tag{4.20}$$

The Dirac equation obtained by decomposing Eqs. (4.11) or, more directly, by varying the decomposed Lagrangian in Eq. (4.20) reads

$$i^*\gamma \wedge D^{\{\}}\Psi + {}^*m\,\Psi - \frac{1}{12}(e_\alpha \rfloor {}^*T^\alpha) \wedge \sigma \wedge \gamma \Psi = 0\,. \tag{4.21}$$

Hence, in a Riemann-Cartan spacetime a spin 1/2 test particle does only feel the axial vector part $e_\alpha \rfloor {}^*T^\alpha$ of the torsion. Of course, in SR torsion vanishes.

Multiply the Dirac equation, Eq. (4.11a), from the left by $\overline{\Psi}$ and the adjoint equation, Eq. (4.11b), from the right by Ψ. Then, by adding these equations, we find that the Dirac Lagrangian is "weakly" zero, even in a spacetime with torsion, i.e.

$$L \simeq 0\,, \tag{4.22}$$

provided the field equations, Eqs. (4.11), are fulfilled.

II.5 Dirac-Maxwell Theory and Gauging of $U(1)$

The Dirac Lagrangian

$$L_D = \frac{i}{2}\left\{\overline{\Psi}{}^*\gamma \wedge D\Psi + \overline{D\Psi}{}^*\gamma \wedge \Psi\right\} + {}^*m\overline{\Psi}\Psi \tag{5.1}$$

is invariant under a global or rigid phase transformation

$$\Psi \to \Psi' = e^{i\theta}\Psi\,, \qquad \theta = \text{const.} \tag{5.2}$$

The phase transformations form the one-dimensional Abelian Lie group $U(1)$ of unitary transformations. The generator of the group is

$$\left(\frac{d}{d\theta}e^{i\theta}\right)(0) = i\,. \tag{5.3}$$

Hence the Lie algebra of $U(1)$ is given by the imaginary numbers. In the following it will be more convenient to consider the infinitesimal transformation

$$\Psi' = (1 + i\theta)\Psi \quad \Leftrightarrow \quad \delta\Psi = i\theta\Psi\,, \qquad \theta \ll 1\,. \tag{5.4}$$

If θ becomes spacetime dependent, we find

$$\delta(D\Psi) = D\delta\Psi = i\theta D\Psi + i(d\theta)\Psi\,. \tag{5.5}$$

The last term would destroy the invariance of L_D under spacetime variations, i.e. *local* phase transformations. In order to restore invariance under these local phase or *gauge transformations*, we introduce a $U(1)$-valued potential 1-form $\mathcal{A}$ and define a new covariant exterior derivative

$$\mathcal{D} = D + \mathcal{A}. \tag{5.6}$$

The requirement of gauge covariance, i.e.

$$\delta(\mathcal{D}\Psi) = i\theta\, \mathcal{D}\Psi\,, \tag{5.7}$$

can now be satisfied, provided the potential transforms under $U(1)$ gauge transformations according to:

$$\delta\mathcal{A} = -id\theta\,. \tag{5.8}$$

The relation to the electromagnetic potential A is given by

$$\mathcal{A} = ieA\,. \tag{5.9}$$

Here e is a coupling constant which stands for the charge of the particle described by the field Ψ.

The commutator of $\mathcal{D}$ with itself does not vanish but yields

$$[\mathcal{D},\mathcal{D}]\Psi = [\mathcal{D}_\mu,\mathcal{D}_\nu]\Psi \wedge \vartheta^\mu \wedge \vartheta^\nu = \frac{i}{4}R^{\alpha\beta}\,\sigma_{\alpha\beta}\wedge\Psi + (D\mathcal{A})\wedge\Psi = \mathcal{F}\wedge\Psi\,, \tag{5.10}$$

since $R^{\alpha\beta} = 0$ in SR. The field strength $\mathcal{F} := D\mathcal{A} = d\mathcal{A}$, which is related to the electromagnetic field strength by $\mathcal{F} = ieF$, is invariant under the local phase transformation in Eq. (5.8). To get the desired coupled Lagrangian, we do not only have to replace D by $\mathcal{D}$, but we also have to add the kinetic term for the field A as in Sect. 3:

$$V = -\frac{1}{2}F\wedge {}^*F = \frac{1}{2e^2}\mathcal{F}\wedge{}^*\mathcal{F}\,. \tag{5.11}$$

In this way we obtain the hermitian gauge-invariant Dirac-Maxwell Lagrangian

$$L = \frac{i}{2}\left\{\overline{\Psi}{}^*\gamma\wedge\mathcal{D}\Psi + \overline{\mathcal{D}\Psi}{}^*\gamma\wedge\Psi\right\} + {}^*m\overline{\Psi}\Psi + \frac{1}{2e^2}\mathcal{F}\wedge{}^*\mathcal{F}\,, \tag{5.12}$$

which can also be rewritten as

$$L = L_D + \mathcal{A}\wedge\mathcal{J} + \frac{1}{2e^2}\mathcal{F}\wedge{}^*\mathcal{F}\,, \tag{5.13}$$

where the 3-forms

$$\mathcal{J} := \frac{\partial L}{\partial\mathcal{A}} = -\overline{\Psi}\,i\,{}^*\gamma\Psi\,, \qquad j := \frac{\partial L}{\partial A} = e\,\overline{\Psi}\,{}^*\gamma\,\Psi = ie\,\mathcal{J} \tag{5.14}$$

are the (antihermitian) $U(1)$-current and the (hermitian) *charge current* of the Dirac field, respectively. In SR, the coupled field equations for matter and gauge fields are, respectively,

$$\frac{\delta L}{\delta\overline{\Psi}} = 0 \quad\Leftrightarrow\quad i\,{}^*\gamma\wedge\mathcal{D}\Psi + {}^*m\,\Psi = 0\,, \tag{5.15}$$

$$\frac{\delta L}{\delta\Psi} = 0 \quad\Leftrightarrow\quad i\,\overline{\mathcal{D}\Psi}\wedge{}^*\gamma + {}^*m\,\overline{\Psi} = 0\,, \tag{5.16}$$

$$\frac{\delta L}{\delta\mathcal{A}} = 0 \quad\Leftrightarrow\quad d\,{}^*\mathcal{F} = -e^2\,\mathcal{J}\,. \tag{5.17}$$

II.6 Noether Theorem

In the following section we will derive a relation between symmetry of the action and conservation of a matter current which is due to Emmy Noether. We will prove the Noether theorem following [9, 124, 140].

Let $\Psi(x)$ be a p-form field and $L = L(x, dx, \Psi, d\Psi)$ a Lagrangian 4-form, then the action W is given by

$$W = \int_N L, \tag{6.1}$$

where N is a subdomain of the Minkowski spacetime M_4. A possible explicit dependence of L on the coordinates x means that Ψ interacts with external sources, i.e. L does not describe a closed system. Let G be a Lie group which acts on Ψ via a representation ρ and which also acts on spacetime, i.e. there exists an associated diffeomorphism ϕ. Let us consider the behavior of W under an infinitesimal variation Δ,

$$\begin{aligned}\Delta W &= \int_{\phi(N)} \hat{L}(\hat{x}, d\hat{x}, \hat{\Psi}, d\hat{\Psi}) - \int_N L(x, dx, \Psi, d\Psi) \\ &= \int_N \phi^* \hat{L}(\phi^*\hat{x}, \phi^* d\hat{x}, \phi^*\hat{\Psi}, d\phi^*\hat{\Psi}) - \int_N L(x, dx, \Psi, d\Psi) \\ &= \int_N \Delta L,\end{aligned} \tag{6.2}$$

$$\text{where} \quad \Delta L := \phi^* \hat{L} - L \tag{6.3}$$

denotes the "total" variation (see Fig. 1). For the definition of ϕ^* and for the transformation theorem, see the Appendix. Because the domain N is arbitrary, we get the relation

$$\Delta W = 0 \qquad \Leftrightarrow \qquad \Delta L = 0\,. \tag{6.4}$$

The total variation yields

$$\begin{aligned}\Delta L =& \Delta x \wedge \frac{\partial L}{\partial x} + \Delta dx \wedge \frac{\partial L}{\partial dx} + \Delta\Psi \wedge \frac{\partial L}{\partial \Psi} + \Delta d\Psi \wedge \frac{\partial L}{\partial d\Psi} \\ =& \Delta x \wedge \left\{\frac{\partial L}{\partial x} - d\frac{\partial L}{\partial dx}\right\} + \Delta\Psi \wedge \left\{\frac{\partial L}{\partial \Psi} - (-1)^p d\frac{\partial L}{\partial d\Psi}\right\} \\ &+ d\left\{\Delta x \wedge \frac{\partial L}{\partial dx} + \Delta\Psi \wedge \frac{\partial L}{\partial d\Psi}\right\}.\end{aligned} \tag{6.5}$$

To find a more appropriate representation for $\delta L/\delta x := \partial L/\partial x - d(\partial L/\partial dx)$, we use the fact that the interior product $\rfloor$ formally has the properties of a derivative, i.e.

$$\xi \rfloor L = \xi^i \partial_i \rfloor L = \xi^i \frac{\partial L}{\partial dx^i} + (\xi \rfloor \Psi) \wedge \frac{\partial L}{\partial \Psi} + (\xi \rfloor d\Psi) \wedge \frac{\partial L}{\partial d\Psi}\,. \tag{6.6}$$

Since the vector ξ is arbitrary, we get

$$\frac{\partial L}{\partial dx^i} = \partial_i \rfloor L - (\partial_i \rfloor \Psi) \wedge \frac{\partial L}{\partial \Psi} - (\partial_i \rfloor d\Psi) \wedge \frac{\partial L}{\partial d\Psi} \,. \tag{6.7}$$

If we introduce the *Lie derivative* for exterior forms by $\ell_\xi = d\xi\rfloor + \xi\rfloor d$, we have

$$\ell_\xi L = d(\xi \rfloor L) \,. \tag{6.8}$$

Here we used $dL = 0$, because L is a 4-form. From Eq. (6.7) we get

$$\ell_\xi x \wedge \frac{\partial L}{\partial dx} + \ell_\xi \Psi \wedge \frac{\partial L}{\partial d\Psi} = \xi \rfloor L - (\xi \rfloor \Psi) \wedge \frac{\delta L}{\delta \Psi} + d\left\{(\xi \rfloor \Psi) \wedge \frac{\partial L}{\partial d\Psi}\right\} , \tag{6.9}$$

and from Eqs. (6.5), (6.8), and (6.9) we obtain, with $\Delta = \ell_\xi$,

$$\frac{\delta L}{\delta x^i} = -(\partial_i \rfloor d\Psi) \wedge \frac{\delta L}{\delta \Psi} - (-1)^p (\partial_i \rfloor \Psi) \wedge d\frac{\delta L}{\delta \Psi} \,. \tag{6.10}$$

By inserting Eq. (6.10) into Eq. (6.5), we have proved a *Noether theorem:*

If the Lagrangian is invariant under the variation Δ and if the field equations are fulfilled, we have a conserved current, i.e.

$$d\left\{\Delta x \wedge \frac{\partial L}{\partial dx} + \Delta \Psi \wedge \frac{\partial L}{\partial d\Psi}\right\} \simeq 0 \,. \tag{6.11}$$

As was noticed by Bessel-Hagen in 1921, the variation of W does not need to vanish in order to find a conserved current. The weaker condition

$$\Delta L = dB_\Delta \tag{6.12}$$

is sufficient, where B_Δ is a 3-form. This leads to a conserved current, too, i.e.

$$d\left\{\Delta x \wedge \frac{\partial L}{\partial dx} + \Delta \Psi \wedge \frac{\partial L}{\partial d\Psi} - B_\Delta\right\} \simeq 0 \,. \tag{6.13}$$

The theorem in the form of Eq. (6.11) or (6.13) is not very convenient for external symmetry groups as, for example, the Poincaré group. The reason is that the Δ variation does not only contain an internal part but also an external part. This external part results from a change of the field Ψ since it is transported to another point. Therefore, let us define, see Fig. 1, the "vertical" variation δ by

$$\delta L := \hat{L} - L \,. \tag{6.14}$$

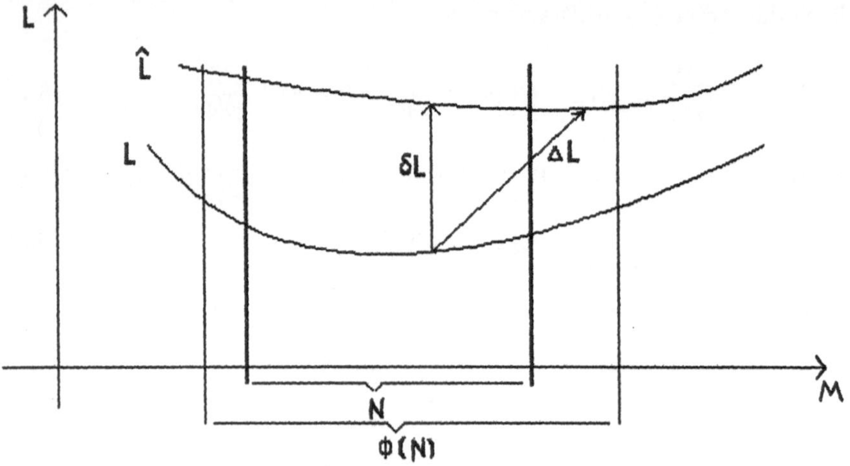

Fig. 1. "Total" and "vertical" variations.

Then we can divide the total variation into two parts:

$$\begin{aligned}\Delta L &= \phi^* \hat{L} - L = \phi^* \hat{L} - \phi^* L + \phi^* L - L = \phi^* \delta L + \phi^* L - L \\ &= \delta L + \phi^* L - L\,. \end{aligned} \qquad (6.15$$

In the case that G is an internal symmetry group, i.e. for $\phi = id$, we hav $\Delta = \delta$. If $\phi = \phi_t$ is a one parameter group, which is generated by a vector fiel ξ', we get

$$\Delta = \delta + \delta t\, \ell_{\xi'} = \delta + \ell_{(\delta t\, \xi')} =: \delta + \ell_\xi\,. \qquad (6.16$$

Then the infinitesimal variation of the Lagrangian reads

$$\Delta L = \delta L + \ell_\xi L = \delta L + d(\xi \rfloor L)\,, \qquad (6.17$$

and we get

$$\Delta L = dB_\Delta \quad \Leftrightarrow \quad \delta L = dB_\delta\,, \qquad \text{where} \quad B_\delta = B_\Delta - \xi \rfloor L\,. \qquad (6.18$$

Thereby we obtain

$$\begin{aligned}\Delta L - dB_\Delta = \delta L - dB_\delta &= \delta\Psi \wedge \frac{\delta L}{\delta \Psi} + \delta x \wedge \frac{\partial L}{\partial x} \\ &+ d\left\{ \delta x \wedge \frac{\partial L}{\partial dx} + \delta \Psi \wedge \frac{\partial L}{\partial d\Psi} - B_\delta \right\}\,. \end{aligned} \qquad (6.19$$

For $\delta L = dB_\delta$, we find the *strong Noether identity*

$$d\left\{ \delta x \wedge \frac{\partial L}{\partial dx} + \delta \Psi \wedge \frac{\partial L}{\partial d\Psi} - B_\delta \right\} = -\delta\Psi \wedge \frac{\delta L}{\delta \Psi} - \delta x \wedge \frac{\delta L}{\delta x}\,. \qquad (6.2C$$

Provided the field equation is fulfilled and $\delta L/\delta x = 0$, the current 3-form

$$j := \delta x \wedge \frac{\partial L}{\partial dx} + \delta \Psi \wedge \frac{\partial L}{\partial d\Psi} - B_\delta \tag{6.21}$$

is *weakly* conserved (*Noether theorem*):

$$dj \simeq 0\,. \tag{6.22}$$

II.7 Noether Current from an Internal Symmetry

As an example let us consider a Lagrangian $L = L(x, dx, \Psi, d\Psi)$ with a q-dimensional internal invariance group G. For the "internal" variation we require

$$\delta x = 0, \qquad \delta\Psi = \epsilon^a I_a \Psi\,, \qquad \text{and} \quad \delta L = 0\,, \tag{7.1}$$

where I_a denotes the infinitesimal generators of G, i.e. they are elements of the corresponding Lie algebra, and ϵ^a are q constant parameters. If the field equation is fulfilled, we find for the currents of the internal charges

$$j_a := I_a \Psi \wedge \frac{\partial L}{\partial d\Psi} \tag{7.2}$$

the weak conservation law

$$d(\epsilon_a j^a) = \epsilon_a dj^a \simeq 0\,. \tag{7.3}$$

Integrate dj_a over a 4-dimensional volume $\Omega := \{\, x^i \,|\, t_0 < x^0 < t_1 \}$, and assume that the currents j_a vanish fast enough for x^1, x^2, $x^3 \to \infty$. Then, by applying Stokes' theorem, we find

$$0 = \int_\Omega dj_a = \int_{\partial\Omega} j_a = \int_{\partial\Omega(x^0=t_0)} j_a + \int_{\partial\Omega(x^0=t_1)} j_a\,. \tag{7.4}$$

Since the two boundaries have opposite orientations, the q charges

$$Q_a := \int_{x^0=t} j_a \tag{7.5}$$

are conserved in time, i.e.

$$\frac{\partial Q_a}{\partial x^0} = 0\,. \tag{7.6}$$

For an internal symmetry group, the conservation law in Eq. (7.3) remains invariant if we add an exact form dY_a to the current:

$$j_a \to \hat{j}_a := j_a + dY_a \qquad \Rightarrow \qquad d\hat{j}_a = dj_a + ddY_a = 0. \tag{7.7}$$

The corresponding charges are

$$\hat{Q}_a = \int_{x^0=t} \hat{j}_a = \int_{x^0=t} j_a + \int_{\partial H_t} Y_a\,. \tag{7.8}$$

Provided the Y_a decrease rapidly enough to zero at the boundary, we obtain

$$\hat{Q}_a = Q_a \,. \tag{7.9}$$

Since we could rearrange the local charge distribution without changing the total charge, we call $\hat{j}_a$ a *relocalized current.*

II.8 Noether Currents from the Poincaré Group

As a second example let us derive the conservation laws resulting from global invariance under the ten-dimensional Poincaré group

$$P = R^4 \times_s SO(1,3)\,, \tag{8.1}$$

which is the *semidirect* product of the four-dimensional translation group and the six-dimensional Lorentz group. Within the canonical formalism, we assume that the matter Lagrangian 4-form depends on the matter field Ψ and, at most, on its first derivative:

$$L = L(g_{\alpha\beta}, \vartheta^\alpha, \Psi, D\Psi)\,. \tag{8.2}$$

In this section, we will be working in a Riemann-Cartan spacetime (U_4) and use the gauge covariant exterior derivative

$$D\Psi := d\Psi + \Gamma_\alpha{}^\beta \wedge I^\alpha{}_\beta \Psi\,, \tag{8.3}$$

in order to allow for curvilinear coordinates and for arbitrary frames which, in this section, are not necessarily orthonormal. Global Poincaré invariance of the M_4 is just a special case of local Poincaré invariance of the U_4 for vanishing torsion and curvature (see the detailed discussion in [46]).

The forms

$$\Sigma_\alpha := \frac{\partial L}{\partial \vartheta^\alpha}\,, \quad \sigma^{\alpha\beta} := 2\,\frac{\partial L}{\partial g_{\alpha\beta}}\,, \quad \text{and} \quad \tau_{\alpha\beta} := \frac{\partial L}{\partial \Gamma^{\alpha\beta}} = I_{\alpha\beta}\Psi \wedge \frac{\partial L}{\partial D\Psi} \tag{8.4}$$

are the canonical *energy-momentum*, the Hilbert stress-energy, and the *spin* currents, respectively. Here and in the following, the partial derivative with respect to the antisymmetric connection 1-form $\Gamma^{\alpha\beta} = -\Gamma^{\beta\alpha}$ is defined by $\delta L = \delta\Gamma^{\alpha\beta} \wedge (\partial L/\partial\Gamma^{\alpha\beta})$. The action

$$W = \int_N L(g_{\alpha\beta}, \vartheta^\alpha, \Psi, D\Psi) \tag{8.5}$$

for the matter Lagrangian is, by construction, invariant under the group *Diff*(M) of coordinate transformations and local frame rotations. In order to obtain a *covariant Noether identity* from invariance of L under a one-parameter group of *local* translations $\subset$ *Diff*(M), we employ the $SO(1,3)$-covariant Lie derivative $\text{Ł}_\xi := \xi\rfloor D + D\xi\rfloor$ on M with respect to an arbitrary vector field ξ (cf. [132, p.160] and [68]). Since $Dg_{\alpha\beta} = 0$, we obtain:

$$\begin{aligned}
Ł_\xi L =&(Ł_\xi\, g_{\alpha\beta})\wedge\frac{\partial L}{\partial g_{\alpha\beta}}+(Ł_\xi\vartheta^\alpha)\wedge\frac{\partial L}{\partial\vartheta^\alpha}\\
&+(Ł_\xi\Psi)\wedge\frac{\partial L}{\partial\Psi}+(Ł_\xi D\Psi)\wedge\frac{\partial L}{\partial D\Psi}\\
=&D\left[(\xi\rfloor\vartheta^\alpha)\wedge\frac{\partial L}{\partial\vartheta^\alpha}+(\xi\rfloor\Psi)\wedge\frac{\partial L}{\partial\Psi}+(\xi\rfloor D\Psi)\wedge\frac{\partial L}{\partial D\Psi}\right]\\
&-(\xi\rfloor\vartheta^\alpha)D\frac{\partial L}{\partial\vartheta^\alpha}+(\xi\rfloor T^\alpha)\wedge\frac{\partial L}{\partial\vartheta^\alpha}+(\xi\rfloor R_\beta{}^\gamma)\wedge I^\beta{}_\gamma\Psi\wedge\frac{\partial L}{\partial D\Psi}\\
&+(\xi\rfloor D\Psi)\wedge\frac{\delta L}{\delta\Psi}+(-1)^p(\xi\rfloor\Psi)\wedge D\frac{\delta L}{\delta\Psi}\,.
\end{aligned}\tag{8.6}$$

Recall that $\xi\rfloor$, which formally acts analogously to a derivative of degree -1, obeys the Leibniz rule. Since the Lagrangian L is a 4-form, its Lie derivative reduces to $Ł_\xi L = D\xi\rfloor L$. Comparing the boundary term, we can read off the identity

$$\xi\rfloor L=(\xi\rfloor\vartheta^\alpha)\wedge\frac{\partial L}{\partial\vartheta^\alpha}+(\xi\rfloor\Psi)\wedge\frac{\partial L}{\partial\Psi}+(\xi\rfloor D\Psi)\wedge\frac{\partial L}{\partial D\Psi}\,.\tag{8.7}$$

Incidentally, the left-hand side is just the term B_Δ, for $B_\delta=0$, which occurred in the non-covariant Noether theorem of Sect. 6, see Eq. (6.18).

Replacing $\xi\to e_\alpha$, Eq. (8.7) yields directly the explicit form of the *canonical energy-momentum current*

$$\Sigma_\alpha=e_\alpha\rfloor L-(e_\alpha\rfloor D\Psi)\wedge\frac{\partial L}{\partial D\Psi}-(e_\alpha\rfloor\Psi)\wedge\frac{\partial L}{\partial\Psi}\,.\tag{8.8}$$

The last term vanishes for a 0-form, as is exemplified by the Dirac field. From the non-divergence part of Eq. (8.6) we can read off the *first Noether identity*

$$\begin{aligned}
D\Sigma_\alpha =&(e_\alpha\rfloor T^\beta)\wedge\Sigma_\beta+(e_\alpha\rfloor R^{\beta\gamma})\wedge\tau_{\beta\gamma}\\
&+(e_\alpha\rfloor D\Psi)\frac{\delta L}{\delta\Psi}+(-1)^p(e_\alpha\rfloor\Psi)\wedge D\frac{\delta L}{\delta\Psi}\\
\simeq&(e_\alpha\rfloor T^\beta)\wedge\Sigma_\beta+(e_\alpha\rfloor R^{\beta\gamma})\wedge\tau_{\beta\gamma}.
\end{aligned}\tag{8.9}$$

Our first result is given in the *strong* form, where no field equation is invoked. A *weak* identities, which is denoted by $\simeq$, holds only provided the matter field equation $\delta L/\delta\Psi=0$ is satisfied.

For the derivation of the Noether identity arising from Lorentz transformations, we apply the "internal variation" of the general Noether procedure:

$$\delta L=\delta g_{\alpha\beta}\wedge\frac{\partial L}{\partial g_{\alpha\beta}}+\delta\vartheta^\alpha\wedge\frac{\partial L}{\partial\vartheta^\alpha}+\delta\Psi\wedge\frac{\partial L}{\partial\Psi}+\delta(D\Psi)\wedge\frac{\partial L}{\partial D\Psi}\,.\tag{8.10}$$

Since $\delta D\Psi=D\delta\Psi+\delta\Gamma_\alpha{}^\beta\wedge I^\alpha{}_\beta\Psi$, this is equivalent to

$$\begin{aligned}
\delta L =&\delta g_{\alpha\beta}\wedge\frac{\partial L}{\partial g_{\alpha\beta}}+\delta\vartheta^\alpha\wedge\frac{\partial L}{\partial\vartheta^\alpha}+\delta\vartheta^\alpha\wedge\frac{\partial L}{\partial\vartheta^\alpha}\\
&+\delta\Gamma_\alpha{}^\beta\wedge I^\alpha{}_\beta\Psi\wedge\frac{\partial L}{\partial D\Psi}+\delta\Psi\wedge\frac{\delta L}{\delta\Psi}+D\left(\delta\Psi\wedge\frac{\partial L}{\partial D\Psi}\right).
\end{aligned}\tag{8.11}$$

Under an infinitesimal Lorentz rotation

$$\epsilon_\alpha{}^\beta(x) := \Lambda_\alpha{}^\beta(x) - \delta_\alpha^\beta \tag{8.12}$$

of the frames, we have

$$\begin{aligned} \delta g_{\alpha\beta} &= 2\epsilon_{(\alpha}{}^\gamma\, g_{\beta)\gamma}, & \delta\vartheta^\alpha &= -\vartheta^\beta\, \epsilon_\beta{}^\alpha, \\ \delta\Gamma_\alpha{}^\beta &= D\epsilon_\alpha{}^\beta, & \delta\Psi &= -\epsilon_\alpha{}^\beta I^\alpha{}_\beta\Psi. \end{aligned} \tag{8.13}$$

Consequently, we get (cf. [50])

$$\delta L = -\epsilon_\alpha{}^\beta\left[-\sigma^{\alpha\gamma}\, g_{\gamma\beta} + \vartheta^\alpha \wedge \frac{\partial L}{\partial\vartheta^\beta} + D\frac{\partial L}{\partial\Gamma_\alpha{}^\beta} + I^\alpha{}_\beta\Psi \wedge \frac{\delta L}{\delta\Psi}\right]. \tag{8.14}$$

Thus, due to the antisymmetry of the U_4-connection, the *second Noether identity* reads

$$D\tau_{\alpha\beta} + \vartheta_{[\alpha} \wedge \Sigma_{\beta]} = -I_{\alpha\beta}\Psi \wedge \frac{\delta L}{\delta\Psi} \simeq 0. \tag{8.15}$$

Again, we distinguish between strong and weak versions.

Let us consider an isolated matter system in SR with the stipulation that the Euler-Lagrange equation $\delta L/\delta\Psi = 0$ for the matter field Ψ is fulfilled. Then $T^\alpha = 0$ and $R^{\alpha\beta} = 0$. Accordingly, *global* (or rigid) Poincaré invariance of SR yields, due to the Noether theorem, the differential identities:

$$D\Sigma_\alpha \simeq 0, \tag{8.16}$$

$$D\tau_{\alpha\beta} + \vartheta_{[\alpha} \wedge \Sigma_{\beta]} \stackrel{*}{=} D(\tau_{\alpha\beta} + x_{[\alpha} \wedge \Sigma_{\beta]}) \simeq 0. \tag{8.17}$$

These 4-form relations represent the 4 plus 6 *conservation laws* of energy-momentum and (total) angular momentum, which, alternatively, could have been obtained from Eq. (6.21). Angular momentum is composed of an intrinsic or *spin* part $\tau_{\alpha\beta} = -\tau_{\beta\alpha}$ and an orbital part $x_{[\alpha} \wedge \Sigma_{\beta]}$, a fact, which is familiar from the second, non-tensorial expression in Eq. (8.17), if we use Cartesian coordinates x^i with $x^\alpha = \delta_i^\alpha\, x^i$.

The conserved Noether currents $(\Sigma_\alpha, \tau_{\alpha\beta})$ provide, in the standard manner, a time-independent total energy-momentum and total angular momentum, respectively:

$$P_\alpha \stackrel{*}{=} \int_{H_t} \Sigma_\alpha, \qquad J_{\alpha\beta} \stackrel{*}{=} \int_{H_t} (\tau_{\alpha\beta} + x_{[\alpha} \wedge \Sigma_{\beta]}). \tag{8.18}$$

The canonical currents $(\Sigma_\alpha, \tau_{\alpha\beta})$ are only determined up to an exact form, as will be described below. Apart from this indeterminacy, these currents represent cornerstones in any interpretation of the dynamic properties of a matter field. Further evidence for their importance is Wigner's successful mass-spin classification of elementary particles [64].

II.9 Symmetric versus Asymmetric Stress

Another theory of electricity, which I prefer, denies action at a distance and attributes electric action to tensions and pressures in an all-pervading medium, these stresses being the same in kind with those familiar to engineers, and the medium being identical with that in which light is supposed to be propagated.

James Clerk Maxwell (1870) [82]

The Maxwell theory is invariant under Poincaré transformations. All we have to do to get the corresponding conservation laws is to calculate Σ_α and $\tau_{\alpha\beta}$. By definition, we obtain from the Maxwell Lagrangian in Eq. (3.15):

$$\begin{aligned} \Sigma_\alpha = e_\alpha \rfloor V - (e_\alpha \rfloor dA) \wedge \frac{\partial V}{\partial(dA)} &= -\frac{1}{2} e_\alpha \rfloor (F \wedge H) + (e_\alpha \rfloor F) \wedge H \\ &= \frac{1}{2}\{(e_\alpha \rfloor F) \wedge H - F \wedge (e_\alpha \rfloor H)\} \\ &= \frac{1}{2}\{(e_\alpha \rfloor F) \wedge {}^*F - F \wedge (e_\alpha \rfloor {}^*F)\}\,. \end{aligned} \tag{9.1}$$

For the spin current we find

$$\tau_{\alpha\beta} = \frac{\partial V}{\partial \Gamma^{\alpha\beta}} = 0\,, \tag{9.2}$$

because Maxwell's equations do not couple to the Lorentz connection. From the second Noether identity we can infer that the energy-momentum current is conserved and symmetric:

$$D\Sigma_\alpha \simeq 0\,, \qquad \vartheta_{[\alpha} \wedge \Sigma_{\beta]} \simeq 0\,. \tag{9.3}$$

For $j \neq 0$ we get the Lorentz force on the right hand side

$$D\Sigma_\alpha \simeq (e_\alpha \rfloor F) \wedge j\,. \tag{9.4}$$

The Hilbert stress-energy 4-form in the Maxwell case is, because of Eq. (9.3b), equivalent to the canonical energy-momentum current in Eq. (9.1). We just have to translate the 3-form Σ_α into the 4-form

$$\sigma_{\alpha\beta} = \vartheta_\alpha \wedge \Sigma_\beta = \vartheta_{(\alpha} \wedge \Sigma_{\beta)}\,. \tag{9.5}$$

In the conventional approach, see [75], the potential A_i is a covariant vector and the spin of the field is defined in accordance with the behavior of A_i under a coordinate transformation. By contrast, here in this lecture, the A is a $U(1)$-valued 1-form and the spin is determined, see Eq. (8.4b), from the response of A to a Lorentz-rotation of the frame. Clearly, spin is linked to the Lorentz group $SO(1,3)$. Therefore the procedure presented here is more appropriate to the physical situation under consideration. It is this reason why we end up directly with the gauge-invariant energy-momentum current in Eq. (9.1), whereas in the conventional procedure the energy-momentum current has to be fixed up in order to become gauge-invariant (and symmetric at the same time).

The energy-momentum current in Eq. (9.1) expands according to $\Sigma_\alpha = \Sigma_\alpha{}^\beta\, \eta_\beta$. Its space-space components $\Sigma_\Xi{}^\Lambda$ represent the Maxwell stress. Like $\Sigma_{\alpha\beta}$, the Maxwell stress is conserved and *symmetric*. Therefore it reminds us of the Euler-Cauchy stress in continuum mechanics "familiar to engineers".

At the end of the last century, it became increasingly clear that the symmetry of the stress in continuum mechanics cannot be deduced from the angular momentum law, as it is so often pretended in textbooks in physics up to this very day, rather an independent axiom is required. The *absence* of hyperstress, in particular of spin moment stress, and of volume torque has to be postulated[6]. The spin current in Eq. (8.4c) has the spin moment stress of continuum mechanics (also called couple stress or torque stress) as its analog. The absence of spin moment stress in the Maxwell field[7], see Eq. (9.2), is the cause of the symmetry of the energy-momentum current (9.1) in vacuum. In media where volume torque may emerge, the symmetry of Eq. (9.1) may be spoiled.

Consequently, the Maxwell field causes symmetric stress in vacuum – and it was Lorentz who had shown that stress can always be understood as a momentum flux density, a concept so beautifully verified by the experiments of Lebedew and Gerlach. Analogously, spin moment stress (introduced by Voigt in 1887) corresponds to a spin flux density. In other words, by the Maxwell field only momentum (and energy) is transported but no (gauge invariant) spin. Therefore the mechanical situtation built up by the Maxwell field is an image of the situation in ordinary continuum mechanics with its symmetry of the stress. And Hilbert's definition

$$\sigma_{\alpha\beta} := 2\frac{\partial L}{\partial g_{\alpha\beta}} \tag{9.6}$$

of the *stress-energy current* as response of the Lagrangian to a change of the metric has the formula stress = δ(elastic energy)/δ(deformation) as its analogue. Clearly the Maxwell stress and its symmetry represent a concept which is deeply rooted in spacetime physics as well as in continuum mechanics.

If we now turn to the Dirac field, we have to expect to loose the symmetry of the energy-momentum current because of the existence of a nontrivial spin angular momentum density. We will consider the currents of the Dirac equation in Minkowski spacetime, but, for later purposes, we will keep all terms involving curvature. Regarding Ψ and $\overline{\Psi}$ as two independent fields in the Noether procedure, the energy-momentum 3-form reads

$$\Sigma_\alpha = e_\alpha \rfloor L_D - (e_\alpha \rfloor D\Psi) \wedge \frac{\partial L_D}{\partial D\Psi} - (e_\alpha \rfloor \overline{D\Psi}) \wedge \frac{\partial L_D}{\partial \overline{D\Psi}} \,. \tag{9.7}$$

---

[6] Compare [137, 62, 32] and the illuminating article of Mindlin [93]. The symmetry of the energy-momentum current and related questions have been put into a historical perspective in [45].

[7] For attempts of Henriot and others to introduce such a concept for the electromagnetic field, see the references in [45].

The terms in Eq. (8.8) which involve $e_\alpha\rfloor\Psi$ and $e_\alpha\rfloor\overline{\Psi}=0$ drop out since Ψ and $\overline{\Psi}$ are 0-forms. In view of $L_D \simeq 0$ we have

$$\Sigma_\alpha \simeq \frac{i}{2}\left\{\overline{\Psi}{}^*\gamma\wedge D_\alpha\Psi - \overline{D_\alpha\Psi}\wedge{}^*\gamma\Psi\right\}, \qquad D_\alpha := e_\alpha\rfloor D\,, \tag{9.8}$$

which is manifestly hermitian. The antisymmetric piece $\vartheta_{[\alpha}\wedge\Sigma_{\beta]}$ does not vanish. Rather, since the canonical spin current of the Dirac field is given by the hermitian 3-form

$$\tau_{\alpha\beta} := \frac{\partial L_D}{\partial\Gamma^{\alpha\beta}} = \frac{1}{8}\overline{\Psi}\left({}^*\gamma\sigma_{\alpha\beta}+\sigma_{\alpha\beta}{}^*\gamma\right)\Psi\,, \tag{9.9}$$

the angular momentum law Eq. (8.17), i.e.

$$D\tau_{\alpha\beta}+\vartheta_{[\alpha}\wedge\Sigma_{\beta]}\simeq 0\,, \tag{9.10}$$

makes the lack of symmetry of the energy-momentum current in Eq. (9.7) manifest. Nevertheless, the canonical spin of the Dirac field is not completely generic but has a specific property. From the anticommutation relation

$$\sigma_{\alpha\beta}\gamma_\gamma+\gamma_\gamma\sigma_{\alpha\beta} = 2i\gamma_{[\alpha}\gamma_\beta\gamma_{\gamma]}\,, \tag{9.11}$$

we can infer that

$$\tau_{\alpha\beta} = \tau_{\alpha\beta\gamma}\,\eta^\gamma = \frac{i}{4}\overline{\Psi}\gamma_{[\alpha}\gamma_\beta\gamma_{\gamma]}\Psi\,\eta^\gamma\,. \tag{9.12}$$

This implies that the components $\tau_{\alpha\beta\mu}=\tau_{[\alpha\beta\mu]}$ of the spin current are totally antisymmetric. Therefore again, as in the Maxwell case,

$$\sigma_{\alpha\beta} = \vartheta_{(\alpha}\wedge\Sigma_{\beta)}\,. \tag{9.13}$$

However, because of Eq. (9.9), there is a genuine antisymmetric part of the energy-momentum current in Eq. (9.7) which is induced by the Dirac spin.

In continuum-mechanical parlance, the Dirac field excites a state of stress *and* of spin moment stress of the vacuum. Continua of this type are known under names such as media with microstructure, micromorphic media, Cosserat continua etc, cf. [72, 32, 41, 62, 65, 73, 93]. One of the simplest of those models is a medium being able to carry stress and spin moment stress obeying momentum and angular momentum laws of the type given in Eqs. (8.16), (8.17). No doubt, the Riemannian spacetime of GR is not the continuum to carry such a state of generalized stress. The existence of the spin of the Dirac field is quite a definite hint that a generalized spacetime with additional local Lorentz-rotational degrees of freedom, capable of carrying spin moment stress, is a desirable "geometrical arena" for a gravitational theory of fermions. The Riemann-Cartan spacetime U_4 fulfills this task. Moreover, the *dilation current* and the *shear current* can also find their proper places in a spacetime continuum with a suitable microstructure, cf. [51,96,97]. Consequently, special-relativistic investigations into the structure of external Dirac currents lead to a deeper understanding of the geometrical properties of the spacetime continuum.

II.10 Relocalization of Energy-Momentum and Spin Currents

Like for internal symmetries, also in the case of the Poincaré group the Noether currents are only determined up to an exact 2-form. This non-uniqueness has troubled physicists already for quite some time. Within gravitational theory, the question of the correct energy-momentum current is as old as GR itself [56, 30, 45]. But only Belinfante [7], in the framework of SR, and Rosenfeld [112], within GR, gave a general prescription of how one can find the "metric" or Hilbert energy-momentum current, which acts as the source on the right hand side of the Einstein field equation, from the canonical or Noether energy-momentum current of an arbitrary matter field Ψ, which is of central importance in canonical field theory. We will now turn our attention to this interrelation between the different energy-momentum currents.

The Noether identity in Eq. (8.16) also holds for an energy-momentum current which is supplemented via $\hat{\Sigma}_\alpha(\mu) = \Sigma_\alpha - D\mu_\alpha$ by an exact form[8]. This does not affect the Noether identity:

$$D\Sigma_\alpha = D\hat{\Sigma}_\alpha + DD\mu_\alpha = D\hat{\Sigma}_\alpha = 0\,. \tag{10.1}$$

If we insert $\Sigma_\alpha = \hat{\Sigma}_\alpha(\mu) + D\mu_\alpha$ into the Noether identity, Eq. (8.17), we find

$$D\tau_{\alpha\beta} + \vartheta_{[\alpha} \wedge \Sigma_{\beta]} = D(\tau_{\alpha\beta} - \vartheta_{[\alpha} \wedge \mu_{\beta]}) + \vartheta_{[\alpha} \wedge \hat{\Sigma}_{\beta]}\,. \tag{10.2}$$

If a relocalized spin $\hat{\tau}_{\alpha\beta}$ is required to fulfill again an identity of the type given in Eq. (8.17), i.e. $D\hat{\tau}_{\alpha\beta} + \vartheta_{[\alpha} \wedge \hat{\Sigma}_{\beta]} = 0$, then

$$\hat{\tau}_{\alpha\beta}(\mu, Y) = \tau_{\alpha\beta} - \vartheta_{[\alpha} \wedge \mu_{\beta]} - DY_{\alpha\beta}\,, \tag{10.3}$$

where $DY_{\alpha\beta}$ is an additional exact form with $Y_{\alpha\beta} = -Y_{\beta\alpha}$. Thus a relocalization of the energy-momentum is, up to an exact form, accompanied by an induced transformation of the canonical spin. Therefore we have the following result:

The canonical currents $(\Sigma_\alpha, \tau_{\alpha\beta})$ fulfill the Noether identities, Eqs. (8.16) and (8.17). Take arbitrary 2-forms μ_α and $Y_{\alpha\beta} = -Y_{\beta\alpha}$ as superpotentials. Then the relocalized currents

$$\Sigma_\alpha \to \hat{\Sigma}_\alpha(\mu) = \Sigma_\alpha - D\mu_\alpha\,, \tag{10.4}$$

$$\tau_{\alpha\beta} \to \hat{\tau}_{\alpha\beta}(\mu, Y) = \tau_{\alpha\beta} - \vartheta_{[\alpha} \wedge \mu_{\beta]} - DY_{\alpha\beta}\,, \tag{10.5}$$

satisfy the same relations

$$D\hat{\Sigma}_\alpha = 0\,, \tag{10.6}$$

$$D\hat{\tau}_{\alpha\beta} + \vartheta_{[\alpha} \wedge \hat{\Sigma}_{\beta]} = 0\,. \tag{10.7}$$

Accordingly, the Noether identities turn out to be invariant under the *relocalization transformation* in Eqs. (10.4) and (10.5). As a consequence, the total

[8] Strictly speaking, $d\mu_\alpha$ is an exact form, but not $D\mu_\alpha$. Since $DD = 0$ in SR, this difference does not matter, however.

energy-momentum P_α and the total angular momentum $J_{\alpha\beta}$ remain invariant up to boundary terms:

$$\hat{P}_\alpha \stackrel{*}{=} P_\alpha - \int_{\partial H_t} \mu_\alpha , \qquad \hat{J}_{\alpha\beta} \stackrel{*}{=} J_{\alpha\beta} - \int_{\partial H_t} (x_{[\alpha} \wedge \mu_{\beta]} + Y_{\alpha\beta}) . \tag{10.8}$$

Provided the superpotentials μ_α and $Y_{\alpha\beta}$ approach zero at asymptotic infinity sufficiently fast, the total quantities are not affected by the relocalization procedure.

II.11 Belinfante-Rosenfeld Symmetrization

A simple way to arrive at a symmetric energy momentum current is to require that the *relocalized spin current vanishes.* This is what the Belinfante-Rosenfeld symmetrization amounts to. Therefore we have for the Belinfante-Rosenfeld energy-momentum current σ_α the following relations:

$$\sigma_\alpha := \hat{\Sigma}_\alpha(\mu) \quad \text{for} \quad \hat{\tau}_{\alpha\beta}(\mu, Y) \stackrel{!}{=} 0 . \tag{11.1}$$

The last equation, together with Eq. (10.5), yields $\tau_{\alpha\beta} = \vartheta_{[\alpha} \wedge \mu_{\beta]} + DY_{\alpha\beta}$, which can be resolved with respect to the superpotential μ^β as follows:

$$\mu^\beta = \mathring{\mu}^\beta - 2e_\gamma \rfloor DY^{\gamma\beta} - \frac{1}{2}\vartheta^\beta \wedge (e_\gamma \rfloor e_\delta \rfloor DY^{\gamma\delta}) . \tag{11.2}$$

Here

$$\mathring{\mu}^\beta = 2e_\gamma \rfloor \tau^{\gamma\beta} + \frac{1}{2}\vartheta^\beta \wedge (e_\gamma \rfloor e_\delta \rfloor \tau^{\gamma\delta}) \tag{11.3}$$

is the proper *spin energy potential.* Observe that μ^β is traceless, i.e. $e_\beta \rfloor \mu^\beta = 0$. Then the new Noether identity in Eq. (8.16) reads alternatively $D\sigma_\alpha \stackrel{*}{=} d\sigma_\alpha = 0$. Let us collect the key formulae for our Belinfante-Rosenfeld current with the superpotential as given in Eqs. (11.2) and (11.3):

$$\sigma_\alpha = \Sigma_\alpha - D\mu_\alpha , \quad \vartheta_{[\alpha} \wedge \sigma_{\beta]} = 0 , \quad D\sigma_\alpha = 0 . \tag{11.4}$$

For $Y_{\alpha\beta} = 0$, these are the familiar Belinfante-Rosenfeld relations [7, 112].

It is remarkable that for a matter field of any spin we can find the relocalized Belinfante-Rosenfeld energy-momentum current σ_α with $D\sigma_\alpha = 0$. If we consider the motion of a "test" field in a Minkowskian spacetime, then our procedure shows that we always can attach to this motion a straight line (geodesic), irrespective of the spin.

It is perhaps surprizing that the relocalization procedure of Sect. 10 can be extended to GR. One has to be careful, however, since the Noether theorem in Eq. (8.16) in a Riemannian spacetime picks up the Mathisson-Papapetrou volume force density on its right hand side, see Eq. (8.9) [125, 45].

Assume that the canonical currents $(\Sigma_\alpha, \tau_{\alpha\beta})$ fulfill the Noether identities

$$D\Sigma_\alpha = (e_\alpha \rfloor R_{\beta\gamma}) \wedge \tau^{\beta\gamma}\,, \qquad (T^\alpha = 0)\,, \tag{11.5}$$

$$D\tau_{\alpha\beta} + \vartheta_{[\alpha} \wedge \Sigma_{\beta]} = 0\,. \tag{11.6}$$

Then the transformed currents in Eqs. (10.4), (10.5), for $Y_{\alpha\beta} = 0$, satisfy the same relations.

Now the Belinfante-Rosenfeld procedure can be executed again. We recover the Eqs. (11.4), this time, however, in the Riemannian spacetime of GR. Observe that then we have

$$D\sigma_\alpha = 0 \quad \Leftrightarrow \quad D\Sigma_\alpha = (e_\alpha \rfloor R_{\beta\gamma}) \wedge \tau^{\beta\gamma}\,. \tag{11.7}$$

Therefore the Belinfante-Rosenfeld current obeys the conventional conservation identity of GR, whereas the canonical current yields the Mathisson-Papapetrou force. This represents perhaps the most succinct derivation of this force.

Here we performed a derivation of the Belinfante-Rosenfeld energy-momentum current σ_α which only takes recourse to special-relativistic concepts. In order to link it up with the conventional approach, we have to show that the current σ_α in Eq. (11.7), as defined in Eq. (11.4a) for $Y_{\alpha\beta} = 0$, is related to the Hilbert current $\sigma_{\alpha\beta} = 2(\partial L/\partial g_{\alpha\beta})$ of Sect. 9 via $\sigma_\alpha = e^\beta \rfloor \sigma_{\alpha\beta}$. That this is, indeed, the case has been shown in a separate paper [70], see also [50].

We have tried to extend the relocalization prescription to the Riemann-Cartan spacetime of the Poincaré gauge theory. Then the 1*st* Noether identity picks up again an additional volume force such that things become more involved, cf. [49, 87]. Similar structures and problems appear in supergravity, see [20].

II.12 Extra Dilation Invariance and Improved Energy-Momentum Current

If we require our matter Lagrangian not only to be Poincaré invariant but, in addition, also to be scale invariant, we obtain for the canonical dilation current Δ the special-relativistic *dilational Noether identity*:

$$D\Delta + \vartheta^\alpha \wedge \Sigma_\alpha \stackrel{*}{=} D(\Delta + x^\alpha \wedge \Sigma_\alpha) = 0\,. \tag{12.1}$$

Here $x^\alpha \wedge \Sigma_\alpha$ is the orbital part of the dilation current in Cartesian coordinates. The intrinsic dilational current is given by

$$\Delta := W\Psi \wedge \frac{\partial L}{\partial(D\Psi)}\,, \tag{12.2}$$

where it is understood that a field with weight W transforms under a scale transformation e^ω according to $\Psi(x) \to \tilde{\Psi}(\tilde{x}) = (e^\omega)^W \Psi(e^\omega x)$.

In SR, the dilational Noether identity plays a crucial role in the construction of the *improved* energy-momentum current [16, 22] which is required to have a "soft", i.e. derivative-free trace for scalar fields. Strictly speaking, the proper meaning of the dilation identity in Eq. (12.1) can be fully expounded only in a Weyl spacetime or a metric-affine spacetime which exhibit local scale invariance [48, 50].

The canonical currents $(\Sigma_\alpha, \tau_{\alpha\beta}, \Delta)$ *fulfill the Noether identities Eqs. (8.16), (8.17), and (12.1). Take arbitrary* 2*-forms* μ_α, $Y_{\alpha\beta} = -Y_{\beta\alpha}$, *and* Z *as superpotentials. Then the relocalized currents in Eqs. (10.4) and (10.5), and*

$$\Delta \to \hat{\Delta}(\mu, Z) = \Delta - \vartheta^\alpha \wedge \mu_\alpha - DZ \tag{12.3}$$

satisfy the same relations Eqs. (10.6), (10.7), and

$$D\hat{\Delta} + \vartheta^\alpha \wedge \hat{\Sigma}_\alpha = 0\,. \tag{12.4}$$

For the *improved* energy-momentum current $\not{\sigma}_\alpha$ we require that, additionally, the intrinsic part of the associated dilation current vanishes, i.e.

$$\not{\sigma}_\alpha := \hat{\Sigma}_\alpha(\mu) \quad \text{for} \quad \hat{\tau}_{\alpha\beta}(\mu, Y) \stackrel{!}{=} 0 \quad \text{and} \quad \hat{\Delta}(\mu, Z) \stackrel{!}{=} 0\,. \tag{12.5}$$

Consequently, the (pseudoscalar) *trace* of our new energy-momentum current vanishes:

$$\vartheta^\alpha \wedge \not{\sigma}_\alpha = \vartheta^\alpha \wedge \Sigma_\alpha + D\Delta - DDZ = 0\,. \tag{12.6}$$

Thus the improved energy-momentum current is symmetric, traceless, and divergence-free:

$$\vartheta_{[\alpha} \wedge \not{\sigma}_{\beta]} = 0\,, \qquad \vartheta^\alpha \wedge \not{\sigma}_\alpha = 0\,, \qquad D\not{\sigma}_\alpha = 0\,. \tag{12.7}$$

In order to find the explicit form of $\not{\sigma}_\alpha$, we split the traceless spin energy potential into its tensor and axial parts [47] according to

$$\mu^\alpha = \widehat{\mu}^\alpha + \frac{1}{3}{}^*\{{}^*(\vartheta^\beta \wedge \mu_\beta) \wedge \vartheta^\alpha\}\,. \tag{12.8}$$

From Eq. (12.5), together with Eq. (12.3), we find

$$\vartheta^\beta \wedge \mu_\beta = \Delta - DZ\,. \tag{12.9}$$

Thus the improved energy momentum current takes the more explicit form, cf. [69, 49]:

$$\begin{aligned} \not{\sigma}_\alpha &= \Sigma_\alpha - D\widehat{\mu}_\alpha - \frac{1}{3}{}^*\{{}^*(D\Delta) \wedge \vartheta_\alpha\} \\ &= \not{\Sigma}_\alpha + \frac{1}{4}{}^*\{{}^*(\Sigma_\gamma \wedge \vartheta^\gamma) \wedge \vartheta_\alpha - D\widehat{\mu}_\alpha - \frac{1}{3}{}^*\{{}^*(D\Delta) \wedge \vartheta_\alpha\} \\ &= \not{\Sigma}_\alpha - D\widehat{\mu}_\alpha - \frac{1}{12}{}^*\{{}^*(D\Delta) \wedge \vartheta_\alpha\}\,. \end{aligned} \tag{12.10}$$

Continuing with SR, one may speculate on the case of a vanishing relocalized energy-momentum current, i.e. $\hat{\Sigma}_\alpha = 0$. Then

$$\Sigma_\alpha = D\mu_\alpha \quad \Rightarrow \quad P_\alpha = \int_{\partial H_t} \mu_\alpha\,, \tag{12.11}$$

and the other Noether identities no longer contain orbital pieces:

$$D\hat{\tau}_{\alpha\beta} = 0, \qquad D\hat{\Delta} = 0\,. \tag{12.12}$$

This would correspond to a dilation-invariant *pure spin system.* A non-vanishing total energy-momentum can only result from a non-trivial boundary term for the spin energy potential.

II.13 Gordon Decomposition of the $U(1)$-Current

The current $\mathcal{J}$ as well as the Lagrangian of the Dirac field can be decomposed into a convective and a polarization part. By definition, the convective part should resemble the current of a Klein-Gordon (scalar) field. Therefore it should not contain pieces originating from the spin of the Dirac particle. Here we rederive this *Gordon decomposition* within the calculus of exterior forms: To this end, we are going to show that *weakly*, i.e. provided that the Dirac equation holds, we have the following relation for the spinor field and its adjoint[9]:

$${}^*\gamma\Psi \simeq \frac{i}{m}\left\{{}^*D\Psi - i\,{}^*\sigma \wedge D\Psi\right\} \quad \Rightarrow \quad \overline{\Psi}\,{}^*\gamma \simeq -\frac{i}{m}\left\{\overline{{}^*D\Psi} + i\overline{D\Psi}\wedge {}^*\sigma\right\}. \tag{13.1}$$

In order to prove this relation, we wedge it from the left with γ and employ the algebraic identities

$$\gamma \wedge {}^*\gamma = 4\,\eta = {}^*4\,, \qquad \gamma \wedge {}^*\sigma = -3i\,{}^*\gamma\,. \tag{13.2}$$

In SR (with $T^\alpha = 0$) we thereby find *weakly*

$$\begin{aligned} &\gamma \wedge {}^*\gamma\Psi - \frac{i}{m}\left\{\gamma \wedge {}^*D\Psi - i\gamma \wedge {}^*\sigma \wedge D\Psi\right\} \\ &= {}^*4\,\Psi - \frac{i}{m}\left\{-{}^*\gamma \wedge D\Psi - 3\,{}^*\gamma \wedge D\Psi\right\} \simeq 0\,. \end{aligned} \tag{13.3}$$

[9] In a Riemann-Cartan spacetime with nonvanishing vector torsion, there occurs the additional term $(i/2m)\,{}^*\gamma \wedge {}^*(D\,{}^*\gamma)\Psi$ on the right-hand side of Eq. (13.1a).

If we insert the weak relation in Eq. (13.1) and its adjoint into the Dirac current $\mathcal{J} = -i\overline{\Psi}{}^*\gamma\Psi$ and write the result in an explicit antihermitian form, we obtain

$$\begin{aligned}\mathcal{J} &\simeq -\frac{1}{2m}\left\{({}^*\overline{\mathcal{D}\Psi} + i\overline{\mathcal{D}\Psi}\wedge{}^*\sigma)\Psi - \overline{\Psi}({}^*\mathcal{D}\Psi - i{}^*\sigma\wedge\mathcal{D}\Psi\right\}\\ &= -\frac{1}{2m}\left\{{}^*\overline{\mathcal{D}\Psi}\Psi - \overline{\Psi}{}^*\mathcal{D}\Psi + i\left(\overline{\mathcal{D}\Psi}\wedge{}^*\sigma\Psi + \overline{\Psi}{}^*\sigma\wedge\mathcal{D}\Psi\right)\right\}\\ &= -\frac{1}{2m}\left\{{}^*\overline{\mathcal{D}\Psi}\Psi - \overline{\Psi}{}^*\mathcal{D}\Psi\right\} - \frac{i}{2m}d\left\{\overline{\Psi}\,{}^*\sigma\Psi\right\}\\ &= \mathcal{J}^c + \mathcal{J}^p\,.\end{aligned} \tag{13.4}$$

In this way, the $U(1)$-current gets decomposed into the convective current

$$\mathcal{J}^c := \frac{1}{2m}\left\{\overline{\Psi}{}^*\mathcal{D}\Psi - {}^*\overline{\mathcal{D}\Psi}\Psi\right\} \qquad (= \frac{1}{ie}j^c) \tag{13.5}$$

and the polarization current

$$\mathcal{J}^p := -\frac{i}{2m}d\{\overline{\Psi}\,{}^*\sigma\Psi\} = -d\mathcal{P} \qquad (= \frac{1}{ie}j^p)\,. \tag{13.6}$$

The polarization current turns out to be the divergence of the 2-form

$$\mathcal{P} := \frac{1}{2m}(\overline{\Psi}\,{}^*\sigma i\Psi) \qquad (= \frac{1}{ie}P) \tag{13.7}$$

of the *magnetic moment density*. In a similar manner, the Lagrangian can also be decomposed into convective and polarization pieces: We insert Eq. (13.1) into the minimally coupled Dirac Lagrangian in Eq. (5.12) (without the Maxwell piece). This yields weakly

$$L_{\mathcal{D}} \simeq \frac{1}{2}L_{\mathcal{D}} \simeq L^c + L^p \simeq 0\,, \tag{13.8}$$

$$\text{where}\quad L^p := \frac{i}{2m}\overline{\mathcal{D}\Psi}\wedge{}^*\sigma\wedge\mathcal{D}\Psi \tag{13.9}$$

is the polarization part, and

$$L^c(\mathcal{D}) := \frac{1}{4m}\left\{{}^*\overline{\mathcal{D}\Psi}\wedge\mathcal{D}\Psi - \overline{\mathcal{D}\Psi}\wedge{}^*\mathcal{D}\Psi + 2{}^*m^2\overline{\Psi}\Psi\right\} \tag{13.10}$$

is a Klein-Gordon type Lagrangian, which is also minimally coupled. The factor 1/2 in Eq. (13.8) takes into account that we have weakly transformed the linear first order Dirac Lagrangian into a quadratic one. Then we can recover the convective and the polarization currents, respectively, by

$$\mathcal{J}^c = \frac{\partial L^c}{\partial\mathcal{A}}\,, \qquad \mathcal{J}^p = \frac{\partial L^p}{\partial\mathcal{A}}\,. \tag{13.11}$$

Using the relation in Eq. (5.10) for the commutator of $\mathcal{D}$ with itself, the polarization part of the Lagrangian can be partly converted into a pure boundary term:

$$L^p = -\mathcal{F} \wedge \mathcal{P} + d\mathcal{M} . \tag{13.12}$$

Here, the 3-form

$$\mathcal{M} := \frac{i}{4m} \left\{ \overline{\Psi}{}^*\sigma \wedge \mathcal{D}\Psi - \overline{\mathcal{D}\Psi} \wedge {}^*\sigma\Psi \right\} , \tag{13.13}$$

is, due to $\mathcal{D} = D + \mathcal{A}$, related to the 2-form of the magnetic moment density via

$$\mathcal{P} = \frac{\partial \mathcal{M}}{\partial \mathcal{A}} . \tag{13.14}$$

Since $\mathcal{A}$ is the gauge potential associated with the Abelian group $U(1)$, we can extract it from the convective part L^c of the Dirac Lagrangian and find eventually

$$\frac{1}{2} L_{\mathcal{D}} \simeq L^c(D) + \mathcal{A} \wedge \mathcal{J}^c - \mathcal{F} \wedge \mathcal{P} + d\mathcal{M} , \tag{13.15}$$

where $L^c(D) := (1/2m)({}^*D\overline{\Psi} \wedge D\Psi + {}^*m^2\overline{\Psi}\Psi)$.

II.14 Gyro-Magnetic Ratio for a Dirac Particle

Next let us investigate the behavior of an electron in a *constant* electromagnetic field F. Under this condition, the Lie derivative of F along the vector field

$$x := x^j \frac{\partial}{\partial x^j} \quad \text{with} \quad \ell_x dx^i = dx^i \tag{14.1}$$

$$\text{yields} \qquad \ell_x F = \frac{1}{2} F_{ij}\, 2(\ell_x dx^i) \wedge dx^j = 2F . \tag{14.2}$$

Note that we have to restrict our analysis to Cartesian coordinates since only then x is guaranteed to be a vector field. Due to the homogeneous Maxwell equation $dF = 0$, Eq. (14.2) is equivalent to

$$\ell_x F = d(x \rfloor F) = 2F = 2dA , \tag{14.3}$$

an equation which can be solved by (cf. [8, 35])

$$A = \frac{1}{2} x \rfloor F . \tag{14.4}$$

Inserting this into Eq. (13.15), we obtain

$$L_{\mathcal{D}} = L^c(D) - \frac{1}{2} F \wedge \left(x \rfloor j^c + 2P \right) + d\mathcal{M} . \tag{14.5}$$

In concordance with the definition $\boldsymbol{L} := \boldsymbol{x} \times \boldsymbol{p}$ of the angular momentum operator and the spin operator $\boldsymbol{S}$, we introduce the 2-forms

$$L = \frac{m}{e}\, x \rfloor j^c = \overline{\Psi}\, x \rfloor i {}^*D\Psi\,, \qquad S = \frac{m}{e}\, P = -\frac{1}{2}\overline{\Psi}\, {}^*\sigma\Psi\,. \tag{14.6}$$

Then the Dirac Lagrangian in a constant field F reads

$$L_{\mathcal{D}} = L^c(D) - \frac{e}{2m} F \wedge (L + 2S) + d\mathcal{M}\,. \tag{14.7}$$

We can read off the gyro-magnetic ratio of the spin relative to the orbital angular momentum as:

$$g_S/g_L = 2\,. \tag{14.8}$$

Our results are in compliance with the definition of the magnetic moment in standard textbooks [1, 6]:

$$\boldsymbol{M} = \boldsymbol{M}_L + \boldsymbol{M}_S = \frac{e}{2m}(g_L \boldsymbol{L} + g_S \boldsymbol{S})\,, \tag{14.9}$$

where g_L and g_S are the gyromagnetic factors of the orbital and spin angular momentum, respectively. (Remember that e is the charge of the particle which for an electron is $e < 0$.)

Note that the structure of Eq. (14.6b) is common to all moment densities which are typically of the form

$$M \sim \overline{\Psi}\, {}^*\sigma I_a \Psi\,. \tag{14.10}$$

Here ${}^*\sigma$ is the spin density indicating that the transport of the quantity defined by $I_a\Psi$ takes place because of the spin flux of the underlying field, and I_a represents the generator(s) of the corresponding gauge group, that is i in the case of $U(1)$.

II.15 Gordon Decomposition of the Energy-Momentum and Spin Currents

It is possible to decompose (weakly) the Poincaré-invariant Dirac Lagrangian and the energy-momentum and spin currents into a convective and a polarization part, respectively, in close analogy to the Gordon decomposition of the $U(1)$-current. The convective part should contain, by definition, only "Schrödinger pieces" but no contributions from spin via ${}^*\sigma$. For the convective Lagrangian we therefore make the ansatz

$$L^c := \frac{1}{4m}\left\{\overline{{}^*D\Psi} \wedge D\Psi - \overline{D\Psi} \wedge {}^*D\Psi + 2\,{}^*m^2\overline{\Psi}\Psi\right\}\,, \tag{15.1}$$

which resembles the Klein-Gordon Lagrangian describing spinless particles. From the Dirac equation we find, analogously to Eq. (13.1), the *weak* relation

$${}^*\gamma\Psi \simeq \frac{i}{m}\{{}^*D\Psi - i\,{}^*\sigma \wedge D\Psi\} \quad \Rightarrow \quad \overline{\Psi}{}^*\gamma \simeq -\frac{i}{m}\{\overline{{}^*D\Psi} + i\overline{D\Psi} \wedge {}^*\sigma\}\,. \tag{15.2}$$

If we insert it into the Dirac Lagrangian, we get

$$\frac{1}{2}L_D \simeq L^c + L^p \simeq 0 \qquad \Rightarrow \qquad L^p \simeq -L^c\,, \tag{15.3}$$

where

$$L^p := \frac{i}{2m}\overline{D\Psi} \wedge {}^*\sigma \wedge D\Psi \tag{15.4}$$

is the polarization part and the factors 1/2 in Eqs. (15.1), (15.3), and (15.4) account for the fact that the Dirac Lagrangian has been transformed weakly into a Lagrangian which is quadratic in the first derivatives.

In the polarization part, we can separate out a boundary term and find, in the case $T^\alpha = 0$, the equivalent form

$$\begin{aligned} L^p &= -R^{\alpha\beta} \wedge M_{\alpha\beta} + \frac{i}{4m} D\left\{\overline{\Psi}{}^*\sigma \wedge D\Psi - \overline{D\Psi} \wedge {}^*\sigma\Psi\right\} \\ &= -R^{\alpha\beta} \wedge M_{\alpha\beta} + dM\,, \end{aligned} \tag{15.5}$$

where we have defined the 3-form

$$M := \frac{i}{4m}\left\{\overline{\Psi}{}^*\sigma \wedge D\Psi - \overline{D\Psi} \wedge {}^*\sigma\Psi\right\}\,, \tag{15.6}$$

and the 2-form

$$M_{\alpha\beta} := \frac{\partial M}{\partial \Gamma^{\alpha\beta}} = -\frac{1}{16m}\overline{\Psi}({}^*\sigma\sigma_{\alpha\beta} + \sigma_{\alpha\beta}{}^*\sigma)\Psi\,. \tag{15.7}$$

Comparing this definition with the general structure of the moment densities in Eq. (14.10), and reminding ourselves that the $\sigma_{\alpha\beta}$ are proportional to the generators of the Lorentz group, it is natural to identify $M_{\alpha\beta}$ with the density of the *rotational gravitational moment.*

Furthermore, we introduce, in analogy to Eq. (14.10) and (15.7), the 2-form

$$M_\alpha := -\frac{i}{4m}\left\{\overline{\Psi}\,{}^*\sigma D_\alpha\Psi - \overline{D_\alpha\Psi}\,{}^*\sigma\Psi\right\}\,. \tag{15.8}$$

Since D_α are the generators of a parallel displacement, one cannot do anything else but to identify M_α with the *translational gravitational moment.* We recognize, however, that the variation of M with respect to the coframe rather yields the modified expression

$$\begin{aligned} \check{M}_\alpha &:= \frac{\partial M}{\partial \vartheta^\alpha} = \frac{i}{4m}\left\{\overline{\Psi}(e_\alpha\rfloor{}^*\sigma) \wedge D\Psi + \overline{D\Psi} \wedge (e_\alpha\rfloor{}^*\sigma) \wedge \Psi\right\} \\ &= M_\alpha + e_\alpha\rfloor M\,. \end{aligned} \tag{15.9}$$

Using the identity $\vartheta^\beta \wedge e_\beta\rfloor\Psi^{(p)} = p\Phi^{(p)}$, we can rewrite the Dirac Lagrangian in the equivalent form

$$\begin{aligned} \frac{1}{2}L_D &\simeq L^c - R^{\alpha\beta} \wedge M_{\alpha\beta} + \frac{1}{2}D(\vartheta^\alpha \wedge \check{M}_\alpha) \\ &= L^c - R^{\alpha\beta} \wedge M_{\alpha\beta} - \frac{1}{2}\vartheta^\alpha \wedge (d\check{M}_\alpha - \Gamma_\alpha{}^\beta \wedge \check{M}_\beta)\,. \end{aligned} \tag{15.10}$$

This result is reminiscent of the $U(1)$-relation in Eq. (13.15). Now the decomposition of the current follows from its canonical definition. Explicitly we get

$$\Sigma^c_\alpha := \frac{\partial L^c}{\partial \vartheta^\alpha} = \frac{1}{2m}\Big\{ e_\alpha \rfloor (\overline{{}^*D\Psi} \wedge D\Psi + {}^*m^2\overline{\Psi}\Psi) \\ + \overline{D_\alpha\Psi} \wedge {}^*D\Psi + \overline{{}^*D\Psi} \wedge D_\alpha\Psi \Big\}, \tag{15.11}$$

$$\Sigma^p_\alpha := \frac{\delta L^p}{\delta \vartheta^\alpha} = -D\check{M}_\alpha . \tag{15.12}$$

Since $\check{M}_\alpha$ itself depends linearly on ϑ^α and $\Gamma^{\alpha\beta}$, the factor 1/2 drops out in the variation. For the convective part of the spin current we have

$$\tau^c_{\alpha\beta} := \frac{\partial L^c}{\partial \Gamma^{\alpha\beta}} = -\frac{i}{8m}\left\{ \overline{{}^*D\Psi}\, \sigma_{\alpha\beta}\Psi - \overline{\Psi}\, \sigma_{\alpha\beta}{}^*D\Psi \right\}, \tag{15.13}$$

whereas the polarization spin current follows now from the variational derivative with respect to the connection. In terms of the moments it reads

$$\tau^p_{\alpha\beta} := \frac{\delta L^p}{\delta \Gamma^{\alpha\beta}} = -\vartheta_{[\alpha} \wedge \check{M}_{\beta]} - DM_{\alpha\beta} . \tag{15.14}$$

The same results can be obtained by decomposing the expressions in Eqs. (9.8) and (9.9) for the currents Σ_α and $\tau_{\alpha\beta}$ directly. In order to recover Eq. (15.11), one has to employ the weak relation $e_\alpha \rfloor (L^c + L^p) \simeq 0$. For the Rarita-Schwinger field a corresponding decomposition has been performed by Seitz [122], see also [123].

Let us summarize our results: Weakly, we have found the following decompositions:

$$\Sigma^c_\alpha \simeq \Sigma_\alpha + D\check{M}_\alpha = \Sigma_\alpha - \Sigma^p_\alpha , \tag{15.15}$$

$$\tau^c_{\alpha\beta} \simeq \tau_{\alpha\beta} + \vartheta_{[\alpha} \wedge \check{M}_{\beta]} + DM_{\alpha\beta} = \tau_{\alpha\beta} - \tau^p_{\alpha\beta} . \tag{15.16}$$

An explicit calculation shows that the convective current in Eq. (15.11) is symmetric, as one would expect for a Schrödinger type energy-momentum current. Consequently, in SR with $R_{\alpha\beta} = 0$, the decomposed currents have the following properties:

$$D\Sigma^c_\alpha \simeq 0 , \quad D\Sigma^p_\alpha \simeq 0 , \\ \vartheta_{[\alpha} \wedge \Sigma^c_{\beta]} \simeq 0 , \quad D\tau^c_{\alpha\beta} \simeq 0 , \quad D\tau^p_{\alpha\beta} + \vartheta_{[\alpha} \wedge \Sigma^p_{\beta]} \simeq 0 . \tag{15.17}$$

Comparing these results with our general relocalization prescription in Sect. 10, we see that a Gordon decomposition in Minkowski spacetime is nothing else but a *specific relocalization* of the currents which is *generated* by the boundary term $dM = (1/2)\, d(\vartheta^\alpha \wedge \check{M}_\alpha)$ in the Dirac Lagrangian. It yields a symmetric energy-momentum current Σ^c_α with a nonvanishing *conserved* spin current $\tau^c_{\alpha\beta}$. The spin tensor of Hilgevoord et al. [57, 58], which was constructed outside of the framework of the Lagrangian formalism, coincides with our convective spin current $\tau^c_{\alpha\beta}$ (up to a factor 2 due to our different conventions).

II.16 Gyro-Gravitational Ratio for a Dirac Particle

Let us investigate, similarly as in Sect. 14, the behavior of an electron in a field of *constant* gravitational curvature $R^{\alpha\beta}$. In order to formulate our conditions invariantly under Lorentz rotations of the anholonomic frame, we use the gauge-covariant Lie derivative $\text{L}_x := Dx\rfloor + x\rfloor D$ along the vector field x. Again, we have to restrict ourselves to Riemannian normal coordinates at a point P. Then we require

$$\text{L}_x R^{\alpha\beta} = \frac{1}{2} R_{ij}{}^{\alpha\beta}\, 2(\ell_x dx^i) \wedge dx^j = 2R^{\alpha\beta}\,. \tag{16.1}$$

Due to the Bianchi identity $DR^{\alpha\beta} = 0$, this is equivalent to

$$\text{L}_x R^{\alpha\beta} = D(x\rfloor R^{\alpha\beta}) = 2R^{\alpha\beta} = 2(d\Gamma^{\alpha\beta} - \Gamma^{\alpha\gamma} \wedge \Gamma_\gamma{}^\beta)\,. \tag{16.2}$$

Similarly as in the $U(1)$-case, this equation can be solved for a weak gravitational field by

$$\Gamma^{\alpha\beta} = \frac{1}{2} x\rfloor R^{\alpha\beta}\,. \tag{16.3}$$

Inserting this into the Gordon-decomposed Dirac Lagrangian in Eq. (15.10), we find

$$\begin{aligned} \frac{1}{2} L_D &\simeq L^c - R^{\alpha\beta} \wedge M_{\alpha\beta} + dM \\ &= L^c(d) + \Gamma^{\alpha\beta} \wedge \tau^c_{\alpha\beta} - R^{\alpha\beta} \wedge M_{\alpha\beta} + dM \\ &= L^c(d) + \frac{1}{2}(x\rfloor R^{\alpha\beta}) \wedge \tau^c_{\alpha\beta} - R^{\alpha\beta} \wedge M_{\alpha\beta} + dM \\ &= L^c(d) - \frac{1}{2} R^{\alpha\beta} \wedge \left(x\rfloor \tau^c_{\alpha\beta} + 2M_{\alpha\beta}\right) + dM\,, \end{aligned} \tag{16.4}$$

where $L^c(d) := (1/2m)({}^*d\overline{\Psi} \wedge d\Psi + {}^*m^2\overline{\Psi}\Psi)$.

According to Eqs. (15.13) and (15.7), both $\tau^c_{\alpha\beta}$ and $M_{\alpha\beta}$ are defined as partial derivatives with respect to $\Gamma^{\alpha\beta}$ such that no disposable factors are involved. In our derivation of the gyro-magnetic ratio in Sect. 14, we took $x\rfloor i^*D$ as the special-relativistic generalization of the angular momentum operator $\boldsymbol{L} := \boldsymbol{x} \times \boldsymbol{p}$. Thus, the 2-forms

$$L_{\alpha\beta} = x\rfloor \tau^c_{\alpha\beta} = \frac{1}{4m} \overline{\Psi}\, \sigma_{\alpha\beta}\, x\rfloor i^*D\Psi \tag{16.5}$$

and
$$M_{\alpha\beta} = -\frac{1}{8m} \overline{\Psi}\, \sigma_{\alpha\beta}{}^*\sigma\Psi \tag{16.6}$$

represent "expectation values" of the orbital and the intrinsic rotational gravitational moments which couple to spacetime curvature.

Then, the Dirac Lagrangian in a field of constant $R^{\alpha\beta}$ finally reads:

$$\frac{1}{2}L_D \simeq \frac{1}{2m}({}^*d\overline{\Psi} \wedge d\Psi + {}^*m^2\overline{\Psi}\Psi) - \frac{1}{2}R^{\alpha\beta} \wedge \left(L_{\alpha\beta} + 2M_{\alpha\beta}\right) + dM \,. \qquad (16.7)$$

Thereby, we can read off the ***gyro-gravitational ratio*** of the spin relative to the orbital moment operators as:

$$g_S/g_L = 2 \,. \qquad (16.8)$$

In an early paper on this issue, Peres [106] argued erroneously that this ratio is zero, although his spin-curvature coupling term[10] gave already hints to our value 2 for g_S/g_L.

Our analysis is still preliminary and needs to be extended in order to include spacetimes with torsion. Then we would expect an additional term in Eq. (16.7) of the type

$$-(1/2)\,T^\alpha \wedge (x\rfloor \Sigma^c_\alpha + 2M_\alpha) \,, \qquad (16.9)$$

but the detailed calculations we have to leave for the future.

Appendix: Some Exterior Calculus

In this appendix, some useful notation and formulae for differential forms are collected, following [47, 86]. For greater detail the reader is referred to Thirring [132], Trautman [136] and Wallner [140]. Here, we are dealing with a four-dimensional spacetime, i.e. a four-dimensional differentiable manifold with a metric g of signature $(+,-,-,-)$ and a metric-compatible connection Γ which, in general, will have nonzero torsion. Such a spacetime will be called a *Riemann-Cartan spacetime*, or a U_4. If the torsion is zero, it is pseudo-Riemannian or a V_4. Flat spacetime of SR is denoted by M_4. A vector basis, or frame, is denoted by $\boldsymbol{e}_\alpha$ ($\alpha = \hat{0}, \hat{1}, \hat{2}, \hat{3}$; $\Xi = \hat{1}, \hat{2}, \hat{3}$). The corresponding dual 1-form basis, or coframe, is given by ϑ^α. For a *holonomic* or coordinate vector basis, there exists a local coordinate system $\{x^i\}$ ($i = 0, 1, 2, 3$; $A = 1, 2, 3$) such that $\boldsymbol{e}_\alpha = \delta^i_\alpha\, \partial/\partial x^i$. Then the coframe is given by $\vartheta^\alpha = \delta^\alpha_i\, dx^i$. For $d\vartheta^\alpha \neq 0$, the basis is said to be *anholonomic*. Of particular relevance is the corresponding orthonormal frame with

$$g_{\alpha\beta} := g(\boldsymbol{e}_\alpha, \boldsymbol{e}_\beta) = \mathrm{diag}(1, -1, -1, -1) \,. \qquad (A.1)$$

Let us stress again that anholonomic indices are always taken from the Greek alphabet and holonomic indices from the Latin alphabet. Moreover, lower case indices are spacetime indices, whereas upper case indices refer only to space. The numbered anholonomic indices are distinguished from the holonomic indices by a hat.

[10] If the gravitational Lagrangian contains a CP violating Chern-Simons term, similarly as in the case of the canonical transformation [92] to Ashtekar's new variables, the gyro-gravitational ratio could even be anomalous (see, however, [66]).

The necessary and sufficient condition for a frame to be holonomic is

$$[e_\alpha, e_\beta] = 0 \quad \text{or, equivalently,} \quad C^\alpha := d\vartheta^\alpha = 0 \,. \tag{A.2}$$

The torsion 2-form T^α and the curvature 2-form $R_\beta{}^\alpha$ are defined by

$$T^\alpha = d\vartheta^\alpha + \Gamma_\beta{}^\alpha \wedge \vartheta^\beta \,, \quad \text{and} \tag{A.3}$$

$$R_\beta{}^\alpha = d\Gamma_\beta{}^\alpha + \Gamma_\gamma{}^\alpha \wedge \Gamma_\beta{}^\gamma \,, \tag{A.4}$$

respectively, where $\Gamma_\beta{}^\alpha$ is the connection 1-form. In terms of the 1-form basis, we obtain the expansions

$$\Gamma_\beta{}^\alpha = \Gamma_{\mu\beta}{}^\alpha \vartheta^\mu \,, \quad T^\alpha = \frac{1}{2} T_{\mu\nu}{}^\alpha \vartheta^\mu \wedge \vartheta^\nu \,, \quad R_\beta{}^\alpha = \frac{1}{2} R_{\mu\nu\beta}{}^\alpha \vartheta^\mu \wedge \vartheta^\nu \,, \tag{A.5}$$

where $\Gamma_{\mu\beta}{}^\alpha$, $T_{\mu\nu}{}^\alpha$, and $R_{\mu\nu\beta}{}^\alpha$ are the tetrad components of the respective forms. Since the connection is metric-compatible, we have

$$Dg_{\alpha\beta} = 0 \,, \tag{A.6}$$

where D is the exterior *covariant* derivative. Provided the tetrad components $g_{\alpha\beta}$ of the metric are constant, the connection 1-form is antisymmetric:

$$\Gamma_{\alpha\beta} = -\Gamma_{\beta\alpha} \,. \tag{A.7}$$

The Bianchi identities,

$$DT^\alpha = R_\beta{}^\alpha \wedge \vartheta^\beta \,, \qquad DR_\alpha{}^\beta = 0 \,, \tag{A.8}$$

follow immediately from Eqs. (A.3) and (A.4).

We denote the volume 4-form by η. Thus, for an arbitrarily oriented basis $\{\vartheta^0, \vartheta^1, \vartheta^2, \vartheta^3\}$, we have

$$\eta = \sqrt{|\det(g_{\mu\nu})|}\; \vartheta^0 \wedge \vartheta^1 \wedge \vartheta^2 \wedge \vartheta^3 \,. \tag{A.9}$$

For an orthonormal basis, $\sqrt{|\det(g_{\mu\nu})|} = 1$. Alternatively, we may write

$$\eta = \frac{1}{4!}\, \eta_{\alpha\beta\gamma\delta}\, \vartheta^\alpha \wedge \vartheta^\beta \wedge \vartheta^\gamma \wedge \vartheta^\delta \,, \tag{A.10}$$

where $\eta_{\alpha\beta\gamma\delta} = \sqrt{|\det(g_{\mu\nu})|}\epsilon_{\alpha\beta\gamma\delta}$ and $\epsilon_{\alpha\beta\gamma\delta}$ is the Levi-Civita permutation symbol with $\epsilon_{\hat{0}\hat{1}\hat{2}\hat{3}} = +1$. In three dimensions we have $\epsilon_{\hat{1}\hat{2}\hat{3}} = +1$. The following forms, together with η, span the algebra of (twisted) exterior forms on spacetime:

$$\begin{aligned}
\eta_\alpha &:= e_\alpha \rfloor \eta = {}^*\vartheta_\alpha = \frac{1}{3!}\eta_{\alpha\mu\nu\rho}\, \vartheta^\mu \wedge \vartheta^\nu \wedge \vartheta^\rho \,, \\
\eta_{\alpha\beta} &:= e_\beta \rfloor \eta_\alpha = {}^*(\vartheta_\alpha \wedge \vartheta_\beta) = \frac{1}{2}\eta_{\alpha\beta\mu\nu}\, \vartheta^\mu \wedge \vartheta^\nu \,, \\
\eta_{\alpha\beta\gamma} &:= e_\gamma \rfloor \eta_{\alpha\beta} = {}^*(\vartheta_\alpha \wedge \vartheta_\beta \wedge \vartheta_\gamma) = \eta_{\alpha\beta\gamma\mu}\, \vartheta^\nu \,, \\
\eta_{\alpha\beta\gamma\delta} &:= e_\delta \rfloor \eta_{\alpha\beta\gamma} = {}^*(\vartheta_\alpha \wedge \vartheta_\beta \wedge \vartheta_\gamma \wedge \vartheta_\delta) \,.
\end{aligned} \tag{A.11}$$

Here, the symbol $\rfloor$ denotes the interior product of a vector with a form and * denotes the Hodge star operator with the property that

$$^{**}\Phi^{(p)} = (-1)^{p(4-p)+1}\Phi^{(p)}. \tag{A.12}$$

The forms defined by Eq. (A.11) satisfy the following relations:

$$\begin{aligned}
\vartheta^\alpha \wedge \eta_\beta &= \delta^\alpha_\beta\, \eta\,, \\
\vartheta^\alpha \wedge \eta_{\beta\gamma} &= \delta^\alpha_\gamma\, \eta_\beta - \delta^\alpha_\beta\, \eta_\gamma\,, \\
\vartheta^\alpha \wedge \eta_{\beta\gamma\mu} &= \delta^\alpha_\mu\, \eta_{\beta\gamma} + \delta^\alpha_\gamma\, \eta_{\mu\beta} + \delta^\alpha_\beta\, \eta_{\gamma\mu}\,, \\
\vartheta^\alpha\, \eta_{\beta\gamma\mu\nu} &= \delta^\alpha_\nu\, \eta_{\beta\gamma\mu} - \delta^\alpha_\mu\, \eta_{\beta\gamma\nu} + \delta^\alpha_\gamma\, \eta_{\beta\mu\nu} - \delta^\alpha_\beta\, \eta_{\gamma\mu\nu}\,.
\end{aligned} \tag{A.13}$$

The following general relations between forms has been of use in the course of our work:

(a) For $\Phi^{(p)}$ and $\Psi^{(p)}$ of the same degree,

$$^*\Phi^{(p)} \wedge \Psi^{(p)} = {}^*\Psi^{(p)} \wedge \Phi^{(p)}\,. \tag{A.14}$$

(b) For a p-form $\Phi^{(p)}$,

$$\vartheta^\alpha \wedge (\mathbf{e}_\alpha \rfloor \Phi^{(p)}) = p\Phi^{(p)}\,, \tag{A.15}$$

$$^*(\Phi^{(p)} \wedge \vartheta_\alpha) = \mathbf{e}_\alpha \rfloor {}^*\Phi^{(p)}\,. \tag{A.16}$$

(c) If the vector-valued 2-form Φ_α and the bivector-valued 1-form $\Psi_{\alpha\beta} = -\Psi_{\beta\alpha}$ are related by

$$\Phi_\alpha = \Psi_{\alpha\beta} \wedge \vartheta^\beta\,, \tag{A.17}$$

then

$$\Psi_{\alpha\beta} = \frac{1}{2}\{\mathbf{e}_\alpha \rfloor \Phi_\beta - \mathbf{e}_\beta \rfloor \Phi_\alpha - (\mathbf{e}_\alpha \rfloor \mathbf{e}_\beta \rfloor \Phi_\gamma)\vartheta^\gamma\} = \frac{1}{2}\big(e_{\{\gamma} \rfloor e_\alpha \rfloor \Phi_{\beta\}}\big)\vartheta^\gamma\,. \tag{A.18}$$

(d) If the vector-valued 2-form Φ_α and the bivector-valued 3-form $\Psi_{\alpha\beta} = -\Psi_{\beta\alpha}$ are related by

$$\vartheta_{[\beta} \wedge \Phi_{\alpha]} = \psi_{\alpha\beta}\,, \tag{A.19}$$

then

$$\Phi_\alpha = 2\mathbf{e}^\mu \rfloor \Psi_{\alpha\mu} - \frac{1}{2}\vartheta_\alpha \wedge (\mathbf{e}_\mu \rfloor \mathbf{e}_\nu \rfloor \Psi^{\mu\nu})\,. \tag{A.20}$$

One reason why exterior forms are so important is that they represent natural objects for the integration over manifolds. In the following we will give some basic results of the corresponding integration theory.

Let M be an n-dimensional manifold with orientation, N a smooth compact p-dimensional submanifold of M with boundary ∂N, ϕ a diffeomorphism which projects N onto N', and Ψ a p-form field over N'. The induced tangential mapping $T\phi$ maps vector fields over N to vector fields over N'. Now define a p-form field $\phi^*\Psi$ over N by

$$\phi^*\Psi := \Psi(T\phi(\,\cdot\,), \ldots, T\phi(\,\cdot\,))\,, \tag{A.21}$$

i.e., if $v_1, \ldots, v_p$ are p vector fields over N, we get

$$\phi^* \Psi(v_1, \ldots, v_p) := \Psi(T\phi(v_1), \ldots, T\phi(v_p)). \tag{A.22}$$

The transformation theorem of integration theory for a p-form reads

$$\int_{N'} \Psi = \int_N \phi^* \Psi, \tag{A.23}$$

whereas for Stokes' theorem

$$\int_N d\Phi = \int_{\partial N} \Phi \tag{A.24}$$

the integrand Φ is a $(p-1)$-form. Now let ϕ_t be a one-parameter subgroup of the diffeomorphism group. Such a one-parameter group is generated by a vector field X, which, in coordinates, is given by

$$X^i := \frac{d\phi_t^i}{dt}|_{t=0}. \tag{A.25}$$

Vice versa, given a smooth vector field X, one can find the associated (local) flow ϕ_t by integrating Eq. (A.25). The Lie derivative ℓ_X is defined by

$$\ell_X \Psi := \frac{d}{dt}(\phi_t^* \Psi)|_{t=0}. \tag{A.26}$$

It can be shown that

$$\ell_X \Psi = X \rfloor d\Psi + d(X \rfloor \Psi). \tag{A.27}$$

Acknowledgments

It is a pleasure for one of us to thank Allen Hirshfeld (Dortmund) for the invitation to present lectures on gravity at his school in Bad Honnef. We thank J.Dermott McCrea (Dublin) for advice and suggestions, in particular on accelerated and rotating reference frames and on exterior calculus. One of the authors (FWH) learnt a lot from Wei-Tou Ni (Hsinchu, Taiwan) on the Dirac electron in non-inertial frames. The collaboration in Hsinchu during a three-months' stay was most enjoyable and fruitful. H.Dehnen (Konstanz), Bahram Mashhoon and Sam Werner (both at Columbia, Missouri) helped us to understand neutrons and their inertial effects. Gerhard Schäfer (Munich) gave hints to some recent literature. We are grateful to all of them.

This work was supported by the German-Israeli Foundation for Scientific Research and Development (GIF), Jerusalem and Munich, by a graduate scholarship of the State of Nordrhein-Westfalen, and by the Deutsche Forschungsgemeinschaft (DFG), Bonn.

Köln, June 1990 FWH, JL, EWM

References

1. M. Alonso and E.J. Finn: ***Fundamental University Physics***, Vol. III (Addison-Wesley, Reading, Mass. 1968).

1a. J. Anandan, Nuovo Cimento **53A**, 221 (1979);
J. Anandan and B. Lesche, Lett. Nuovo Cimento **37**, 391 (1983).

2. D.K. Atwood, M.A. Horne, C.G. Shull, and J. Arthur, Phys. Rev. Lett. **52**, 1673 (1984).
3. J. Audretsch and C. Lämmerzahl, J. Phys. **A16**, 2457 (1983).
4. J. Audretsch and C. Lämmerzahl, Ann. Phys. (Leipzig) **44**, 145 (1987).

4a. J. Bailey et al., Nature **268**, 301 (1977).

5. V.G. Baryshevskii and S.V. Cherepitsa, Class. Quantum Grav. **3**, 713 (1986).
6. L.F. Bates: ***Modern Magnetism***, 3rd edition (Cambridge University Press, Cambridge 1951).
7. F. J. Belinfante, ***Physica* 6**, 887 (1939); **7**, 449 (1940).
8. J.D. Bjorken and S. Drell: ***Relativistic Quantum Mechanics*** (McGraw-Hill, New York 1964).
9. G.W. Bluman and S. Kumei: ***Symmetries and Differential Equations*** (Springer, New York 1989).
10. U. Bonse and A. Rumpf, Phys. Rev. Lett. **56**, 2441 (1986).
11. U. Bonse and T. Wroblewski, Phys. Rev. Lett. **51**, 1401 (1983).
12. M. Born: ***Zur Statistischen Deutung der Quantentheorie*** (Battenberg Verlag, Stuttgart 1962).
13. Th. Bührke, Phys. Blätter **45**, 339 (1989);
Frankfurter Allgemeine Zeitung, 02-May (1990).
14. W.L. Burke: ***Applied differential geometry*** (Cambridge University Press, Cambridge 1985).
15. E.R. Caianiello, Lett. Nuovo Cimento **41**, 370 (1984).
16. C. Callan, S. Coleman, and R. Jackiw, Ann. Phys. (N.Y.) **59**, 42 (1970).
17. E. Cartan: ***On Manifolds with an Affine Connection and the Theory of General Relativity***, transl. from the French (Bibliopolis, Napoli 1986).
18. Y. Choquet-Bruhat, C. DeWitt-Morette, and M. Dillard-Bleick: ***Analysis, Manifolds and Physics***, revised edition (North-Holland, Amsterdam 1982).
19. A. Cimmino, G.I. Opat, A.G. Klein, H. Kaiser, S.A. Werner, M. Arif, and R. Clothier, Phys. Rev. Lett. **63**, 380 (1989).
20. T.E. Clark and S.T. Love, Phys. Rev. **D39**, 2391 (1989).
21. R. Colella, A.W. Overhauser, and S.A. Werner, Phys. Rev. Lett. **34**, 1472 (1975).
22. S. Coleman, ***Aspects of Symmetry*** (Cambridge University Press, Cambridge 1985) p. 67.

22a. R. Collier, Czech. J. Phys. **B27**, 991 (1977).

23. E.M. Corson: ***Introduction to Tensors, Spinors, and Relativistic Wave-Equations*** (Blackie, London 1953).
24. J.W.T. Dabbs, J.A. Harvey, D. Paya, and H. Horstmann, Phys. Rev. **B139**, 756 (1965).
25. R. Dandoloff, Phys. Lett. **A139**, 19 (1989).
26. B. DeFacio, P.W. Dennis, and D.G. Retzloff, Phys. Rev. **D18**, 2813 (1978).
27. H. Dehnen, in: ***Philosophie und Physik der Raum-Zeit***, J. Audretsch und K. Mainzer (eds.) (Bibliographisches Institut, Mannheim 1988) p. 182.
28. T. Dray, R. Kulkarni, and J. Samuel, J. Math. Phys. **30**, 1306 (1989).
29. R.W.P. Drever, Phil. Mag. **6**, 683 (1961).
30. A. Einstein, Sitzber. Preuss. Akad. Wiss. Berlin, p. 1111 (1916).
31. A.M. Eisele, Helv. Phys. Acta **60**, 1024 (1987).
32. A.C. Eringen and C.B. Kafadar, in: ***Continuum Physics, Vol. IV***, A.C. Eringen (ed.) (Academic Press, New York 1976) pp. 1-73.
33. E. Fischbach, D. Sudarsky, A. Szafer, C. Talmadge, and A. Aronson, Phys. Rev. Lett. **56**, 3 (1985).
34. V.L. Fitch, Rev. Mod. Phys. **53**, 367 (1981);
I.R. Kenyon, Phys. Lett. **B237**, 274 (1990).

34a. G. Gabrielse et al., Phys. Rev. Lett. **65**, 1317 (1990).

35. S. Gasiorowicz: ***Quantum Physics*** (J. Wiley and Sons, New York 1974).

36. I.M. Gelfand and S.V. Fomin: *Calculus of Variations* (Prentice Hall, Englewood Cliffs 1963).
37. M. Göckeler and T. Schücker: *Differential Geometry, Gauge Theories, and Gravity* (Cambridge University Press, Cambridge 1987).
38. T. Goldman, R.J. Hughes, and M.M. Nieto, Scientific American **258**, 32 (March 1988); H. Koch, Phys. Bl. **44**, 411 (1988); B. Schwarzschild, Physics Today **43**, 17 (July 1990).
39. M. Good, Phys. Rev. **121**, 311 (1961).
40. A.Gorbatsievich, Experimentelle Technik der Physik **27**, 529 (1979).
41. A. E. Green and R. S. Rivlin, Arch. Rat. Mech. Anal. **17**, 113 (1964).
42. D.M. Greenberger, Rev. Mod. Phys. **55**, 875 (1983).
43. D.M. Greenberger and A.W. Overhauser, Rev. Mod. Phys. **51**, 43 (1979).
44. D.M. Greenberger and A.W. Overhauser, Scientific American **242**, 54 (May 1980).
45. F.W. Hehl, Reports on Math. Phys. (Torun) **9**, 55 (1976).
46. F.W. Hehl, in: *Cosmology and Gravitation*, P.G. Bergmann and V. de Sabbata (eds.) (Plenum, New York 1980) p. 5.
47. F.W. Hehl and J.D. McCrea, Found. Phys. **16**, 265 (1986), reprinted in: *Between Quantum and Cosmos. Studies and Essays in Honor of John Archibald Wheeler*, W.H. Zurek, A. van der Merwe, and W.A. Miller (eds.) (Princeton University Press, Princeton 1988).
48. F.W. Hehl, J.D. McCrea, and E.W. Mielke, in: *Exact Sciences and their Philosophical Foundations – Vorträge des Internationalen Hermann-Weyl-Kongresses, Kiel 1985*, W. Deppert et al. (eds.) (P. Lang Verlag, Frankfurt a.M. 1988) p. 241.
49. F.W. Hehl and E.W. Mielke, Wiss. Zeitschr. Friedrich-Schiller-Universität Jena (Festschrift for Schmutzer) **39**, 58 (1990); E.W. Mielke, F.W. Hehl, and J.D. McCrea, Phys. Lett. **A104**, 368 (1989).
50. F.W. Hehl, J.D. McCrea, E.W. Mielke, and Y. Ne'eman, Found. Phys. **19**, 1075 (1989).
51. F.W. Hehl and Y. Ne'eman: "Spacetime as a continuum with microstructure and metric-affine gravity". Preprint TAUP N202-90, Tel Aviv University (Jan. 1990), (Festschrift for Ivanenko, to be published).
52. F.W. Hehl and W.-T. Ni, Phys. Rev. **D42**, 2045 (1990).
53. F.W. Hehl, P. von der Heyde, G.D. Kerlick, and J.M. Nester, Rev. Mod. Phys. **48**, 393 (1976).
54. H. Heintzmann und P. Mittelstaedt, Springer Tracts Mod. Phys. **47**, 185 (1968).
55. W. Heisenberg und H. Euler, Z. Phys. **98**, 714 (1936).
56. D. Hilbert, *Königl. Gesellsch. d. Wiss. Göttingen*, Nachr. Math.-Phys. Kl., p. 395 (1915).
57. J. Hilgevoord and E.A. De Kerf, Physica **31**, 1002 (1965).
58. J. Hilgevoord and S.A.Wouthuysen, Nucl. Phys. **40**, 1 (1963).
59. C.-H. Hsieh, P.-Y. Jen, K.-L. Ko, K.-Y. Li, W.-T. Ni, S.-s. Pan, Y.-H. Shih and R.-J. Tyan, Mod. Phys. Lett. **A4**, 1597 (1989); Y. Chou, W.-T. Ni, and S.-L. Wang, Mod. Phys. Lett. **A5**, 2297 (1990).
60. R. Huber: "Foldy-Wouthuysen-Transformation eines Spin-0 und eines Spin-$\frac{1}{2}$ Teilchens in einem Weyl-Cartan-Raum." Ph. D. thesis, University of Konstanz (1988).
61. B.R. Iyer, Phys. Rev. **D26**, 1900 (1982).
62. W. Jaunzemis: *Continuum Mechanics* (Macmillan, New York 1967).
63. J. Kalckar, J. Lindhard, and O. Ulfbeck, Mat. Fys. Medd. Dan. Vid. Sels. **40** (1982) #11.
64. Y.S. Kim and M.E. Noz: *Theory and Applications of the Poincaré Group* (Reidel, Dordrecht 1986).
65. H. Kleinert: *Gauge Fields in Condensed Matter, Vol. 2: Stresses and Defects* (World Scientific, Singapore 1989).
66. I.Yu. Kobzarev, and L.B. Okun, Sov. Phys. JETP **16**, 1343 (1963).
67. L. Koester, Phys. Rev. **D14**, 907 (1976).
68. W. Kopczyński, J.Phys. **A15**, 493 (1982).
69. W. Kopczyński, J.D. McCrea, and F.W. Hehl, Phys. Lett. **A128**, 313 (1988).
70. W. Kopczyński, J.D. McCrea, and F.W. Hehl, Phys. Lett. **A135**, 89 (1989).
71. M. Kretzschmar and W. Fugmann, Nuovo Cimento **103B**, 389 (1989).
72. E. Kröner (ed.): *Mechanics of Generalized Continua, IUTAM Symposium* (Springer, Berlin 1968).

73. E. Kröner, in ***Physics of Defects***, Les Houches, Session XXXV, 1980. R. Balian et al. (eds.) (North-Holland, Amsterdam 1981) p. 215.
74. S.K. Lamoreaux, J.P. Jacobs, B.R. Heckel, F.J. Raab, and E.N. Fortson, Phys. Rev. Lett. **57**, 3125 (1986).
75. L.D. Landau and E.M. Lifshitz: ***The Classical Theory of Fields*** (Pergamon Press, Oxford 1962).
76. L. Lange, Ber. Königl. Ges. Wiss., Math.-Phys. Kl., 333 (1885); cf. M. von Laue, Naturwiss. **35**, 193 (1948).
77. W.-Q. Li and W.-T. Ni, Chinese J. Phys. **16**, 214 (1978).
78. W.-Q. Li and W.-T. Ni, J. Math. Phys. **20**, 1473 (1979).
79. D. Lovelock, J. Math. Phys. **12**, 498 (1971).
80. B. Mashhoon, Phys. Rev. Lett. **61**, 2639 (1988).
81. B. Mashhoon, Phys. Lett. **A143**, 176 (1990); **A145**, 147 (1990).
82. J.C. Maxwell: ***The Scientific Papers of James Clerk Maxwell***, W.D.Niven (ed.), Vol. II, p. 228 (Dover, New York 1965).
83. J.D. McCrea, in: ***Proceedings of the 14th Int. Conference on Differential Geometric Methods in Mathematical Physics***, Salamanca 1985, P.L. García and A. Pérez-Rendón (eds.), Lecture Notes in Mathmatics, Vol. **1251** (Springer, Berlin 1987) p. 222.
84. J.D. McCrea: "REDUCE in General Relativity and Poincaré Gauge Theory", lectures given in Rio de Janeiro, to be published (Cambridge University Press, Cambridge 1991).
85. J.D. McCrea, private communication.
86. J.D. McCrea, "Beyond General Relativity: Theories of Gravitation with Dynamic Torsion", Int. J. Modern Phys. **A**, in preparation.
87. J.D. McCrea, F.W. Hehl, and E.W. Mielke, Int. J. Theor. Phys. **29**, 1185 (1990).
88. K. Meetz und W.L. Engl: ***Elektromagnetische Felder*** (Springer, Berlin 1980).
89. E.W. Mielke, Gen. Rel. Grav. J. **8**, 321 (1977).
90. E.W. Mielke, Z. Naturforschung **41a**, 777 (1986).
91. E.W. Mielke: ***Geometrodynamics of Gauge Fields – On the Geometry of Yang-Mills and Gravitational Gauge Theories*** (Akademie-Verlag, Berlin 1987).
92. E.W. Mielke, Phys. Rev. **D42**, 3388 (1990).
93. R.D. Mindlin, Arch. Rat. Mech. Anal. **16**, 51 (1964).
94. C.W. Misner, K.S. Thorne, and J.A. Wheeler: ***Gravitation*** (Freeman, San Francisco 1973).
95. O. Nachtmann, Acta Phys. Aust., Suppl. **6**, 485 (1969).
96. Y. Ne'eman and Dj. Šijački, Phys. Rev. **D37**, 3267 (1988).
97. Y. Ne'eman and Dj. Šijački, Phys. Lett. **B200**, 489 (1988).
98. R.A. Nelson, J. Math. Phys. **28**, 2379 (1987).
99. R.A. Nelson, Gen. Rel. Grav. J. **22**, 431 (1990).
100. W.-T. Ni, Chinese J. Phys. **15**, 51 (1977).
101. W.-T. Ni, Physica **B165** & **166** 157 (1990); R.C. Ritter, C.E. Goldblum, W.-T. Ni, G.T. Gillies, and C.C. Speake, Phys. Rev. **D42**, 977 (1990).
102. W.-T. Ni, private communication.
103. T.M. Niebauer, M.P. Hugh, and J.E. Faller, Phys. Rev. Lett. **59**, 609 (1987).
104. A. Page, Phys. Rev. Lett. **35**, 543 (1975).
104a. S. Pakvasa, W.A. Simmons, and T.J. Weiler, Phys. Rev. **D39**, 1761 (1989).
105. W. Paul, Phys. Bl. **46**, 227 (1990).
106. A. Peres, Nuovo Cimento **28**, 1091 (1963).
107. E.J. Post: ***Formal Structure of Electromagnetics*** (North-Holland, Amsterdam 1962).
108. E.J. Post, Rev. Mod. Phys. **39**, 475 (1967).
109. E.J. Post, Found. Phys. **9**, 619 (1979).
110. G. de Rham: ***Differentiable Manifolds, Forms, Currents, Harmonic Forms.*** Transl. from the French by F.R. Smith (Springer, Berlin 1984).
111. W. Rindler: ***Essential Relativity – Special, General, and Cosmology***, 2nd edition (Springer, New York 1977).
112. L. Rosenfeld, Mém. Acad. Roy. Belgique, cl. sc., **18**, fasc. 6 (1940).
113. R.K. Sachs and H. Wu: ***General Relativity for Mathematicians*** (Springer, New York 1977).
114. L.I. Schiff, Proc. Nat. Acad. Sci. (USA) **45**, 69 (1959).

115. L.I. Schiff and M.V. Barnhill, Phys. Rev. **151**, 1067 (1966).
116. E. Schmutzer and J. Plebański, Fortschr. Phys. **25**, 37 (1977).
117. J.A. Schouten: *Ricci Calculus*, 2nd ed. (Springer, Berlin 1954).
118. E. Schrödinger: *Die Struktur der Raum-Zeit*, transl. by J. Audretsch (Wiss. Buchgesellschaft, Darmstadt 1987).
119. V. Schroth: "Nicht-Inertialsysteme, das Dirac-Elektron und die Anwendung des Äquivalenzprinzips". Diploma thesis, University of Cologne (1984).
120. E. Schrüfer, F.W. Hehl, and J.D. McCrea, Gen. Rel. Grav. J. **19**, 197 (1987).
121. B. Schutz: *Geometrical Methods of Mathematical Physics* (Cambridge University Press, Cambridge 1980).
122. M. Seitz, Ann. Phys. (Leipzig) **41**, 280 (1984).
123. M. Seitz, Class. Quantum Grav. **3**, 175 (1986).
124. R.U. Sexl and H.K. Urbantke: *Relativität, Gruppen, Teilchen* (Springer, Wien 1976).
125. R.U. Sexl and H.K. Urbantke: *Gravitation und Kosmologie* (Bibliographisches Institut, Mannheim 1983).
126. I.I. Shapiro, R.D. Reasenberg, J.F. Chandler, and R.W. Babcock, Phys. Rev. Lett. **61**, 2643 (1988).
127. *Sky & Telescope*, p. 127 (Feb. 1990).
128. D. Stauffer, F.W. Hehl, V. Winkelmann, and J.G. Zabolitzky: *Computer Simulation and Computer Algebra*, 2nd edition (Springer, Berlin 1989).
129. S. Sternberg, Lecture Notes in Mathematics, Vol. **676**, 1 (Springer, Berlin 1978).
129a. L. Stodolsky, Gen. Rel. Grav. J. **11**, 391 (1979).
130. N. Straumann: *Allgemeine Relativitätstheorie und relativistische Astrophysik*, 2.Auflage, Lecture Notes in Physics **150** (Springer, Berlin 1988).
131. C.W. Stubbs, E.G. Adelberger, B.R. Heckel, W.F. Rogers, H.E. Swanson, R. Watanabe, J.H. Gundlach, and F.J. Raab, Phys. Rev. Lett. **62**, 609 (1989); E.G. Adelberger et al., Phys. Rev. **D42**, 3267 (1990).
132. W. Thirring: *Lehrbuch der Mathematischen Physik*, Bd.1 & Bd.2 (Springer, Wien 1977, 1990); see also the English transl. of these books (Springer, New York).
133. M. Toller, Int. J. Theor. Phys. **29**, 963 (1990).
134. A. Trautman, Sov. Phys. Uspekhi **9**, 319 (1967).
135. A. Trautman, in: *Differential Geometry, Symposia Matematica* **12**, p. 139 (Academic Press, London 1973).
136. A. Trautman: *Differential Geometry for Physicists* (Bibliopolis, Napoli 1984).
136a. H.-J. Treder, Ann. Phys. (Leipzig) **39**, 265 (1982).
137. C. Truesdell and R.A. Toupin: *The Classical Field Theories*, in: *Handbuch der Physik* (S. Flügge ed.), Vol. III/1, p. 226 (Springer, Berlin 1960).
138. P. von der Heyde, Lett. Nuovo Cimento **14**, 250 (1975).
139. K. Vonlanthen: "Supergravitation und Velo-Zwanziger Phänomene". Diploma thesis, University of Zürich (1978).
140. R. P. Wallner, "Feldtheorie im Formenkalkül". Ph.D. thesis, University of Vienna (1982).
141. S.A. Werner, J.L. Staudemann, and R. Colella, Phys. Rev. Lett. **42**, 1103 (1979).
142. S.A. Werner and H. Kaiser: "Neutron interferometry – macroscopic manifestations of quantum mechanics". In: *Quantum Mechanics in Curved Space-Time*, 11th Course of the Internat. School of Cosmology and Gravitation. V. deSabbata and J. Audretsch (eds.) (Plenum Press, New York, to appear 1990/91).
143. J.A. Wheeler: *A Journey into Gravity and Spacetime* (Freeman, New York 1990).
144. F.C. Witteborn and W.M. Fairbank, Rev. Sci. Instrum. **48**, 1 (1977).
145. Y.L. Wu, Phys. Lett. **A131**, 419 (1988).
146. C.Q. Xia and Y.L. Wu, Phys. Lett. **A141**, 251 (1989).

Braid Group Statistics[1]

K.-H. Rehren

Instituut voor Theoretische Fysica
P.O.B. 80.006, NL-3508 TA Utrecht,
Netherlands

Abstract

The present lectures give an introduction to the Doplicher-Haag-Roberts theory of superselection sectors and statistics [1, 2] with special emphasis on the occurrence of braid group statistics in low-dimensional space-time [3]. We sketch the Doplicher-Roberts theorem on the relation between statistics and gauge symmetry valid for permutation group statistics [4], and terminate with two-dimensional conformal field theories providing examples of braid group statistics with particularly simple kinematics.

The DHR theory itself is an example for the effectivity of the algebraic Haag-Kastler framework [5] as an approach to the understanding of structural foundations of quantum field theory. While it proved very flexible with respect to physically motivated modifications of the basic axioms [6, 7], our presentation will be restricted to the easiest case: charges localizable within *bounded* regions of space-time ("particles"). The issue of statistics of a local quantum field theory may be regarded as an "invariant", i.e. independent of the choice of a description in terms of unobservable charged fields, characterization of local algebras.

1. Introduction

In conventional approaches, a quantum field theory is defined in terms of a field algebra $\mathcal{F}$ represented in an appropriate Hilbert space $\mathcal{H}$. In general, $\mathcal{F}$ will contain unobservable fields (such as fermion multiplets $\psi_i(x)$) transforming under some *gauge symmetry*. There is a subalgebra $\mathcal{A}$ of observables, consisting of the gauge invariant operators in $\mathcal{F}$, such as currents $\bar{\psi}\gamma_\mu\psi$. With respect to the observables, the full Hilbert space $\mathcal{H}$ will decompose into irreducible subspaces $\mathcal{H}_\alpha$, transforming differently under the gauge group. Consequently, any non-vanishing interference between states in different such subspaces would

[1] based on a review by D.Kastler, M.Mebkhout, K.-H. Rehren (ref.[10])

provide a "frame" to experimentally fix the gauge. Thus, the very concept of gauge symmetry forbids such interference. The inequivalent subspaces $\mathcal{H}_\alpha$ are called *superselection sectors*. Only charged (gauge non-invariant) fields can interpolate among different sectors.

It is known from examples, that mathematically inequivalent field algebras may contain the same subalgebra of observables and generate the same superselection sectors: they describe the same physics. In choosing the "correct" field algebra, one is guided by convenience, e.g., if one wants to study perturbations of the symmetry.

The following questions arise. Given "the physics", i.e. the algebra of observables and its relevant representations. To what extent are the non-observable fields creating superselection charges, and their algebraic properties determined? Is there an algebra at all consisting of objects which deserve the name $\psi(x)$, i.e. commute with observables $\varphi(y)$ at space-like distance $y - x$, and describing unphysical operations like charge creation?

In order to answer questions like these, it is necessary to study the algebra $\mathcal{A}$ and its representations $\mathcal{H}_\alpha$ intrinsically. In particular, one has to find a composition prescription for its representations which reflects the observable content of the conventional prescription using charged fields, namely if ψ_α and ψ_β generate representations $\mathcal{H}_\alpha = \mathcal{A}\psi_\alpha\Omega$ and $\mathcal{H}_\beta = \mathcal{A}\psi_\beta\Omega$ from the vacuum state Ω, then the composite representation is the one generated by the product of fields: $\mathcal{H}_{\alpha\times\beta} = \mathcal{A}\psi_\alpha\psi_\beta\Omega$. Conversely, once this prescription has been found, it will provide information about the multiplicative properties of the would-be fields ψ.

The starting point for the theory of superselection sectors is the basic observation [1], that representations with "particle-like" properties possess a natural such composition law. Namely, one can equivalently describe *representations* in terms of *endomorphisms* of the algebra. Then, in a nutshell, one has the following statement.

Let $\mathcal{A}$ be an algebra, $\pi_0 : \mathcal{A} \to \mathcal{B}(\mathcal{H}_0)$ some representation by bounded operators in the Hilbert space $\mathcal{H}_0$. If $\rho : \mathcal{A} \to \mathcal{A}$ is an algebra endomorphism then $\pi := \pi_0 \circ \rho : \mathcal{A} \to \mathcal{B}(\mathcal{H}_0)$ is another representation in the same Hilbert space, but not necessarily equivalent to π_0. If π_0 is faithful, then there is a natural composition of representations of this type: namely, for $\pi_i = \pi_0 \circ \rho_i$ define

$$\pi_1 \times \pi_2 := \pi_0 \circ (\rho_1 \circ \rho_2).$$

The theory of *statistics* is the study of this composition law (in particular its non-commutativity) under appropriate additional physically motivated assumptions on the algebra $\mathcal{A}$ and its representations of interest. It provides detailed control over the superselection structure of the theory.

Let us in the first place give some physical justification for the assumptions on the algebra and its representations. The essential assumptions on $\mathcal{A}$ are its *local net* structure [5] expressing Einstein causality of physical measurements (observables), and a technical *maximality* property known under the name of Haag duality [8].

The assumption that there is a faithful vacuum representation π_0 reflects the idea that no information accessible by *local* measurements is lost if a charged state is mimiced by another state in the vacuum representation consisting of the charged particles "in the laboratory" and their anti-particles (compensating charges) "behind the moon" (Haag). In particular, one may then identify the abstract algebra $\mathcal{A}$ with its vacuum representation $\pi_0(\mathcal{A}) \subset \mathcal{B}(\mathcal{H}_0)$ (and omit the symbol π_0), and $\mathcal{A}$ equipped with the norm of $\mathcal{B}(\mathcal{H}_0)$ is a C^*-algebra.

Poincaré invariance of the vacuum representation and the positivity of the energy spectrum are needed only in the form of a technical "Property B" (Borchers property, [9]). The latter follows from covariance and spectrum condition, but is certainly a weaker assumption.

Among the host of representations of the C^* local net $\mathcal{A}$ one selects a class of physically relevant representations in the sense that they are locally equivalent to the vacuum representation. This certainly excludes temperature $T > 0$ representations, but fits the conception of asymptotic states containing only finitely many particles [1]. Actually, even in the absence of scattering states the criterion applies to non-trivial situations, e.g. in the case of conformal light-cone theories.

With these assumptions one shows that every such representation is equivalent to a representation of the type

$$\pi = \pi_0 \circ \rho \equiv \rho,$$

where ρ is a *localized* and *transportable* (C^*-endo-)morphism of $\mathcal{A}$. The physical connotation is that situations with global charges can be prepared with all these charges localized within some finite "laboratory" volume – undetectable by measurements in the causal complement – and vacuum fluctuations outside, and that the position of the laboratory in space-time is irrelevant for the equivalence class of the representation. In fact, transportability of a localized morphism is a weaker assumption than translation covariance, but is sufficient for our purposes.

Equivalence classes of endomorphisms $\rho : \mathcal{A} \to \mathcal{A}$ with the described properties are called superselection sectors, or less stringently, charges.

Now we are in the situation described in the beginning. What is then the physical interpretation of the composition of representations $\pi_1 \times \pi_2$, or equivalently, $\rho_1 \circ \rho_2$? If π_i describe superselection sectors containing, say, asymptotic one-particle state vectors $|p_i\rangle$ then $\pi = \pi_1 \times \dots \pi_n$ describes a superselection sector containing the asymptotic n-particle state vector $|\underline{p}\rangle \equiv |p_1, \dots, p_n\rangle$, i.e. the composition is a physical "addition of charges".

Now, every permutation of the order of the factors $\pi_1, \ldots, \pi_n$ of π will yield another n-particle *state vector* $|\underline{p}'\rangle$ with permuted entries, which however describes the same *state*. The unitary operators taking these permuted state vectors into each other characterize the statistics of the corresponding particles.

The point is that these statistics operators can be determined abstractly in terms of the transportable morphisms alone without any reference to asymptotic states; consequently, statistics is a concept of local quantum field theory well-defined also if there is no scattering theory, as in the case of conformal quantum field theory.

The statistics operators may be regarded as an intrinsic characterization of the local algebra $\mathcal{A}$ and its superselection structure ("charge or particle content"). The evaluation of the important structural consequences of the initial physical assumptions thus offer a starting point for an ambitious classification program of local quantum field theory. In fact, the admissible statistics has been classified in high (e.g. four) space-time dimensions: there can occur only Fermi or Bose parastatistics of order $N = 1, 2, \ldots, \infty$. While $N = \infty$ is considered as pathological, finite order parastatistics is equivalent to the existence of multiplet fields in an N-dimensional representation of some compact (global) gauge group. Moreover, the superselection rules for the composition and reduction of irreducible charges (reflecting "interactions") are isomorphic to the unitary representation theory of that symmetry group.

More or less surprisingly, the statistics of low-dimensional quantum field theory turns out to define representations of the braid group instead of the permutation group (in terms of the above asymptotic particle picture: the statistics operator describing the transposition of the factors π_i and π_{i+1} depends on the relative space-like position of the in- or out-going wave packets, or equivalently, on the ordering of the momenta).

No complete classification of braid group statistics exists sofar. But there is the powerful tool of "left-inverses", closely related to the existence of conjugate charges (anti-particles). With the help of left-inverses one can construct Markov traces (i.e. maps from the infinite braid group B_∞ into $\mathbb{C}$ with an invariance under certain knot-theoretical manipulations on braids). The class of Markov traces obtained this way ("field theoretical Markov traces") has a number of additional, quite restrictive non-linear properties. They are expected to encode all relevant information necessary to classify the admissible braid group statistics; but this – most fascinating – part of the program is far from being settled at this moment.

In order to emphasize the physical relevance and the mathematical logic of the sequence of definitions and propositions, which constitute the theory of superselection sectors, we sent all the proofs to the appendix of this article (or referred to the literature). One might sometimes wonder, why certain statements *have to* be proven, while they are so physically convincing at first sight; but after a closer view one will find, that the more surprising fact is that things *can* be proven, and have not to be built into the theory from outside.

2. Axioms

Let us begin with some space-time geometry. We consider quantum field theory in $\mathbb{R}^d$, $d = 1, 2, 3, \ldots$, where the meaning of quantum field theory in $d = 1$ will be explained later (Sect. 7). For $d > 2$, $\mathbb{R}^d$ is equipped with the Minkowski metric $(+, (-)^{d-1})$.

A double cone is a non-empty intersection $\mathcal{O}$ of a forward and a backward open light-cone if $d \geq 2$, and a bounded open interval if $d = 1$. The causal complement is the set

$$\mathcal{O}' := \{y \in \mathbb{R}^d \,|\, (x-y)^2 \leq 0 \; \forall x \in \mathcal{O}\}$$

if $d \geq 2$, and the complement $\mathcal{O}' := \mathbb{R} \setminus \mathcal{O}$ if $d = 1$. We write

$$\mathcal{O}_1 \between \mathcal{O}_2 \;\text{ iff }\; \mathcal{O}_1 \subset \mathcal{O}_2' \;(\Leftrightarrow \mathcal{O}_2 \subset \mathcal{O}_1').$$

For $d \leq 2$ observe that $\mathcal{O}'$ has two connected components. Then we write $\mathcal{O}_1 < \mathcal{O}_2$, if $\mathcal{O}_2$ lies to the right of $\mathcal{O}_1$ (with respect to the spatial coordinate).

By $\mathcal{O}_1 \vee \mathcal{O}_2$ we denote the "smallest" double cone which contains $\mathcal{O}_1$ and $\mathcal{O}_2$. For $d \leq 2$ every other double cone which contains $\mathcal{O}_1$ and $\mathcal{O}_2$, also contains $\mathcal{O}_1 \vee \mathcal{O}_2$. This is not true for $d > 2$, where $\mathcal{O}_1 \vee \mathcal{O}_2$ is chosen the smallest by volume; but all statements about $\mathcal{O}_1 \vee \mathcal{O}_2$ in fact hold true for every other double cone which contains $\mathcal{O}_1$ and $\mathcal{O}_2$. Care has to be taken in non-simply-connected space-time, cf. Sect. 7.

2.1. Local observables. To every double cone $\mathcal{O}$ is associated a C^*-algebra $\mathcal{A}(\mathcal{O}) \subset \mathcal{B}(\mathcal{H}_0)$, realized as bounded operators in a Hilbert space $\mathcal{H}_0$. (We shall refer to this realization as the "vacuum representation" $\pi_0 : \mathcal{A} \to \mathcal{B}(\mathcal{H}_0)$, and to $\mathcal{H}_0$ as the "vacuum Hilbert space", since the technical axioms 2.4. and 2.5. below can be traced back [8, 9] to more fundamental physical assumptions which justify this name. However, other representations may be as good for our purposes.) π_0 is assumed to be irreducible; in other words, $\pi_0(\mathcal{A})$ has trivial center.

Operators $A \in \mathcal{A}(\mathcal{O})$ are called "observables localized in $\mathcal{O}$". We introduce the C^*-algebras[2]

$$\mathcal{A}(\mathcal{O}') := \overline{\bigcup_{\mathcal{Q}:\, \mathcal{Q} \subset \mathcal{O}'} \mathcal{A}(\mathcal{Q})}, \qquad \mathcal{A} := \overline{\bigcup_{\mathcal{O}} \mathcal{A}(\mathcal{O})}.$$

2.2. Net property. If $\mathcal{O}_1 \subset \mathcal{O}_2$ then $\mathcal{A}(\mathcal{O}_1) \subset \mathcal{A}(\mathcal{O}_2)$.

2.3. Locality. If $\mathcal{O}_1 \between \mathcal{O}_2$ then $[\mathcal{A}(\mathcal{O}_1), \mathcal{A}(\mathcal{O}_2)] = 0$.

2.4. Haag duality. $\pi_0(\mathcal{A}(\mathcal{O})) = \pi_0(\mathcal{A}(\mathcal{O}'))'$.

2.5. Borchers property. Let the closure of a double cone $\mathcal{O}$ be contained in some larger double cone $\mathcal{O}_1$. Then every projection $E \in \mathcal{A}(\mathcal{O})$ is of the form $E = WW^*$, where $W \in \mathcal{A}(\mathcal{O}_1)$ is an isometry: $W^*W = 1$, localized in $\mathcal{O}_1$.

[2] For a subalgebra $\mathcal{B} \subset \mathcal{B}(\mathcal{H}_0)$, the norm-closure is denoted by $\overline{\mathcal{B}}$, the commutant by $\mathcal{B}'$.

Whenever we refer to covariance properties with respect to some covariance group G (translation group, Poincaré group, conformal group), we assume in addition

2.6. Covariance. The covariance group G is represented by automorphisms α_g, $g \in G$, of $\mathcal{A}$ such that

$$\alpha_g(\mathcal{A}(\mathcal{O})) = \mathcal{A}(g\mathcal{O}).$$

There is a unitary implementation of G by operators $\mathcal{U}_0(g)$ in the vacuum Hilbert space such that

$$\pi_0(\alpha_g(A)) = \mathcal{U}_0(g)\pi_0(A)\mathcal{U}_0(g)^*,$$

there is an invariant cyclic vacuum state $\mathcal{U}_0(g)\Omega = \Omega$, and the spectrum of the translation operators is contained in the forward light-cone.

3. Representations and Localized Morphisms

In this section we introduce the class of representations of physical relevance and show that this class can be equivalently studied in terms of endomorphisms of the algebra of observables. In particular, this class is closed under taking direct sums and subrepresentations. We introduce the notion of intertwiners between endomorphisms with common subrepresentations.

Once the appropriate definitions have been made, the subsequent conclusions are technically straightforward (the use of Haag duality is essential to prove that certain operators of interest are in fact local operators in 3.3., 3.5., and 3.8., and the Borchers property in 3.6.). We content ourselves here to take notice of the facts. Proofs can be found in [1, Sect. 1 and 2].

3.1. Definition:

A representation $\pi : \mathcal{A} \to \mathcal{B}(\mathcal{H}_\pi)$ is called *locally equivalent* to the vacuum representation, if its restrictions to the causal complements of all translates of some double cone are unitarily equivalent to the vacuum representation, i.e. if there are a double cone $\mathcal{O}$ and unitaries $V_x : \mathcal{H}_0 \to \mathcal{H}_\pi$ such that

$$\pi(A)V_x = V_x\pi_0(A) \quad \forall A \in \mathcal{A}((\mathcal{O} + x)').$$

3.2. Definition:

(i) An endomorphism $\rho : \mathcal{A} \to \mathcal{A}$ is called *localized* in the double cone $\mathcal{O}$ (or a *localized morphism*) if

$$\rho(A) = A \quad \forall A \in \mathcal{A}(\mathcal{O}').$$

The set of all localized morphisms is denoted by Δ_l.

(ii) A morphism ρ localized in $\mathcal{O}$ is called *transportable* if for every $x \in \mathbb{R}^d$ there is a unitarily equivalent morphism $\rho_x \cong \rho$ localized in the translate double cone $\mathcal{O} + x$. The set of all transportable morphisms is denoted by Δ_t.

(iii) The set of all *inner automorphisms* σ_U of $\mathcal{A}$ defined by $\sigma_U(A) = UAU^*$, U unitary in some $\mathcal{A}(\mathcal{O})$, is denoted by Δ_i. Obviously, Δ_i is a group contained in Δ_t.

3.3. Proposition:

A representation $\pi : \mathcal{A} \to \mathcal{B}(\mathcal{H}_\pi)$ is locally equivalent to the vacuum representation, if and only if it is equivalent to a representation of the form[3] $\pi_0 \circ \rho$, where $\rho : \mathcal{A} \to \mathcal{A}$ is a transportable morphism.

3.4. Definition and Lemma:

A representation π of $\mathcal{A}$ is *covariant*, if there is a continuous representation of the universal covering group $\tilde{G}$ of G by unitary operators $\mathcal{U}_\pi(\gamma) \in \mathcal{B}(\mathcal{H}_\pi)$ such that for $\gamma \in \tilde{G}$

$$\pi(\alpha_\gamma(A)) = \mathcal{U}_\pi(\gamma)\pi(A)\mathcal{U}_\pi(\gamma)^*.$$

A covariant representation π is a *positive energy representation*, if the subgroup of translations is represented by $\mathcal{U}_\pi(x)$ with spectrum in the (closed) forward light-cone. For representations of the form $\pi = \pi_0 \circ \rho$, $\rho \in \Delta_t$, we shall write $\mathcal{U}_\pi \equiv \mathcal{U}_\rho$.

(i) If π is irreducible and $\gamma \in \tilde{G}$ projects onto id $\in G$, then $\mathcal{U}_\pi(\gamma)$ is a scalar multiple $\zeta_\pi(\gamma)$ of $\mathbf{1}$.

(ii) For a covariant morphism ρ, $\gamma \in \tilde{G}$, put $U = \mathcal{U}_0(\gamma)\mathcal{U}_\rho(\gamma)^*$. If ρ is localized in $\mathcal{O}$, then $U \in \mathcal{A}(\mathcal{O} \vee \gamma(\mathcal{O}))$ and $\sigma_U \circ \rho = \alpha_\gamma \circ \rho \circ \alpha_\gamma^{-1}$ is localized in $\gamma(\mathcal{O})$.

---

[3] By virtue of 3.1., the kernel of π_0 is contained in the kernel of every representation π which is locally equivalent to π_0. Thus, this class of representations actually represents the algebra $\mathcal{A}/ker(\pi_0)$ for which π_0 is injective. Consequently, for our purposes we are free to assume π_0 injective and use it as an *identification* of the algebra $\mathcal{A}$ with its image in the bounded operators $\mathcal{B}(\mathcal{H}_0)$, hence $\pi_0 \equiv$ id. Nevertheless, we often prefer to write $\pi_0 \circ \rho$ in order to distinguish the endomorphism ρ from the corresponding representation.

3.5. Lemma:

If ρ_1, ρ_2 are unitarily equivalent morphisms localized in $\mathcal{O}_1$, $\mathcal{O}_2$:

$$\rho_1(A) = U\rho_2(A)U^* \quad \forall A \in \mathcal{A}, \quad U \in \mathcal{B}(\mathcal{H}_0) \text{ unitary},$$

then they differ by an *inner* automorphism: $\rho_1 = \sigma_U \circ \rho_2$. Namely

$$U \in \mathcal{A}((\mathcal{O}_1 \vee \mathcal{O}_2)')' \equiv \mathcal{A}(\mathcal{O}_1 \vee \mathcal{O}_2) \subset \mathcal{A};$$

in other words: unitary equivalence of localized morphisms is always implemented by *local* operators. In particular, the operators $\mathcal{U}_0(\gamma)\mathcal{U}_\rho(\gamma)^*$ in 3.4(ii) are local "charge transporting" operators, unitarily relating a localized morphism with an equivalent one localized in the shifted double cone.

3.6. Proposition:

Let $\rho_i \in \Delta_t$, $\pi_i = \pi_0 \circ \rho_i$, $i = 1, 2$. Then $\pi_1 \oplus \pi_2$ is locally equivalent to the vacuum representation. If π is a subrepresentation of π_1, then π is locally equivalent to the vacuum representation.

More specifically, the projection $E \in \rho_1(\mathcal{A})'$ on the subrepresentation π can be obtained in the form $E = WW^*$ with an isometry W contained in some local algebra $\mathcal{A}(\mathcal{O})$, $\mathcal{O}$ containing the closure of the localization of ρ_1, and

$$\rho(A) := W^* \rho_1(A) W$$

defines a transportable morphism with $\pi_0 \circ \rho$ equivalent to π.

3.7. Corollary:

Let $\rho_i \in \Delta_t$, $i = 1, 2$, such that $\pi_0 \circ \rho_i$ possess a common (up to equivalence) non-trivial subrepresentation. Then there is a partial isometry $0 \neq T \in \mathcal{A}$ (i.e.: TT^* and T^*T are projections in $\rho_1(\mathcal{A})'$ and $\rho_2(\mathcal{A})'$ resp.) satisfying

$$\rho_1(A)T = T\rho_2(A) \quad \forall A \in \mathcal{A}.$$

3.8. Definition and Proposition:

For $\rho_1, \rho_2 \in \Delta_l$ the *set of intertwiners* T from ρ_2 to ρ_1 is defined by

$$(\rho_1|\rho_2) := \{T \in \mathcal{B}(\mathcal{H}_0) |\ \rho_1(A)T = T\rho_2(A) \ \forall A \in \mathcal{A}\}.$$

In particular, $(\rho|\rho) \equiv \rho(\mathcal{A})'$. Intertwiners between localized morphisms are local operators. More specifically, if ρ_i are localized in $\mathcal{O}_i$, then $(\rho_1|\rho_2) \subset \mathcal{A}(\mathcal{O}_1 \vee \mathcal{O}_2)$.

4. Statistics

We now turn to the study of the (in general non-commutative) composition law $\rho_1\rho_2 \equiv \rho_1 \circ \rho_2$ of transportable morphisms, leading to the intrinsic definition of "statistics". The latter arises in the form of homomorphisms of the (infinite) braid or permutation group – depending on the number of space-time dimensions – into the unitary operators of $\mathcal{A}$. We shall see in Sect. 6, how statistics may be converted into commutation relations for charged fields. It is important to note that the existence of charges with more general commutation relations than bosonic local commutativity 2.3. is an intrinsic property of $\mathcal{A}$ itself and its representation theory.

In covariant theories, statistics operators can be computed in terms of translation operators. However, as a concept, statistics is independent of covariance properties. Some of the laws concerning the covariance of composite representations have to be modified when non-simply-connected space-time is considered, e.g. for conformal covariance as discussed in Sect. 7. For proofs of the less evident statements and some useful formulae see the appendix.

4.1. Lemma:

Let $\rho_i \in \Delta_t$ be localized in $\mathcal{O}_i$, $i = 1, 2$.

(i) Then $\rho_1 \circ \rho_2$ is transportable and localized in $\mathcal{O}_1 \vee \mathcal{O}_2$. Thus Δ_t is a semi-group.

(ii) Let $\rho_i \cong \tilde{\rho}_i \in \Delta_t$, $i = 1, 2$. Then $\rho_1 \circ \rho_2 \cong \tilde{\rho}_1 \circ \tilde{\rho}_2$. Thus the set of equivalence classes of transportable morphisms (superselection sectors) $\widehat{\Delta}_t := \Delta_t/\Delta_i$ is a semi-group.

4.2. Proposition:

Let $\rho_i \in \Delta_t$ be localized in $\mathcal{O}_i$, $i = 1, 2$, at space-like distance: $\mathcal{O}_1 \times \mathcal{O}_2$. Then they commute: $\rho_1\rho_2 = \rho_2\rho_1$.

4.3. Corollary:

The composition of transportable morphisms is commutative up to unitary equivalence. In other words: $\widehat{\Delta}_t$ is an abelian semi-group.

4.4. Proposition:

If $\rho_i \in \Delta_t$ are covariant, then $\rho_1\rho_2$ is covariant. More specifically, one has

$$\mathcal{U}_{12}(\gamma) = \mathcal{U}_1(\gamma)\rho_1(\mathcal{U}_0(\gamma)^*\mathcal{U}_2(\gamma)) = \rho_1(\mathcal{U}_2(\gamma)\mathcal{U}_0(\gamma)^*)\mathcal{U}_1(\gamma).$$

If γ projects onto the identity in G, then for irreducible ρ_i one obtains from this formula $\mathcal{U}_{12}(\gamma) = \zeta_1(\gamma)\zeta_2(\gamma)$, e.g. the usual additivity of spins up to integers in more than two dimensions.

4.5. Proposition:

If the morphisms ρ_i are covariant with respect to one of the usual[4] covariance groups G, and if $\rho_1\rho_2$ is finitely reducible (this will be the case for "proper" ρ_i discussed below), then every subrepresentation of $\rho_1\rho_2$ is covariant, and the corresponding intertwiners intertwine also the representations of $\tilde{G}$:

$$T \in (\rho_1\rho_2|\rho_3) \Rightarrow \mathcal{U}_{12}(\gamma)T = T\mathcal{U}_3(\gamma).$$

4.6. Theorem:

There is a collection of unitary statistics operators for every pair $\rho_1, \rho_2 \in \Delta_t$, implementing the unitary equivalence between $\rho_1\rho_2$ and $\rho_2\rho_1$:

$$\varepsilon(\rho_1, \rho_2) \in (\rho_2\rho_1|\rho_1\rho_2),$$

which is uniquely determined by the properties

(i)
$$\varepsilon(\rho_1\rho_2, \rho_3) = \varepsilon(\rho_1, \rho_3)\rho_1(\varepsilon(\rho_2, \rho_3)),$$
$$\varepsilon(\rho_3, \rho_1\rho_2) = \rho_1(\varepsilon(\rho_3, \rho_2))\varepsilon(\rho_3, \rho_1),$$

(ii) for $T \in (\rho_2|\rho_1)$:
$$\rho_3(T)\varepsilon(\rho_1, \rho_3) = \varepsilon(\rho_2, \rho_3)T,$$
$$\rho_3(T)\varepsilon(\rho_3, \rho_1)^* = \varepsilon(\rho_3, \rho_2)^*T,$$

and the "initial values"

(iii) $\varepsilon(\rho_1, \rho_2) = 1$, if ρ_i are localized in $\mathcal{O}_i$ with $\begin{cases} \mathcal{O}_1 \times \mathcal{O}_2 & \text{for } d > 2 \\ \mathcal{O}_2 < \mathcal{O}_1 & \text{for } d \leq 2. \end{cases}$

These operators satisfy in addition:

(iv) $\rho_3(\varepsilon(\rho_1, \rho_2))\varepsilon(\rho_1, \rho_3)\rho_1(\varepsilon(\rho_2, \rho_3)) = \varepsilon(\rho_2, \rho_3)\rho_2(\varepsilon(\rho_1, \rho_3))\varepsilon(\rho_1, \rho_2)$, and
(v) $\varepsilon(\rho_1, \rho_2)\varepsilon(\rho_2, \rho_1) = 1$, if $d > 2$.

We construct the statistics operators explicitly in the proof in the appendix, vividly exploiting the transportability of morphisms and locality properties. In particular, if ρ_i are translation covariant, the statistics operators can be expressed in terms of covariance operators. Namely, denote by $U_{ix} = \mathcal{U}_0(x)\mathcal{U}_i(x)^*$ the local operators (cf. 3.4. and 3.5.) shifting the charges ρ_i localized in $\mathcal{O}_i$ into $\tilde{\rho}_i$ localized in $\mathcal{O}_i + x$. Then

$$\varepsilon(\rho_1, \rho_2) = \begin{cases} \rho_2(U_{1x}^*)U_{1x}, & x > 0 \text{ sufficiently large,} \\ U_{2x}^*\rho_1(U_{2x}), & x < 0 \text{ sufficiently small.} \end{cases}$$

[4] which share the property that all its unitary finite-dimensional ray representations are trivial; the translation group alone would not be sufficient.

4.7. Corollary:

(i) $\varepsilon_\rho := \varepsilon(\rho,\rho) \in (\rho^2|\rho^2) \equiv \rho^2(\mathcal{A})'$.
(ii) $\varepsilon_\rho \rho(\varepsilon_\rho)\varepsilon_\rho = \rho(\varepsilon_\rho)\varepsilon_\rho \rho(\varepsilon_\rho)$.
(iii) If $d > 2$ then $\varepsilon_\rho^2 = \mathbf{1}$.

4.8. Corollary:

Let B_n denote the braid group generated by the symbols σ_i, $i = 1, \dots, n-1$, with the relations

$$\sigma_i\sigma_j = \sigma_j\sigma_i \quad \text{if } |i-j| \geq 2,$$
$$\sigma_i\sigma_{i+1}\sigma_i = \sigma_{i+1}\sigma_i\sigma_{i+1}.$$

Let $\rho \in \Delta_t$. Then the mapping $\varepsilon_\rho^{(n)} : \sigma_i \mapsto \rho^{i-1}(\varepsilon_\rho)$ defines a homomorphism of the braid group B_n into the unitary operators in $(\rho^n|\rho^n) = \rho^n(\mathcal{A})' \subset \mathcal{A}$. The family $\{\varepsilon_\rho^{(n)} \mid n \in \mathbb{N}\}$ is compatible with the inclusions $B_{n-1} \subset B_n$ and $(\rho^{n-1}|\rho^{n-1}) \subset (\rho^n|\rho^n)$, and thus extends to a homomorphism $\varepsilon_\rho^{(\infty)}$ of the infinite braid group B_∞ into $\bigcup_n (\rho^n|\rho^n) \subset \mathcal{A}$.

If $d > 2$, then $\varepsilon_\rho^{(n)}$ respect also the additional relation $\sigma_i^2 = 1$, hence define homomorphisms of the permutation groups S_n into the unitaries in $(\rho^n|\rho^n)$.

(There is a natural generalization to the groupoid of "colored braids", involving the statistics operators $\varepsilon(\rho_j, \rho_k)$ instead of $\varepsilon_\rho = \varepsilon(\rho,\rho)$.)

In the following section we shall define characters (Markov traces) for these representations of the braid group, providing information about the equivalence classes of representations present in the theory. Speaking in terms of particles, these are "quantum numbers" of the latter, generalizing the fermion number and the degeneracy as a multiplet with respect to some gauge symmetry. However, for braid group statistics the range of these quantum numbers is much richer than it is for permutation group statistics. In more abstract terms, the Markov traces intrinsically *characterize* the local algebra of observables itself, to some extent comparable with the characterization of Lie groups by their Dynkin diagrams (rather than their basis-dependent structure constants). The restrictive properties of field theoretic Markov traces also provide a starting point to *classify* the possible statistics.

5. Conjugates and Markov Traces

In general, the endomorphisms $\rho \in \Delta_t$ need not to be invertible. (See, however, 5.7. For the case of permutation group statistics it is well known that the occurrence of non-invertible morphisms is related to the existence of a non-abelian gauge group [11, 4].) The best one may expect is that for $\rho \in \Delta_t$ there is $\bar{\rho} \in \Delta_t$, such that $\bar{\rho}\rho$ contains the vacuum sector as a subrepresentation, i.e.

$$(\bar{\rho}\rho|\,\mathrm{id}) \neq \{0\}.$$

(ρ and $\bar{\rho}$ will be called *conjugates* if they satisfy some more detailed conditions.) We shall next introduce the concept of left-inverses, which may be regarded as a tool to study the "degree of non-invertibility" of transportable morphisms, and – as we shall see – to control the reducibility of product morphisms ("superselection structure" or "fusion rules"). We shall here, however, not discuss the powerful general theory of left-inverses and its implication on the existence of conjugates, since we are only interested in the "proper" situation described in Prop. 5.2., which is believed to be the only non-pathological one. (Covariant superselection sectors with non-negative energy spectrum are proper.) The superselection structure for proper sectors is particularly well-behaved, as expressed by the propositions below.

If ρ *is* invertible, then $\rho^{-1} \in \Delta_t$ is a conjugate, and if $\bar{\rho}\rho = \sigma_U$ is equivalent to id, then $U \in (\bar{\rho}\rho|\,\mathrm{id})$, and $\sigma_{U^*}\bar{\rho}$ inverts ρ. This observation generalizes to the following.

5.1. Proposition and Definition:

Let $R \in (\bar{\rho}\rho|\,\mathrm{id})$ be an isometry. Then the mapping

$$\phi_R : \mathcal{A} \to \mathcal{A}, \quad A \mapsto R^*\bar{\rho}(A)R$$

is a *left-inverse* of ρ. By definition, left-inverses of ρ are positive C^*-linear mappings $\phi : \mathcal{A} \to \mathcal{A}$ satisfying

$$\phi(1) = 1,$$
$$\phi(\ \rho(A)B\rho(C)\ ) = A\,\phi(B)\,C$$

(in particular $\phi \circ \rho = \mathrm{id}$, whence the name).

5.2. Lemma:

Let $\rho, \rho_i \in \Delta_t$, and ϕ any left-inverse of ρ. Then

$$\phi((\rho\rho_1|\rho\rho_2)) \subset (\rho_1|\rho_2).$$

In particular, if ρ is irreducible, then $\phi(\varepsilon_\rho) \in \phi((\rho^2|\rho^2)) \subset (\rho|\rho) = \mathbb{C}$ is a scalar $\lambda \in \mathbb{C}$.

5.3. Proposition and Definition:

Let $\rho \in \Delta_t$ be irreducible. We call an irreducible $\bar{\rho} \in \Delta_t$ *conjugate* to ρ, if there is an isometry $R \in (\bar{\rho}\rho|\mathrm{id})$ and if the corresponding left-inverse ϕ_R as in 5.1. satisfies $\phi_R(\varepsilon_\rho) \neq 0$. Then

(i) every other irreducible conjugate of ρ is unitarily equivalent to $\bar{\rho}$.
(ii) $\bar{\rho}\rho$ contains the vacuum sector precisely once, i.e. $\dim(\bar{\rho}\rho|\mathrm{id}) = 1$.
(iii) $\phi = \phi_R$ is the unique left-inverse of ρ. In particular it is independent of the choice of the isometry R.
(iv) ρ is a conjugate of $\bar{\rho}$. $\bar{\rho}$ has a unique left-inverse $\bar{\phi}$, and

$$\lambda_{\bar{\rho}} \equiv \bar{\phi}(\varepsilon_{\bar{\rho}}) = \lambda_\rho \equiv \phi(\varepsilon_\rho).$$

In this case, ρ is called *proper*. If ρ is proper, then $\bar{\rho}$ is proper.

λ_ρ is the statistics parameter of ρ, $d(\rho) := |\lambda_\rho|^{-1}$ the statistical dimension, and $\omega(\rho) := \lambda_\rho/|\lambda_\rho|$ the statistics phase. The statistics parameter depends only on the equivalence class of $\rho \in \Delta_t$. The statistical dimension takes values $d \geq 1$.

5.4. Theorem:

(i) If ρ_i are proper, then $\rho_1\rho_2$ is equivalent to a direct sum of finitely many proper morphisms $\rho^{(j)}$

$$\pi_0 \circ \rho_1\rho_2 \cong \bigoplus_j \pi_0 \circ \rho^{(j)}.$$

(ii) The spectrum of the monodromy operator $\varepsilon_M = \varepsilon(\rho_2,\rho_1)\varepsilon(\rho_1,\rho_2)$ consists of the eigenvalues $\omega(\rho^{(j)})/\omega(\rho_1)\omega(\rho_2)$. The statistical dimensions satisfy the "sum rule"

$$d(\rho_1)d(\rho_2) = \sum_j d(\rho^{(j)}).$$

(iii) All irreducible components of $\rho_1\rho_2$ have statistical dimensions in the interval

$$\max(\frac{d_1}{d_2},\frac{d_2}{d_1}) \leq d \leq d_1 d_2, \quad d_i = d(\rho_i).$$

We have introduced the statistics parameters $\lambda = \omega/d$ as class-invariants of proper morphisms, and the previous theorem yields class-invariant fusion matrices $N_{\alpha\beta}^\gamma$ describing the superselection structure, defined as the multiplicity of the equivalence class $[\rho_\gamma]$ occurring in the decomposition of $\rho_\alpha\rho_\beta$; see 5.6. below. In general these numbers will not be sufficient as complete characterizations of equivalence classes. We shall now introduce huge class-invariant functionals ("field-theoretical Markov traces") which generically encode much more information (actually, for permutation group statistics the Markov traces are completely determined by the statistics parameters, and do *not* provide additional information). On the other hand the functional properties of field-theoretical Markov traces are very restrictive and (at least in various important cases) lead to a quantization of the admissible values of the parameters.

5.5. Theorem:

Let ρ be proper, ϕ its left-inverse. The mapping

$$\mathrm{tr}_\rho^{(n)} := \phi^n \circ \varepsilon_\rho^{(n)} : B_n \to \mathbb{C}$$

extends to a normalized positive strong Markov trace tr_ρ on the group algebra of B_n; in other words, the following hold:

(i) For $b \in B_n \subset B_{n+1}$ one has $\mathrm{tr}_\rho^{(n+1)}(b) = \mathrm{tr}_\rho^{(n)}(b)$.
(ii) $\mathrm{tr}_\rho(e) = 1$, $\mathrm{tr}_\rho(b^{-1}) = \mathrm{tr}_\rho(b)^*$.
(iii) $\mathrm{tr}_\rho(b_1 b_2) = \mathrm{tr}_\rho(b_2 b_1)$.
(iv) $\mathrm{tr}_\rho(b_1 b_2) = \mathrm{tr}_\rho(b_1)\ \mathrm{tr}_\rho(b_2)$, if b_1 is a word in the generators $\sigma_i^{\pm 1}$, $i < k$, and b_2 is a word in the generators $\sigma_j^{\pm 1}$, $j \geq k$, for some k.
(v) $\mathrm{tr}_\rho(xx^*) \geq 0$, where $*$ is the anti-linear extension of the inverse on B_∞ to the group algebra.

The trace tr_ρ depends only on the equivalence class of ρ, and

(vi) $\mathrm{tr}_{\bar\rho}(\bar b) = \mathrm{tr}_\rho(b)$, where if $b = s_1 \dots s_r$ is a word in the symbols $\sigma_i^{\pm 1}$ then $\bar b = s_r \dots s_1$ is the reversed word.

To give an idea how to exploit this proposition, we shall sketch the quantization of statistics parameters in the "Hecke case" [1, 3, 12]. The Hecke case is characterized by the validity of a quadratic equation for the statistics operator ε_ρ implying that ε_ρ has at most two different eigenvalues:

$$(\varepsilon_\rho - \mu_1)(\varepsilon_\rho - \mu_2) = 0.$$

This is guaranteed, e.g., in the case of permutation group statistics: $\varepsilon_\rho^2 = 1$, or if ρ^2 has at most two irreducible components: then $\rho^2(\mathcal{A})'$ is at most two-dimensional, hence $1, \varepsilon_\rho, \varepsilon_\rho^2 \in \rho^2(\mathcal{A})'$ are linearly dependent.

Using the quadratic equation in addition to 4.8. and 5.5., one can compute $\mathrm{tr}_\rho(b)$ recursively for every braid b, as a function of the parameters μ_1, μ_2, and $\lambda = \mathrm{tr}_\rho(\sigma_1)$. On the other hand, one can find projectors $e_T = e_T^* = e_T^2$ (generalizing Young tableaux of S_n) in the Hecke algebra which is the group algebra of B_n divided by the ideal generated by $(\sigma_i - \mu_1)(\sigma_i - \mu_2) = 0$. Hence one obtains infinitely many inequalities $\mathrm{tr}_\rho(e_T) \geq 0$ from the positivity of the trace. Solving these inequalities for μ_1, μ_2, λ one finds only a discrete series of admissible values for the statistical dimensions:

$$d = |\lambda|^{-1} = \frac{\sin \frac{N}{N+L}\pi}{\sin \frac{1}{N+L}\pi}, \quad N, L = 1, 2, 3, \dots, \infty,$$

and relations among the phases. The special case of permutation group statistics, i.e. $\mu_1, \mu_2 = \pm 1$ reduces to $L = \infty$ and

$$d = N = 1, 2, 3, \dots, \quad \omega = \pm 1,$$

where by a powerful analysis [4], N acquires the interpretation of the dimension of a representation (associated with the sector) of some compact gauge group, and $\omega = \pm 1$ distinguishes bosons from fermions. We shall give some details of this in the next section.

A similar analysis has been performed [13, 14] for the case that ρ^2 has three irreducible components one of which is an automorphism. This case reduces to the analysis of traces on Birman-Wenzl algebras.

While 5.5. refers to every sector separately, the generalization to Markov traces on the groupoid of "colored braids" provides more class-invariants related to the superselection structure. For an example we refer to [15].

As by-products of the general analysis of the reducibility of products of proper morphisms we quote the following results.

Let $\widehat{\Delta}$ denote the set of equivalence classes $[\rho]$ of proper morphisms. Then by 4.1., 4.3., 5.4., the equivalence classes of the (proper) subrepresentations of $\rho_\alpha\rho_\beta$ are independent of the representatives $\rho_\alpha \in [\rho_\alpha]$, $\rho_\beta \in [\rho_\beta]$, and coincide with those of $\rho_\beta\rho_\alpha$. Thus it is justified to introduce the class-invariant multiplicities

$$N_{\alpha\beta}^{\gamma} = \dim(\rho_\alpha\rho_\beta|\rho_\gamma)$$

of $[\rho_\gamma]$ in $[\rho_\alpha][\rho_\beta]$, which may be organized into "fusion matrices" $(N_\alpha)_\beta^\gamma \equiv N_{\alpha\beta}^{\gamma}$.

5.6. Proposition:

For $\alpha \in \widehat{\Delta}$ denote by $\bar{\alpha}$ the conjugate class $[\rho_{\bar{\alpha}}] = [\bar{\rho}_\alpha]$ (cf. 5.2(i)), and by $[0]$ the class of the identity. Then

(i) $N_{\alpha\beta}^{\gamma} = N_{\beta\alpha}^{\gamma} = N_{\bar{\alpha}\gamma}^{\beta} = N_{\bar{\alpha}\bar{\beta}}^{\bar{\gamma}}$, $\quad N_{0\beta}^{\gamma} = \delta_{\beta\gamma}$, $\quad N_{\bar{\alpha}} = N_\alpha^T$.
(ii) $N_\alpha N_\beta = N_\beta N_\alpha = \sum_\gamma N_{\alpha\beta}^{\gamma} N_\gamma$.
(iii) $N_\alpha \cdot \underline{d} = d_\alpha\ \underline{d}$, where $\underline{d}$ is the vector with components $d_\beta = d(\rho_\beta)$.

5.7. Proposition:

Let $\rho \in \Delta_t$ be proper. The following statements are equivalent:

(i) ρ is an automorphism, i.e. ρ is invertible.
(ii) ρ^2 is irreducible.
(iii) ε_ρ is a scalar.
(iv) $d(\rho) = 1$.

Equivalence classes of proper automorphisms are called *simple* sectors. Their statistics provide abelian representations of the braid group. By 5.4., the composition of an automorphism with a proper morphism is irreducible (and thus again proper).

6. Charged Fields

We have so far studied representations of the algebra of observables $\mathcal{A}$. It is now desirable to have "charged field operators" which interpolate among inequivalent representations of $\mathcal{A}$, and in particular create charged sectors from the vacuum sector. Charged fields are in general physically unobservable (like $\mathbb{Z}_2$-charged Fermi fields). Correspondingly there are different possibilities to choose the field algebras, although they all describe the same physics. We shall in this section introduce three types of field algebras, which differ by the multiplicities of equivalent charged sectors featuring in the full Hilbert space.

The first one, the gauge covariant field algebra [4], consists of finite field multiplets. These transform under some compact gauge group (determined by the superselection structure of Δ_t) and are represented in a Hilbert space which contains as many copies of a proper sector as its statistical dimension indicates. In particular, the duality relation between statistics and symmetry is most clearly exhibited. This construction is valid only for permutation group statistics, since the triviality of the monodromy operators and the quantization of statistical dimensions $d \in \mathbb{N}$ are crucial. The generalization to braid group statistics has been attempted recently in special cases [16] with a "quantum group" [17] in the place of the compact symmetry group.

The second one, the "field bundle" [2], exists for any type of statistics, being essentially the crossed product of the semigroup Δ_t with the algebra $\mathcal{A}$. It is an extremely redundant description entailing infinite multiplicities of equivalent sectors, and misses the aspect of symmetry.

In the third type of a field algebra, the "reduced field bundle" [3], the infinitely many copies of equivalent sectors of the field bundle are by means of charge reducing intertwiners transferred into the direct sum of precisely one "reference copy" of every proper sector. Among its algebra relations are the exchange algebra and a finite operator product expansion for bounded operators. Precisely the same type of relations have been encountered in models of two-dimensional conformal field theory. The present derivation of the reduced field bundle from local and spectral properties alone shows that not the conformal exchange algebra itself, but only its particular kinematical simplicity is due to conformal covariance and light-cone factorization. We sketch the fitting of conformal field theories into the present framework in Sect. 7.

The gauge covariant field algebra [4]. The gauge covariant field algebra (which exists only for permutation group statistics) is best suited to illustrate how to make contact with the common description of $d \geq 3$-dimensional quantum field theory in terms of (unobservable) tensor fields.

Let us assume for simplicity that ρ is a proper morphism with trivial determinant[5], $d = N \in \mathbb{N}$ its statistical dimension, $\omega = \pm 1$. By definition of the determinant there is an isometry $T_{\text{det}} \in (\rho^N \,|\, \text{id})$. Let $\mathcal{H}_0$ denote the vacuum Hilbert space of $\mathcal{A}$. Our presentation can be only sketchy.

For the construction of the gauge covariant field algebra one considers the Cuntz-algebra C_N, which is the $SU(N)$-covariant C^*-algebra generated by a multiplet of N isometries ψ_i, $i = 1, \ldots, N$, satisfying the relations

$$\psi_i^* \psi_j = \delta_{ij}, \quad \sum_i \psi_i \psi_i^* = 1.$$

The main theorem [4] states that there is a unique subgroup $G \subset SU(N)$ such that there is an isomorphism between the G-invariant polynomials in C_N of the form $\underbrace{\psi \ldots \psi}_{n} \underbrace{\psi^* \ldots \psi^*}_{m}$ and the local operators in $(\rho^n | \rho^m) \subset \mathcal{A}$. A crucial condition for this isomorphism is the fact that the observables $\varepsilon_\rho^{(n)}(\pi) \in (\rho^n | \rho^n)$ and the Cuntz-algebra operators $\Psi^{(n)}(\pi) = \sum \psi_{i_{\pi(1)}} \cdots \psi_{i_{\pi(n)}} \psi_{i_n}^* \cdots \psi_{i_1}^*$ have the same spectra; namely the Markov trace state ϕ^n on $\varepsilon_\rho^{(n)}(\pi)$ coincides with the positive trace state φ^n on $\Psi^{(n)}(\pi)$ defined by

$$\varphi(x) = N^{-1} \sum_i \psi_i^* x \psi_i, \quad x \in C_N.$$

The gauge covariant field algebra $\mathcal{F}_{\text{DR}}$ is generated by the algebras $\mathcal{A}$ and C_N by identifying the corresponding subalgebras, in particular (for $\omega(\rho) = \pm 1$)

$$(\rho^2 | \rho^2) \ni \varepsilon_\rho = \pm \textstyle\sum_{ij} \psi_i \psi_j \psi_i^* \psi_j^*,$$
$$(\rho^N \,|\, \text{id}) \ni T_{\text{det}} = (N!)^{-1/2} \textstyle\sum_{\pi \in S_N} \text{sig}(\pi) \psi_{\pi(1)} \cdots \psi_{\pi(N)},$$

and imposing the relations

$$\rho(A) = \sum_i \psi_i A \psi_i^* \;\Leftrightarrow\; \psi_i A = \rho(A) \psi_i.$$

With these identifications one finds that for $T \in (\rho^n | \rho_\alpha)$ the subspace of $\mathcal{F}_{\text{DR}}$ spanned by $T^* \psi_{i_1} \cdots \psi_{i_n}$ depends only on ρ_α (but neither on n nor on the choice of the isometry), and is a $d(\rho_\alpha)$-dimensional subspace of isometries, transforming irreducibly under G. Thus, the equivalence classes $[\rho_\alpha]$ of proper morphisms contained in powers of ρ are in 1:1 correspondence with the equivalence classes

[5] The determinant of a proper morphism ρ with permutation group statistics is defined as follows. Let e be the completely antisymmetrizing Young projection in the group algebra of S_N, $N = d(\rho)$. Then $E = \varepsilon_\rho^{(N)}(e)$ is a minimal projection in $\rho^N(\mathcal{A})'$ with $\phi^N(E) \equiv \text{tr}_\rho(e) = N^{-N}$, hence by 5.4. and A.4., $E = TT^*$, $T \in (\rho^N | \tau)$ with τ an automorphism. The equivalence class of τ is called the determinant of ρ. If $\det(\rho) \neq [\text{id}]$, then one has either to deal with additional $U(1)$ charges, or one may pass to consider a morphism equivalent to the direct sum of ρ and $\bar{\rho}$ (with appropriate extension of the relevant techniques to reducible morphisms), which amounts to consider $U(N)$ as a subgroup of $SO(2N) \subset SU(2N)$.

$[r_\alpha]$ of unitary representations of G contained in powers of the vector representation r of G as a subgroup of $SU(N)$ acting on ψ_i^*, and $d(\rho_\alpha) = \dim(r_\alpha)$. Moreover, the algebra of observables $\mathcal{A}$ coincides with the subalgebra of G-invariants in $\mathcal{F}_{\rm DR}$, justifying the notion "gauge group" for G. Operators ψ^{ρ_α} in the spaces spanned by $T^*\psi_{i_1}\ldots\psi_{i_n}$ satisfy

$$\psi^{\rho_\alpha}A = \rho_\alpha(A)\psi^{\rho_\alpha},$$

and can be moved to any localization by

$$U\cdot\psi^{\rho_\alpha} = \psi^{\sigma_U\circ\rho_\alpha}.$$

Typical gauge covariant charged operators in $\mathcal{F}_{\rm DR}$ are of the form

$$(\psi^{\rho_\alpha})^*\cdot C,$$

transforming in the $d(\rho_\alpha)$-dimensional representation $[r_\alpha]$. They create the charge ρ_α in the sense that $\mathcal{A}$ acts on the Hilbert spaces $(\psi^{\rho_\alpha})^*\mathcal{H}_0$ in the representation $\rho_\alpha \equiv \pi_0\circ\rho_\alpha$:

$$A\cdot(\psi^{\rho_\alpha})^*\Phi = (\psi^{\rho_\alpha})^*\rho_\alpha(A)\Phi.$$

They are localized in $\mathcal{O}$ (that is: they commute with $\mathcal{A}(\mathcal{O}')\subset\mathcal{F}_{\rm DR}$) if $C\in\mathcal{A}(\mathcal{O})$ and ρ_α is localized in $\mathcal{O}$.

If $(\psi_1^{\rho_\alpha})^*C_1$ and $(\psi_2^{\rho_\beta})^*C_2$ are two such operators localized at space-like distance, then one has the ordinary graded commutation relations of fermionic or bosonic tensor fields[6]

$$(\psi_1^{\rho_\alpha})^*C_1\cdot(\psi_2^{\rho_\beta})^*C_2 = \pm(\psi_2^{\rho_\beta})^*C_2\cdot(\psi_1^{\rho_\alpha})^*C_1,$$

with the $-$ sign if both ρ_α and ρ_β are fermionic sectors.

To make contact with the usual description in terms of Wightman fields, observe that the latter are singular and unbounded point-like fields. Thus our field operators should be regarded as bounded functions of Wightman fields smeared with test functions. Their symmetry and localization degrees of freedom are carried by ψ^{ρ_α}, while C contains all remaining dynamical degrees of freedom.

The field bundle. Let us now turn to the field bundle as an intermediate step to the reduced field bundle. Recall that by the very construction of localized morphisms, the *same* Hilbert space $\mathcal{H}_0$ serves as a representation space for all representations $\pi = \pi_0\circ\rho$. Thus it is reasonable to make a notational distinction, writing

$$\imath(A)(\rho,\Phi) = (\rho,\rho(A)\Phi)\in(\rho,\mathcal{H}_0)\equiv\mathcal{H}_\rho,$$

where $\imath(A)$ is the action of $A\in\mathcal{A}$ on the direct sum of Hilbert spaces $\mathcal{H}_\rho$, $\rho\in\Delta_t$.

---

6 The check of these commutation relations from the previous formulae is an instructive exercise.

Introducing charged fields as pairs (ρ, A), $\rho \in \Delta_t$, $A \in \mathcal{A}$, acting on $\mathcal{H}_\rho$ by

$$(\rho_2, A)(\rho_1, \Phi) := (\rho_1\rho_2, \rho_1(A)\Phi),$$

(in particular $(\mathrm{id}, A) \equiv \imath(A)$), one obtains the algebra product

$$(\rho_1, A_1)(\rho_2, A_2) = (\rho_2\rho_1, \rho_2(A_1)A_2),$$

which is mathematically a crossed product of Δ_t with $\mathcal{A}$. Physically, a charged field (ρ, A) adds its charge ρ to the charge of the state. By definition, (e, A) is said to be localized in $\mathcal{O}$ if (e, A) commutes with $\imath(\mathcal{A}(\mathcal{O}'))$, which turns out to be equivalent to the condition that A is of the form $A = U^*C$ with some unitary U such that $C \in \mathcal{A}(\mathcal{O})$ and $\sigma_U \circ \rho$ is localized in $\mathcal{O}$. One finds the commutation relations

$$F_1F_2 = {}^{\varepsilon(\rho_1,\rho_2)}(F_2F_1) \quad \text{resp.} \quad F_1F_2 = {}^{\varepsilon(\rho_2,\rho_1)^*}(F_2F_1),$$

if F_i are localized in $\mathcal{O}_i$ with $\mathcal{O}_2 < \mathcal{O}_1$ resp. $\mathcal{O}_1 < \mathcal{O}_2$, and where the *exterior* action of intertwiners (here the statistics operator) on states or fields is given by ${}^T(\rho_2, \Phi) := (\rho_1, T\Phi)$ resp. ${}^T(\rho_2, A) := (\rho_1, TA)$ if $T \in (\rho_1|\rho_2)$.

The reduced field bundle. In general it is impossible to find a finite collection of transportable morphisms per superselection sector (equivalence class) such that the corresponding charged field bundle operators act within the corresponding direct sum of Hilbert spaces. However, by 5.4. and exploiting the exterior action of intertwiners it is possible to map every subrepresentation of a reducible representation space, say $\mathcal{H}_{12} = (\rho_1\rho_2, \mathcal{H}_0)$, unitarily onto an equivalent "reference" representation space, say $\mathcal{H}_3 = (\rho_3, \mathcal{H}_0)$:

$$\begin{gathered} {}^{T^*}(\rho_1\rho_2, E\mathcal{H}_0) = (\rho_3, T^*\mathcal{H}_0) \equiv (\rho_3, \mathcal{H}_0), \\ E = TT^* \in \rho_1\rho_2(\mathcal{A})', \quad T \in (\rho_1\rho_2|\rho_3). \end{gathered}$$

Including this "transfer operation" with T^* into the action of the field operator as given above, one obtains a closed algebra of fields carrying charge ρ_2 and interpolating from the sector with charge ρ_1 to the sector with charge ρ_3. For definiteness, one fixes a collection ∇ of proper reference morphisms, one per equivalence class, and fixes orthonormal bases T_e of the intertwiner spaces $(\rho_\alpha\rho_\beta|\rho_\gamma)$ according to 5.4. and A.4. The collective label e (the "superselection channel") stands for the triple $\rho_\alpha, \rho_\beta, \rho_\gamma \in \nabla$ as well as for a multiplicity index ranging from 1 to $N^\gamma_{\alpha\beta}$. It will turn out convenient to call ρ_α the source $s(e)$, ρ_γ the range $r(e)$, and ρ_β the charge $c(e)$. Then by definition, the charged fields (e, A) act on states $(\rho_\delta, \Phi) \in \mathcal{H}_\delta$ by

$$(e, A)(\rho_\delta, \Phi) = \delta_{\alpha\delta}(\rho_\gamma, T_e^*\rho_\alpha(A)\Phi).$$

A reduced field bundle operator (e, A) is again by definition localized in $\mathcal{O}$ if it commutes with $\imath(\mathcal{A}(\mathcal{O}'))$ acting as before, which turns out equivalent to $(\rho, A) = (c(e), A)$ (in the field bundle) being localized in $\mathcal{O}$.

From this, and using the commutation relation of the field bundle in the form $\rho_1(A_2)A_1 = \varepsilon(\rho_2,\rho_1)\cdot\rho_2(A_1)A_2$ if $\mathcal{O}_1 < \mathcal{O}_2$, one derives the multiplication law for reduced field bundle operators as a finite operator product expansion, as well as commutation relations for reduced field bundle operators (e_i, A_i) localized in $\mathcal{O}_i$ at space-like distance. In particular, the reduced field bundle operators span an *algebra* $\mathcal{F}_{\text{red}}$ of bounded operators in $\mathcal{H}_{\text{red}} = \bigoplus_{\alpha\in\nabla}\mathcal{H}_\alpha$, and the field bundle operators localized in $\mathcal{O}$ span subalgebras denoted by $\mathcal{F}_{\text{red}}(\mathcal{O})$:

$$(e_2, A_2)(e_1, A_1) = \delta_{s(e_2)r(e_1)} \sum_{e,f} D_{e_1\circ e_2;f,e} \quad (e, A_f),$$

$$(e_2, A_2)(e_1, A_1) = \sum_{e_2'\circ e_1'} R^{(\pm)}_{e_1\circ e_2;e_2'\circ e_1'} (e_1', A_1)(e_2', A_2) \quad \text{if} \begin{cases} \mathcal{O}_1 < \mathcal{O}_2 \\ \mathcal{O}_2 < \mathcal{O}_1 \end{cases}$$

where $A_f = T_f^*\rho_1(A_2)A_1$, and the scalar coefficients D and R are matrix elements of the charge reducing intertwiners resp. statistics operators in the following sense. By A.2(iii), intertwiners map appropriate intertwiner spaces into intertwiner spaces by left multiplication, namely

$$\rho_\alpha(\varepsilon(\rho_2,\rho_1)) : (\rho_\alpha\rho_2\rho_1|\rho_\gamma) \to (\rho_\alpha\rho_1\rho_2|\rho_\gamma) \quad \text{and}$$
$$\rho_\alpha(T_f) : (\rho_\alpha\rho|\rho_\gamma) \to (\rho_\alpha\rho_1\rho_2|\rho_\gamma).$$

By A.4., the intertwiner spaces $(\rho_\alpha\rho_1\ldots\rho_n|\rho_\gamma)$ have orthonormal bases $T_\xi = T_{e_1}\ldots T_{e_n}$, where T_{e_i} are a chain of basis isometries with charges $c(e_i) = \rho_i \in \nabla$, $s(e_{i+1}) = r(e_i)$, $s(e_1) = \rho_\alpha$, and $r(e_n) = \rho_\gamma$. In these bases,

$$D_{e_1\circ e_2;f,e} = T_{e_2}^* T_{e_1}^* \cdot (\rho_\alpha(T_f)) \cdot T_e \in (\rho_\gamma|\rho_\gamma) = \mathbb{C},$$

$$R^{(\pm)}_{e_1\circ e_2;e_2'\circ e_1'} := T_{e_2}^* T_{e_1}^* \cdot \begin{pmatrix} \rho_\alpha(\varepsilon(\rho_2,\rho_1)) \\ \rho_\alpha(\varepsilon(\rho_1,\rho_2)^*) \end{pmatrix} \cdot T_{e_2}T_{e_1} \in (\rho_\gamma|\rho_\gamma) = \mathbb{C}.$$

The following proposition is easily verified by direct calculation.

6.1. Proposition:

In the reduced field bundle constructed with covariant proper reference morphisms $\rho_\alpha \in \nabla$, denote by $\mathcal{U} := \bigoplus_\alpha \mathcal{U}_\alpha$ the representation of $\tilde{G}$ in $\oplus_\alpha\mathcal{H}_\alpha$. Then $\mathrm{Ad}_{\mathcal{U}(\gamma)}(\mathcal{F}_{\text{red}}(\mathcal{O})) = \mathcal{F}_{\text{red}}(\gamma\mathcal{O})$, and with $\rho = c(e)$

$$\mathrm{Ad}_{\mathcal{U}(\gamma)}((e, A)) = (e, \mathcal{U}_\rho(\gamma)\mathcal{U}_0(\gamma)^*\alpha_\gamma(A)),$$

i.e. the operators of the reduced field bundle transform covariantly.

7. Conformal Covariance

A distinctive feature of two-dimensional *conformal* quantum field theories is the existence of *local light-cone fields*, such as currents or the stress-energy tensor, possessing representations with positive spectrum for light-cone translations. The reason is that the conformal covariance group of $\mathbb{M}^2$ factorizes into two Möbius groups transforming either light-cone separately. *Globally* covariant fields, which commute at space-like distances, thus also commute at time-like distances, and globally covariant local fields, which transform in the trivial representation of one of the Möbius factor groups (i.e. are independent of the corresponding light-cone coordinate), may be regarded as fields living on the opposite light-cone with local light-cone commutativity, i.e. $[A, B] = 0$ if A, B are localized in disjoint intervals of the light-cone.

The light-cone transformed by the Möbius group is not the naive light-cone coordinate axis $\mathbb{R}$ in $\mathbb{M}^2$, but rather its compactification S^1. Hence, if we want to apply the general theory of statistics based on local light-cone fields, we have to consider a local net over S^1, which necessarily contain the subalgebra of *global* operators $\mathcal{A}(S^1)$. But the argument for the injectivity of the vacuum representation π_0 fails for global operators, because S^1 cannot be shifted to its complement, and the laws 4.4., 6.1. hold only in restriction to appropriate simply connected neighbourhoods of the origin of $\tilde{G}$.[7]

On the other hand, a conformally covariant local net over S^1 has the advantages [18] that it satisfies Haag duality 2.4. automatically (which must not be postulated as an axiom), and that all its positive energy representations are locally equivalent to each other. In particular, the selection criterion 3.1. covers all positive energy representations.

Let us in a first step avoid all technical problems with global operators and precisely recover the situation of the previous sections by "brute force". We de-compactify the light-cone by omitting one "point at infinity" called $\xi \in S^1$, and restrict ourselves to the local net $\mathcal{A}_\xi$ over $S_\xi = S^1 \setminus \{\xi\} \equiv \mathbb{R}$, with the stabilizer subgroup G_ξ of ξ in the Möbius group G as covariance group. Then all the results of the previous sections hold for $\mathcal{A}_\xi$, with the ordering of disjoint intervals on the circle (distinguishing between ε_ρ and ε_ρ^*) determined with respect to the point ξ. The covariance group G_ξ is implemented in the reduced field bundle $\mathcal{F}_{\mathrm{red}\,\xi}$ acting on $\mathcal{H}_{\mathrm{red}} = \bigoplus_\alpha \mathcal{H}_\alpha$ by $\mathcal{U}(\gamma) = \bigoplus_\alpha \mathcal{U}_\alpha(\gamma)$, and one verifies that 4.4. and 6.1. hold within $\mathcal{F}_{\mathrm{red}\,\xi}$. In particular, by 6.1. one may consider pointlike interpolating (Wightman) fields $\phi_e(x)$ transforming covariantly under G_ξ as scaling limits $\lim_{\lambda\to 0} \mathcal{N}(\lambda) \cdot \mathrm{Ad}_{\mathcal{U}(x)\mathcal{U}(\lambda)}(e, A)$ (with appropriate scaling functions $\mathcal{N}(\lambda)$) localized at $\lim(\lambda I + x) = \{x\}$. Without entering into any technical details of the scaling limits, but only remarking that due to 6.1. the scaling functions may be chosen independently of $s(e)$ and $r(e)$, we conclude that after the limit the exchange algebra remains valid in the form

$$\phi_{1e_1}(x)\phi_{2e_2}(y) = \sum R^{(sig(x-y))}_{e_2 \circ e_1; e_1' \circ e_2'} \, \phi_{2e_2'}(y)\phi_{1e_1'}(x).$$

---

7 Thus the conclusion in 4.4., concerning the additivity of spin, will fail. Instead there will be a modification by the insertion of a monodromy operator ε_M, see below.)

In contrast, the operator product expansion turns into a more complicated expansion involving infinite series of "secondary fields" (a linear basis of Wightman fields). The coefficients of the dominating (most singular) contribution to these expansions will be the matrices D.

In the Wightman framework, the above exchange commutation relations imply monodromy properties for the analytic continuations of n-point functions

$$\langle \Omega, \phi_{ne_n}(x_n) \ldots \phi_{1e_1}(x_1)\Omega \rangle$$

with $s(e_1) = r(e_n) = [0]$, $s(e_{i+1}) = r(e_i)$ to multivalued functions in the set $\mathbb{C}^n \setminus \{z_i = z_j\}$. These are well-known under the name of "conformal block functions" [19, 20]. Although the light-cone fields ϕ_e themselves are non-local, they can be paired into two-dimensional local fields with single-valued correlation functions. For this construction, a TPC-symmetry of the structure constants D and R is crucial.

Actually, hidden in these sketchy arguments we used the fact that the true covariance group is G and not only G_ξ. In fact, there is a family of reduced field bundles, one for every $\xi \in S^1$, acting in the same Hilbert space, and transformed into each other by the full conformal group G. They are subalgebras of $\mathcal{B}(\mathcal{H}_{\text{red}})$ with large intersections (but eventually with different notions of localization, since the criterion "(e, A) commutes with $\mathcal{A}(S_\xi \backslash I)$" depends on ξ). 6.1. holds within $\mathcal{F}_{\text{red}\,\xi}$ for γ in a neighbourhood of id $\in \tilde{G}$ (strictly larger than G_ξ). One has to carefully distinguish observables in $\mathcal{A}$ from operators in $\mathcal{B}(\mathcal{H}_0)$; e.g. in 6.1. the operator $\mathcal{U}_\rho(\gamma)\mathcal{U}_0(\gamma)^* \in \mathcal{B}(\mathcal{H}_0)$ has to be replaced by $U_\gamma \in \mathcal{A}_\xi$ such that $\pi_0(U_\gamma) = \mathcal{U}_\rho(\gamma)\mathcal{U}_0(\gamma)^*$.

We shall now give an example for the close interplay of covariance with statistics, leading to a spin-statistics theorem. Although the following argument provides only a weaker form of this relation, we present it instead of the stronger one, since it most clearly exhibits the role of charge transporting operators as the link between covariance and statistics.

Consider the geometrical situation in Fig.1a. We assume two conformally covariant representations π_1, π_2 to be localized in $I \subset S^1$. We study rigid rotations $\varphi \in \tilde{G}$ of S^1 by an angle φ, and are specially interested in the central elements $\mathcal{U}(2\pi) = \zeta$ in irreducible representations.

The operator $\mathcal{U}_0(+\pi)\mathcal{U}_2(-\pi)$ takes π_2 into a representation localized in $\tilde{I}$ and commutes with $\pi_0(\mathcal{A}(I_\xi))$, where I_ξ is a small interval around ξ. Hence there is a unitary $U_+ \in \mathcal{A}_\xi$ such that $\pi_0(U_+) = \mathcal{U}_0(+\pi)\mathcal{U}_2(-\pi)$. According to 4.4. and 4.6. (since $I < \tilde{I}$ with respect to ξ) we have

$$\mathcal{U}_{12}(-\pi) = \mathcal{U}_1(-\pi)\pi_1(U_+), \quad \varepsilon(\pi_2, \pi_1) = \pi_1(U_+^*)\pi_0(U_+).$$

Arguing similarly with Fig.1b, there is a unitary $U_- \in \mathcal{A}_\zeta$ such that

$$\pi_0(U_-) = \mathcal{U}_0(-\pi)\mathcal{U}_2(+\pi) \quad \text{and}$$
$$\mathcal{U}_{12}(+\pi) = \mathcal{U}_1(+\pi)\pi_1(U_-), \quad \varepsilon(\pi_1, \pi_2) = \pi_0(U_-^*)\pi_1(U_-).$$

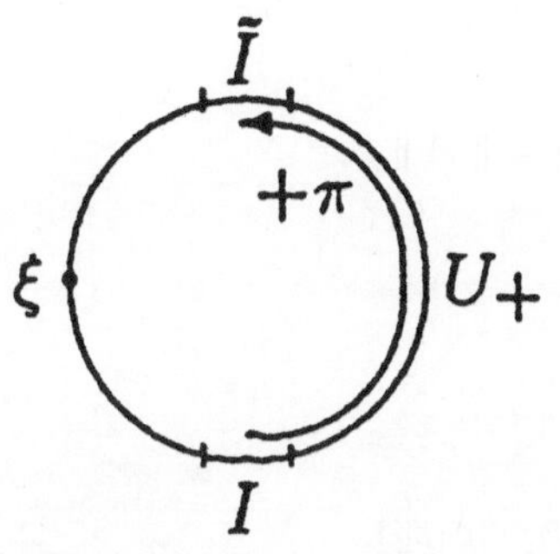

Fig. 1a.

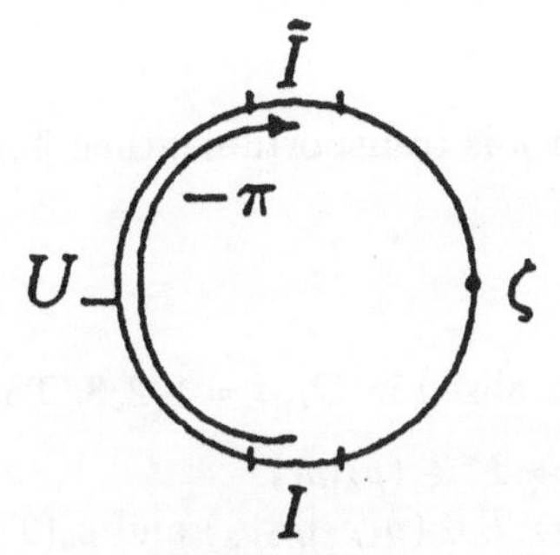

Fig. 1b.

With these formulae we compute the central element of the conformal group in the composite representation:

$$\begin{aligned}\mathcal{U}_{12}(2\pi) &= \varepsilon(\pi_2,\pi_1)\mathcal{U}_2(+\pi)\mathcal{U}_0(-\pi)\mathcal{U}_1(2\pi)\mathcal{U}_0(-\pi)\mathcal{U}_2(+\pi)\varepsilon(\pi_1,\pi_2) = \\ &= \zeta_1\zeta_2\cdot\varepsilon(\pi_2,\pi_1)\varepsilon(\pi_1,\pi_2),\end{aligned}$$

and with 5.4. and 3.4. we obtain the central element in an irreducible subrepresentation π_3 by

$$\frac{\zeta_3}{\zeta_1\zeta_2} = \frac{\omega_3}{\omega_1\omega_2}.$$

The additivity of spin is corrected by statistics phases due to the monodromy operator. In fact, the stronger result $\zeta = \omega$ can also be proven.

Appendix

A.1. Lemma: If ρ is transportable, then $\| \rho(A) \| = \| A \|$.

A.2. Lemma:

Let $\rho_i \in \Delta_t$ be localized in $\mathcal{O}_i$, $i = 1, 2, 3$. Then

(i) $T \in (\rho_1|\rho_2) \Rightarrow T^* \in (\rho_2|\rho_1)$.
(ii) $T \in (\rho_1|\rho_2) \Rightarrow T \in (\rho_1\rho_3|\rho_2\rho_3)$ and $\rho_3(T) \in (\rho_3\rho_1|\rho_3\rho_2)$.
(iii) $T \in (\rho_1|\rho_2)$ and $S \in (\rho_2|\rho_3) \Rightarrow TS \in (\rho_1|\rho_3)$.

A.3. Lemma:

Let $\rho_1, \rho_2 \in \Delta_t$, ρ_2 irreducible. Then

(i) $(\rho_1|\rho_2) \subset \mathcal{A}$ is a Hilbert space with inner product

$$\langle S, T \rangle := S^*T \in (\rho_2|\rho_2) \equiv \rho_2(\mathcal{A})' \equiv \mathbb{C}.$$

(ii) If T is a unit vector in $(\rho_1|\rho_2)$, then T is an isometry and $E = TT^* \in \rho_1(\mathcal{A})'$ is a projection onto a subrepresentation of ρ_1 unitarily equivalent to $\pi_0 \circ \rho_2$.

Proof of 4.2.: Let $A \in \mathcal{A}(\mathcal{O})$. It is possible to choose $\tilde{\mathcal{O}}_i \subset \mathcal{O}'$, $i = 1, 2$, and unitary "charge transporting operators" $U_i \in \mathcal{A}(\mathcal{O}_i \vee \tilde{\mathcal{O}}_i)$, such that $(\mathcal{O}_1 \vee \tilde{\mathcal{O}}_1) \times (\mathcal{O}_2 \vee \tilde{\mathcal{O}}_2)$ and $\tilde{\rho}_i = \sigma_{U_i} \circ \rho_i$ are localized in $\tilde{\mathcal{O}}_i$. Then with 3.2. and 2.3. one computes

$$\rho_1\rho_2(A) = U_1^*\tilde{\rho}_1(U_2^*\tilde{\rho}_2(A)U_2)U_1 = U_1^*U_2^*AU_2U_1 =$$
$$= U_2^*U_1^*AU_1U_2 = U_2^*\tilde{\rho}_2(U_1^*\tilde{\rho}_1(A)U_1)U_2 = \rho_2\rho_1(A).$$

Proofs of 4.4., 4.5. can be found in [2, *Sect.*2].

Proof of 4.6.: By the required properties (i) – (iii) of the theorem, which imply $\varepsilon(\rho, \mathrm{id}) = \varepsilon(\mathrm{id}, \rho) = \mathbf{1}$ and $\varepsilon(\sigma_U, \rho) = \rho(U)U^*, \varepsilon(\rho, \sigma_U) = U\rho(U^*)$, we can give a formula for the statistics operator. Namely, let ρ_i be localized in $\mathcal{O}_i$, $i = 1, 2$. It is possible to choose auxiliary double cones $\tilde{\mathcal{O}}_i$ such that $\tilde{\mathcal{O}}_1 \times \tilde{\mathcal{O}}_2$ for $d > 2$ resp. $\tilde{\mathcal{O}}_2 < \tilde{\mathcal{O}}_1$ for $d \leq 2$, and unitary charge transporters $U_i \in \mathcal{A}(\mathcal{O}_i \vee \tilde{\mathcal{O}}_i)$ such that $\tilde{\rho}_i = \sigma_{U_i} \circ \rho_i$ are localized in $\tilde{\mathcal{O}}_i$. Then by (i) – (iii)

$$\varepsilon(\rho_1, \rho_2) \stackrel{!}{=} \varepsilon := \rho_2(U_1^*)U_2^*U_1\rho_1(U_2) \in (\rho_2\rho_1|\rho_1\rho_2) \subset \mathcal{A}(\mathcal{O}_1 \vee \mathcal{O}_2),$$

where the stated intertwining property of ε is verified from $U_i \in (\tilde{\rho}_i|\rho_i)$, using 3.8., A.2, 4.1., 4.2. It remains, however, to show that ε is independent of the choice of the auxiliary double cones $\mathcal{O}_i$ and charge transporters U_i: First, for a given pair of auxiliary double cones $\tilde{\mathcal{O}}_i$, ε is independent of the choice of the unitaries U_i as long as $\sigma_{U_i} \circ \rho_i$ remain localized in $\tilde{\mathcal{O}}_i$; namely if $U_i^!$ is another such pair of unitaries, then $V_i := U_i^!U_i^* \in (\tilde{\rho}_i^!|\tilde{\rho}_i) \subset \mathcal{A}(\tilde{\mathcal{O}}_i)$, hence for $i \neq j = 1, 2$

$$U_i^! \rho_i(U_j^!) = V_i U_i \rho_i(V_j U_j) = V_i \tilde{\rho}_i(V_j) U_i \rho_i(U_j) = V_i V_j \cdot U_i \rho_i(U_j),$$

which implies the claim since $V_1 V_2 = V_2 V_1$. Second, let $\tilde{\mathcal{Q}}_i$ be another pair of auxiliary double cones sufficiently close to $\tilde{\mathcal{O}}_i$ such that

$$(\tilde{\mathcal{O}}_1 \vee \tilde{\mathcal{Q}}_1) \big\backslash\!\big/ (\tilde{\mathcal{O}}_2 \vee \tilde{\mathcal{Q}}_2) \quad \text{resp.} \quad (\tilde{\mathcal{O}}_2 \vee \tilde{\mathcal{Q}}_2) < (\tilde{\mathcal{O}}_1 \vee \tilde{\mathcal{Q}}_1),$$

and $U_i^!$ a pair of unitary charge transporters into $\tilde{\mathcal{Q}}_i$. Then both U_i and $U_i^!$ are also charge transporters into the auxiliary double cones $\tilde{\mathcal{O}}_i \vee \tilde{\mathcal{Q}}_i$, and by the preceding argument ε defined by U_i and ε' defined by $U_i^!$ coincide. Repeating the argument one finds that ε remains unchanged if the pair of auxiliary double cones is iteratively changed by a chain of moves of the above type, keeping the separation space-like. Now, in $d > 2$ any such pair can be changed into every other such pair by a chain of moves of the above type, hence ε is completely independent of the pair of auxiliary double cones chosen. (v) is then a consequence of the asymmetry under $(1 \leftrightarrow 2)$ of the definition of ε. In $d \leq 2$, a chain of moves of the above type can take any pair $\tilde{\mathcal{O}}_2 < \tilde{\mathcal{O}}_1$ into every other pair $\tilde{\mathcal{Q}}_2 < \tilde{\mathcal{Q}}_1$, hence again ε is independent of the auxiliary double cones $\tilde{\mathcal{O}}_2 < \tilde{\mathcal{O}}_1$ chosen.

In particular, $\varepsilon(\rho_1, \rho_2)$ is uniquely determined by the properties (i) – (iii). (Had one chosen auxiliary double cones $\tilde{\mathcal{O}}_1 < \tilde{\mathcal{O}}_2$ instead, one would have obtained $\varepsilon' = \varepsilon(\rho_2, \rho_1)^*$ instead of $\varepsilon(\rho_1, \rho_2)$, but since the ordering cannot be changed continuously while keeping the distance space-like, there is no reason why the two operators should coincide.)

To prove the first equations of (i) and (ii), one may choose for simplicity $\tilde{\mathcal{O}}_3 < (\mathcal{O}_1 \vee \mathcal{O}_2)$, $\tilde{\mathcal{O}}_1 = \mathcal{O}_1$, $\tilde{\mathcal{O}}_2 = \mathcal{O}_2$, and $U_1 = U_2 = \mathbf{1}$. Then

$$\varepsilon(\rho_1, \rho_3) \rho_1(\varepsilon(\rho_2, \rho_3)) = U_3^* \rho_1(U_3) \rho_1(U_3^* \rho_2(U_3)) = U_3^* \rho_1 \rho_2(U_3) = \varepsilon(\rho_1 \rho_2, \rho_3),$$

$$\begin{aligned} \rho_3(T)\varepsilon(\rho_1, \rho_3) &= \rho_3(T) U_3^* \rho_1(U_3) = U_3^* \tilde{\rho}_3(T) \rho_1(U_3) = \\ &= U_3^* T \rho_1(U_3) = U_3^* \rho_2(U_3) T = \varepsilon(\rho_2, \rho_3) T, \end{aligned}$$

where we used 4.1(i) resp. 3.2(i) and 3.8. The second equations are obtained by choosing $(\mathcal{O}_1 \vee \mathcal{O}_2) < \tilde{\mathcal{O}}_3$ instead. Finally, (iv) is proven by the substitutions $\rho_2 \to \rho_2 \rho_1, \rho_1 \to \rho_1 \rho_2$ and $T = \varepsilon(\rho_1, \rho_2)$ in (ii), using (i).

Proof of 4.8.: 4.7(i) and (ii) imply that $\varepsilon_\rho^{(n)}(\sigma_i)$ satisfy the defining relations of the braid group. For $d > 2$ the additional relation defining the permutation group is implied by 4.7(iii). The remaining statements are evident.

Proof of 5.3.: From 4.6(ii) with $\rho_1 = \text{id}$, $\rho_2 = \bar{\rho}\rho$ one checks the identity

$$\begin{aligned} \bar{\rho}(\varepsilon(\rho, \rho)) R &= \varepsilon(\bar{\rho}, \rho)^* \rho(R), \quad \text{hence} \\ \lambda \equiv \phi_R(\varepsilon_\rho) &= R^* \varepsilon(\bar{\rho}, \rho)^* \rho(R) = \bar{R}^* \rho(R), \end{aligned}$$

where we called $\bar{R}$ the isometry $\bar{R} = \varepsilon(\bar{\rho}, \rho) R \in (\rho\bar{\rho} | \text{id})$. Since by assumption the scalar λ is different from zero, one can rewrite every $A \in \mathcal{A}$ in the form

$$A = |\lambda|^{-2} \rho(R^*) \bar{R} A \bar{R}^* \rho(R) = |\lambda|^{-2} \rho(R^*) \bar{R} \bar{R}^* \rho(\bar{\rho}(A) R),$$

hence for every left-inverse ϕ of ρ one gets $\phi(A) = |\lambda|^{-2} R^* \phi(\bar{R}\bar{R}^*)\bar{\rho}(A)R$. Now, $\phi(\bar{R}\bar{R}^*) \in (\bar{\rho}|\bar{\rho})$ is a scalar, and putting $A = 1$ one finds $\phi(\bar{R}\bar{R}^*) = |\lambda|^2$, hence $\phi(A) = \phi_R(A)$ implying (iii). Next, let $\bar{\rho}' \in \Delta_t$ be another irreducible morphism for which there is an isometry $\bar{R}' \in (\bar{\rho}'\rho|\,\mathrm{id})$. Then $\bar{\rho}'(\bar{R}^*)R' \in (\bar{\rho}'|\bar{\rho})$ does not vanish, since $\bar{\rho}'(\bar{R}^*)R'R = \bar{\rho}'(\bar{R}^*\rho(R))R' = \lambda \cdot R' \neq 0$, and is a multiple of a unitary intertwiner since $\bar{\rho}, \bar{\rho}'$ are both irreducible. This implies (i).

For $R_1, R_2 \in (\bar{\rho}\rho|\,\mathrm{id})$ isometries one has

$$\lambda R_2 = \bar{\rho}(R_1^* \varepsilon(\bar{\rho},\rho)^* \rho(R_1))R_2 = \bar{\rho}(R_1^*\varepsilon(\bar{\rho},\rho)^*)R_2 \cdot R_1.$$

Since $\bar{\rho}(R_1^*\varepsilon(\bar{\rho},\rho)^*)R_2 \in (\bar{\rho}|\bar{\rho}) = \mathbb{C}$, R_2 is a scalar multiple of R_1, hence (ii). Finally, again with 4.6(ii) one finds $\bar{R}^*\rho(\varepsilon(\bar{\rho},\bar{\rho})) = \bar{\rho}(\bar{R}^*)\varepsilon(\bar{\rho},\rho)^*$, and thus

$$\mathbb{C} \ni \bar{\phi}_R(\varepsilon_{\bar{\rho}}) = \bar{\rho}(\bar{R}^*)R = R^*\bar{\rho}(\bar{R}^*)RR = \phi(\bar{R}^*)R = \phi(\bar{R}^*\rho(R)) = \phi(\lambda) = \lambda \neq 0.$$

Applying the results (i) – (iii) to $\bar{\rho}$ instead of ρ, we prove (iv).

To prove the class-invariance of the statistics parameter, let $\rho' = \sigma_U \rho$ be equivalent to the proper morphism ρ. Then $R' := \bar{\rho}(U)R$ is an isometry in $(\bar{\rho}\rho'|\,\mathrm{id})$. In particular, $\bar{\rho}$ is conjugate to ρ', and by 4.6(i) one finds $\varepsilon_{\rho'} = U\rho(U)\varepsilon_\rho \rho(U^*)U^*$. Hence

$$\phi_{R'}(\varepsilon_{\rho'}) = R^*\bar{\rho}(\rho(U)\varepsilon_\rho\rho(U^*))R = U\phi(\varepsilon_\rho)U^* = \phi(\varepsilon_\rho).$$

Finally, $d \geq 1$ will become evident in the following proof.

Proof of 5.4.: Let $E \in \rho_1\rho_2(\mathcal{A})'$ be a projection onto a subrepresentation of $\rho_1\rho_2$. We claim that $\phi_1(E) \geq d_1^{-2}$. (Since positivity of left-inverses implies $\phi_1(1-E) \geq 0$, we conclude in particular $d_1 \geq 1$.) Namely, by 4.1(i) and 3.8., if ρ_i are localized in $\mathcal{O}_i$ then $E \in \mathcal{A}(\mathcal{O}_1 \vee \mathcal{O}_2)$, and we may choose $\tilde{\mathcal{O}}_1 < (\mathcal{O}_1 \vee \mathcal{O}_2)$ and U such that $\tilde{\rho}_1 = \sigma_U\rho_1$ is localized in $\tilde{\mathcal{O}}_1$. Then $\varepsilon_{\rho_1} = U^*\rho_1(U)$, hence

$$EU = \tilde{\rho}_1(E)U = U\rho_1(E) = \rho_1(U)\varepsilon_{\rho_1}^*\rho_1(E),$$

hence $\phi_1(EU) = \lambda_1^*\, UE$, $\phi_1(U^*E) = \lambda_1\, EU^*$. Then from the estimate

$$\begin{aligned} \phi_1(E) &= R_1^*\bar{\rho}_1(EUU^*E)R_1 \geq \\ &\geq R_1^*\bar{\rho}_1(EU)\cdot R_1R_1^* \cdot \bar{\rho}_1(U^*E)R_1 = \phi_1(EU)\phi_1(U^*E) = d_1^{-2}UEU^*, \end{aligned}$$

by taking norms and noting that $\phi_1(E) \in \rho_2(\mathcal{A})'$ is a positive scalar, we get the claim. Consequently every partition of unity into orthogonal projections $E \in \rho_1\rho_2(\mathcal{A})'$ can have at most d_1^2 elements, and there is a *finite* partition of unity into orthogonal *minimal* projections $E^{(j)} \in \rho_1\rho_2(\mathcal{A})'$.

Since $\varepsilon_M := \varepsilon(\rho_2,\rho_1)\varepsilon(\rho_1,\rho_2) \in \rho_1\rho_2(\mathcal{A})'$, we may assume that $E^{(j)}$ are dominated by spectral projections of ε_M, i.e. $\varepsilon_M E^{(j)} = \mu^{(j)}E^{(j)}$.

By 3.6., let $E^{(j)} = T^{(j)}T^{(j)*}$ with local isometries $T^{(j)}$. Then by virtue of the minimality of $E^{(j)}$, the transportable morphisms

$$\rho^{(j)}(A) = T^{(j)*}\rho_1\rho_2(A)T^{(j)}$$

are irreducible. We have to show that $\rho^{(j)}$ are proper.

First, pick any of the irreducible components $\rho^{(j)} =: \rho$ (and omit the label (j) for quantities referring to ρ). Observe that

$$\phi(A) := \phi_2\phi_1(TT^*)^{-1} \cdot \phi_2\phi_1(TAT^*)$$

defines a left-inverse of ρ. By 4.6(i), (ii), and (iv) one finds

$$\begin{aligned}\phi_2\phi_1(TT^*) \cdot \phi(\varepsilon_\rho) &= \phi_2\phi_1(T\varepsilon(\rho,\rho)T^*) = \phi_2\phi_1(\rho_1\rho_2(T^*)\varepsilon(\rho_1\rho_2,\rho_1\rho_2)\rho_1\rho_2(T)) \\ &= \lambda_1\lambda_2 \cdot \phi_2\phi_1(T^*\varepsilon_M T) = \mu\ \lambda_1\lambda_2 =: \phi_2\phi_1(TT^*) \cdot \lambda \neq 0.\end{aligned}$$

Second, since $\bar\rho_i$ are also proper, let $E_{(i)} = T_{(i)}T^*_{(i)}$ be projections (with the respective properties) onto subrepresentations of $\bar\rho_2\bar\rho_1$ equivalent to $\rho_{(i)}$, and denote $W_i := T^*_{(i)}\bar\rho_2\bar\rho_1(T^*)\bar\rho_2(R_1)R_2 \in (\rho_{(i)}\rho|\,\mathrm{id})$. These cannot all vanish, since one finds

$$\sum_i W_i^*\rho_{(i)}(\varepsilon_\rho)W_i = \phi_2\phi_1(T\varepsilon_\rho T^*) = \phi_2\phi_1(TT^*) \cdot \phi(\varepsilon_\rho) \neq 0.$$

In particular, at least one among the W_i is a non-vanishing multiple of an isometry $R \in (\rho_{(i)}\rho|\,\mathrm{id})$, for which $\phi_R(\varepsilon_\rho) \neq 0$. Hence ρ is proper with the corresponding $\rho_{(i)}$ as a conjugate $\bar\rho$, and by 5.3., $\phi = \phi_R$. This completes the proof of (i).

Comparing phases and moduli in the above equation for $\lambda = \phi(\varepsilon_\rho)$, we find the spectrum of the monodromy operator: $\mu = \omega(\rho)/\omega(\rho_1)\omega(\rho_2)$, as well as $\phi_2\phi_1(TT^*) = d(\rho)/d(\rho_1)d(\rho_2)$. This, and $\sum_j \phi_2\phi_1(E^{(j)}) = 1$, proves the "sum rule" and the upper bound for $d(\rho)$, while the lower bound follows from the previous estimate $\phi_1(TT^*) \geq d_1^{-2}$, together with the symmetry $1 \leftrightarrow 2$, cf. 4.3.

We retain some useful formula in

A.4. Lemma:

(i) The product $\rho_1\rho_2$ of two proper morphisms is decomposed into irreducible components $\rho^{(j)}$ by isometries $T^{(j)} \in (\rho_1\rho_2|\rho^{(j)})$ satisfying

$$T^{(j)*}T^{(k)} = \delta_{jk}, \quad \sum_j T^{(j)}T^{(j)*} = 1,$$

$$\varepsilon(\rho_2,\rho_1)\varepsilon(\rho_1,\rho_2)\ T^{(j)} = \frac{\omega(\rho^{(j)})}{\omega(\rho_1)\omega(\rho_2)}\ T^{(j)}.$$

(ii) One may assume that any two of the $\rho^{(j)}$ are either equal or inequivalent. Then the collection of intertwiners $T^{(j)} \in (\rho_1\rho_2|\rho^{(j)})$ with common $\rho^{(j)}$ provides an orthonormal basis of $(\rho_1\rho_2|\rho^{(j)})$ in the sense of A.3., and if $\rho^{(j)} = \rho^{(k)}$, then

$$\phi_1(T^{(j)}T^{(k)*}) \equiv \phi_2\phi_1(T^{(j)}T^{(k)*}) = \frac{d(\rho^{(j)})}{d(\rho_1)d(\rho_2)}\delta_{jk}.$$

(iii) More generally, let $\rho_1, \ldots, \rho_n$ be proper. There are local isometries T_ξ such that

$$T_\xi^* T_{\xi'} = \delta_{\xi\xi'}, \quad \sum_\xi T_\xi T_\xi^* = \mathbf{1},$$

and $E_\xi = T_\xi T_\xi^*$ are orthogonal minimal projections in $\rho_1 \ldots \rho_n(\mathcal{A})'$. T_ξ may be chosen such that the proper morphisms $\rho_\xi(A) = T_\xi^* \rho_1 \ldots \rho_n(A) T_\xi$ are either inequivalent or equal. All T_ξ with common ρ_ξ provide an orthonormal basis of $(\rho_1 \ldots \rho_n | \rho_\xi)$. If $\rho_\xi = \rho_{\xi'}$, then

$$\phi_n \ldots \phi_1(T_{\xi'} T_\xi^*) = \frac{d(\rho_\xi)}{d(\rho_1) \ldots d(\rho_n)} \delta_{\xi\xi'}.$$

Proof: (i) The orthonormality and completeness statements for the isometry $T^{(j)} \in (\rho_1\rho_2|\rho^{(j)})$ are equivalent to the orthogonality and completeness of the projections $E^{(j)}$ in the proof of 5.4., where also the eigenvalues of ε_M have been computed. (ii) Observe that if $\rho^{(k)} = \sigma_U \circ \rho^{(j)}$ is equivalent to $\rho^{(j)}$, then with the choice $T^{(k)'} = T^{(k)} U \in (\rho_1\rho_2|\rho^{(j)})$ all of the formulae in the proof of 5.4. remain valid, with $\rho^{(k)'} = \rho^{(j)}$ instead of $\rho^{(k)}$. Thus we may choose $T^{(j)}$ such that $\rho^{(j)} = \rho^{(k)}$ whenever they are equivalent. Now let $\rho^{(j)} = \rho^{(k)} = \rho$, and let $\bar{R} \in (\rho\bar{\rho}|\,\mathrm{id})$ be as in the proof of 5.3. Observe that $\rho(\bar{R}^*)\varepsilon_\rho\rho(\bar{R}) \in (\rho|\rho)$ is a scalar, which by application of the left-inverse of ρ turns out to be the statistics parameter λ of ρ. Hence

$$\begin{aligned}\phi_2\phi_1(T^{(j)}T^{(k)*}) &= \lambda^{-1}\ \phi_2\phi_1(T^{(j)}\rho(\bar{R}^*)\varepsilon_\rho\rho(\bar{R})T^{(k)*}) = \\ &= \lambda^{-1}\lambda_1\lambda_2\ \bar{R}^* T^{(k)*}\varepsilon_M T^{(j)}\bar{R} = (d/d_1 d_2)\ \delta_{jk}.\end{aligned}$$

The proof of (iii) is a straightforward generalization.

Proof of 5.5.: $\mathrm{tr}_\rho^{(n)}(b)$ is a scalar by 5.2., since $\varepsilon_\rho^{(n)}(b) \in (\rho^n|\rho^n)$. With $\varepsilon_\rho^{(n)}(b) = \varepsilon_\rho^{(n+1)}(b)$ and $\phi^n(\varepsilon_\rho^{(n)}(b)) \in \mathbb{C}$, one gets (i). (ii) follows from unitarity of ε_ρ and normalization and C^*-linearity of ϕ, (v) from positivity of ϕ. For the strong Markov property (iv) observe that $\varepsilon_\rho^{(n)}(b_1 b_2) = \varepsilon_\rho^{(n)}(b_1)\varepsilon_\rho^{(n)}(b_2)$ with $\varepsilon_\rho^{(n)}(b_1) \in (\rho^k|\rho^k)$ and $\varepsilon_\rho^{(n)}(b_2) = \rho^{k-1}(u)$ for some $u \in \mathcal{A}$. Thus

$$\phi^{k-1}(\varepsilon_\rho^{(n)}(b_1 b_2)) = \phi^{k-1}(\varepsilon_\rho^{(n)}(b_1)) \cdot u = \mathrm{tr}_\rho(b_1) \cdot u,$$

since $\phi^{k-1}(\varepsilon_\rho^{(n)}(b_1)) \in (\rho|\rho)$ is already a scalar. Finally, with $u = \phi^{k-1}(\varepsilon_\rho^{(n)}(b_2))$, one gets (iv) by further application of ϕ. To prove the trace property (iii), choose intertwiners $T_\xi \in (\rho^n|\rho_\xi)$ as in A.4(iii). Then for every $m \in (\rho^n|\rho^n)$, in particular for $\varepsilon_\rho^{(n)}(b)$,

$$\phi^n(m) = \sum_{\xi\xi'} \phi^n(T_{\xi'}T_{\xi'}^*\ m\ T_\xi T_\xi^*) = \sum_{\xi\xi'} m_{\xi'\xi}\phi^n(T_{\xi'}T_\xi^*),$$

where $m_{\xi'\xi} = T_{\xi'}^* m T_\xi \in (\rho_{\xi'}|\rho_\xi)$ are scalars which vanish unless $\rho_{\xi'} = \rho_\xi$. But then A.4(iii) applies and yields

$$\phi^n(m) = \sum_\alpha \frac{d(\rho_\alpha)}{d(\rho)^n} \sum_{\xi:\ \rho_\xi = \rho_\alpha} m_{\xi\xi},$$

where ρ_α runs over all the values ρ_ξ can take. This is manifestly a trace, since $m_{\xi'\xi}$ are the matrix elements with respect to the orthonormal bases T_ξ of the linear actions (by left multiplication) of m in $(\rho^n|\rho_\alpha)$.

Proof of 5.6.: The first equality of (i) is 4.3. The second equality is proven as follows. By virtue of the proof of 5.3(iv), the linear mappings

$$X : (\rho_\alpha\rho_\beta|\rho_\gamma) \to (\rho_\beta|\bar{\rho}_\alpha\rho_\gamma), \quad T \mapsto R^*_\alpha\bar{\rho}_\alpha(T)$$
$$Y : (\rho_\beta|\bar{\rho}_\alpha\rho_\gamma) \to (\rho_\alpha|\rho_\beta\rho_\gamma), \quad S \mapsto \rho_\alpha(S)\bar{R}_\alpha$$

invert each other: $X \circ Y = \lambda(\rho)\ \mathrm{id} = Y \circ X$. Hence the two spaces of intertwiners have equal dimensions, and $\dim(\rho_\alpha\rho_\beta|\rho_\gamma) = \dim(\rho_\beta|\bar{\rho}_\alpha\rho_\gamma)$ by A.2(i). The third equality is obtained by iterating the first and second one. The rest of (i) is obvious. (iii) is just the "sum rule" in 5.4.

It remains to prove (ii). First observe, that $(N_\alpha N_\beta)^\varepsilon_\delta = \dim(\rho_\delta\rho_\alpha\rho_\beta|\rho_\varepsilon)$. Namely, for ζ ranging over all inequivalent subrepresentations of $\rho_\delta\rho_\alpha$ let $T_e \in (\rho_\delta\rho_\alpha|\rho_\zeta)$ and $T_f \in (\rho_\zeta\rho_\beta|\rho_\varepsilon)$ be respective orthonormal intertwiner bases. Then the $\sum_\zeta N^\zeta_{\alpha\delta}N^\varepsilon_{\beta\zeta} = (N_\alpha N_\beta)^\varepsilon_\delta$ operators T_eT_f provide a basis of $(\rho_\delta\rho_\alpha\rho_\beta|\rho_\varepsilon)$: by A.2., they are contained in these intertwiner spaces, and by A.4(i), they are orthonormal and every intertwiner $T \in (\rho_\delta\rho_\alpha\rho_\beta|\rho_\varepsilon)$ can be written as $T = \sum_{\zeta,e} T_eT^*_eT$, where $T^*_eT \in (\rho_\zeta\rho_\beta|\rho_\varepsilon)$ is a linear combination of T_f. Similarly, one verifies $\sum_\gamma N^\gamma_{\alpha\beta}N^\varepsilon_{\gamma\delta} = \dim(\rho_\alpha\rho_\beta\rho_\delta|\rho_\varepsilon)$. Finally, the dimensions of the two intertwiner spaces coincide, since again by A.2., the former is unitarily mapped onto the latter by the statistics operator $\varepsilon(\rho_\alpha\rho_\beta, \rho_\delta)$.

Proof of 5.7.: (i) implies (ii) implies (iii) implies (iv) trivially by use of the definitions. By the sum rule for statistical dimensions in 5.4., and $d \geq 1$, ρ^2 can contain only one subrepresentation if $d(\rho) = 1$, hence (iv) implies (ii). We finally show that (iii) implies (i). Let $A \in \mathcal{A}(\mathcal{Q})$, and ρ be localized in $\mathcal{O}$. Choose $\tilde{\rho} = \sigma_U \circ \rho$ localized in $\tilde{\mathcal{O}} < (\mathcal{O} \vee \mathcal{Q})$. Then $\varepsilon_\rho = U^*\rho(U)$, and $A = \tilde{\rho}(A) = U\rho(A)U^* = \rho(U)\rho(A)\rho(U^*) \in \rho(\mathcal{A})$. Thus all local algebras are in $\rho(\mathcal{A})$, and taking the norm closure of their union, $\mathcal{A}$ itself is in $\rho(\mathcal{A})$. Thus ρ is surjective. Since, by A.1., transportable morphisms are always injective, ρ is invertible.

Acknowledgements: I am particularly indebted to D.Kastler who has the merit (besides being one of the founders of the theory, and having contributed to its development with unceasing enthusiasm) of having organized the bulk of material into a "linearly presentable" scheme. This led to the recent review [10], along the lines of which the major part of the present lectures is oriented. My own contributions to the theory arose from joint work with K.Fredenhagen and B.Schroer, to whom I am also much indebted.

References

1. S.Doplicher, R.Haag, J.E.Roberts: Commun.Math.Phys. **23**, 199 (1971).
2. S.Doplicher, R.Haag, J.E.Roberts: Commun.Math.Phys. **35**, 49 (1974).
3. K.Fredenhagen, K.-H.Rehren, B.Schroer: Commun.Math.Phys. **125**, 201 (1989); and Part II in preparation.
4. S.Doplicher, J.E.Roberts: in *Proceedings of the* VIIIth *Intern. Congress on Math. Phys.* (Marseille 1986), p.489; eds. M.Mebkhout, R.Sénéor (Singapore 1987); Bull.Am.Math.Soc. (New Series) **11**, 333 (1984); Invent.Math. **98**, 157 (1989).
5. R.Haag, D.Kastler: J.Math.Phys. **5**, 848 (1964); H.J.Borchers: Commun.Math.Phys. **1**, 281 (1965).
6. D.Buchholz, K.Fredenhagen: Commun.Math.Phys. **84**, 1 (1982).
7. K.Fredenhagen: "Structure of Superselection Sectors in Low-Dimensional Quantum Field Theory", Proceedings Lake Tahoe 1989, and Schladming lectures 1990; J.Fröhlich, F.Gabbiani, P.-A.Marchetti: "Superselection Structure and Statistics in Three-Dimensional Local Quantum Theory" and "Braid Statistics in Three-Dimensional Local Quantum Field Theory", ETH Zürich preprints ETH-TH/89-22 and 89-36.
8. J.J.Bisognano, E.H.Wichmann: Journ.Math.Phys. **16**, 985 (1975) and **17**, 303 (1976).
9. H.J.Borchers: Commun.Math.Phys. **4**, 315 (1967).
10. D.Kastler, M.Mebkhout, K.-H.Rehren: "Introduction to the Algebraic Theory of Superselection Sectors", in *Algebraic Theory of Superselection Sectors and Field Theory*, ed. D.Kastler; to be published by World Scientific.
11. S.Doplicher, R.Haag, J.E.Roberts: Commun.Math.Phys. **15**, 173 (1969).
12. H.Wenzl: Invent.Math. **92**, 349 (1988).
13. R.Longo: Commun.Math.Phys. **126**, 217 (1989) and **130**, 285 (1990).
14. J.Birman, H.Wenzl: Trans.Am.Math.Soc. **313**, 249 (1989); H.Wenzl: "Quantum Groups and Subfactors of Lie Type B, C, and D", preprint San Diego 1989.
15. K.-H.Rehren: "Braid Group Statistics and their Superselection Rules", in *Algebraic Theory of Superselection Sectors and Field Theory*, ed. D.Kastler; to be published by World Scientific; and "Markov Traces as Characters for Local Algebras", in *Proceedings of the* 4th *Annecy Meeting on Theoretical Physics 1990*, eds. P.Binetruy et.al.; to be published as Nucl.Phys.B (Proc.Suppl.).
16. G.Mack, V.Schomerus: "Conformal Field Algebras with Quantum Symmetry from the Theory of Superselection Sectors", Hamburg preprint 1989; and "Endomorphisms and Quantum Symmetry of the Conformal Ising Model", in *Algebraic Theory of Superselection Sectors and Field Theory*, ed. D.Kastler; to be published by World Scientific.
17. V.G.Drinfel'd: in *Proceedings of the Intern. Congress of Mathematicians* (Berkeley 1986), p.799; ed. A.Gleason (Berkeley 1987).
18. D.Buchholz, G.Mack, I.Todorov: Nucl.Phys.B (Proc.Suppl.) **5B**, 20 (1988).
19. A.A.Belavin, A.M.Polyakov, A.B.Zamolodchikov: Nucl.Phys. **B241**, 333 (1984).
20. K.-H.Rehren, B.Schroer: Nucl.Phys. **B312**, 715 (1989); J.Fröhlich: in *Proceedings Cargèse 1987*, p.71; eds. G.'tHooft et al. (New York 1988); *Proceedings Como 1987*, p.173; eds. K.Bleuler et al. (Dordrecht 1988); Nucl.Phys.B (Proc.Suppl.) **5B**, 110 (1988).

Infinite Dimensional Algebras and (2+1)-Dimensional Field Theories: Yet Another View of gl(∞); Some New Algebras

Jens Hoppe

Institut für Theoretische Physik der Universität
Postfach 6380, D(W)-7500 Karlsruhe,
Federal Republic of Germany

Abstract

Over the past years, associative algebras have come to play a major rôle in several areas of theoretical physics. Firstly, it has been realized that Yang-Baxter algebras [1] constitute the relevant structure underlying (1+1)-dimensional integrable models; in addition, their relation to braid groups, the theory of knots and links, and the exchange algebras of (1+1)-dimensional conformal field theories [2] is by now well understood. Secondly, deformations of Poisson structures which appeared in (2+1)-dimensional field theories as infinite dimensional symmetry algebras possess underlying associative structures, which have also been studied in some detail (concerning higher spin theories see, e.g., [3,4] and references therein, concerning the enveloping algebra of $sl(2,c)$ see, e.g., [5], concerning deformations of $diff_A T^2$ – the Lie algebra of infinitesimal area preserving diffeomorphisms of the torus – see [6 – 9]). Ideas on how both investigations could eventually converge (i.e. a relation between (2+1)- and (1+1)-dimensions) have e.g. been expressed in [10].

As indicated by the two subtitles, there are two parts to my talk: The first one presents a view on something I met long ago [11], and recently got interested in again [5,7,9,12], while the second part introduces some algebraic structures that seem to be interesting and possibly new.

1. Yet Another View of gl(∞)

This part of my talk is based on work done in collaboration with Martin Bordemann, Peter Schaller and Martin Schlichenmaier [12,9,5]. The material is mostly elementary, but worthwhile noting, as quite a lot of confusion exists in the literature concerning $su(N \to \infty)$.

Let us start with the canonical basis E_{ij} $(i,j = 1,\ldots,N)$ of $gl(N,\mathbb{C})$ satisfying

$$[E_{ij},E_{kl}] = \delta_{jk}E_{il} - \delta_{li}E_{kj} \qquad (i,j,k,l = 1,\ldots N) \tag{1.1}$$

In the fundamental representation E_{ij} is an $N \times N$ matrix whose entries are all zero except the one at position (i,j), which is $+1$. In the limit $N \to \infty$ each (arbitrary but fixed) E_{ij} just gets more and more rows and columns that are identically zero. One thus arrives at the standard $gl_{(+)}(\infty)$:

$$[E_{ij},E_{kl}] = \delta_{jk}E_{il} - \delta_{li}E_{kj} \qquad (i,j,k,l = 1,\ldots \infty) \tag{1.2}$$

consisting of all infinite dimensional complex matrices that have finitely many non-zero elements (the subscript $(+)$ denotes the fact that $i,j,k,l \in \mathbb{N} = \mathbb{Z}^+$ rather than $\mathbb{Z}$. As mere Lie algebras $gl_+(\infty)$ and $gl(\infty)$ are isomorphic to each other, the isomorphism being induced by any bijective map $\varphi : \mathbb{N} \to \mathbb{Z}$; but as this isomorphism is not canonical, and would also not respect the standard grading, they are usually distinguished. In this paper, however, I will use $gl(\infty)$ and $gl_+(\infty)$ as synonyms).

Now consider an *arbitrary* basis of $gl(N,\mathbb{C})$, i.e., generators T_a, where a takes values between 1 and N^2, that are related to the E_{ij} by an invertible basis transformation $\Gamma = \Gamma^{(N)} = (\gamma_{a,ij})$:

$$T_a = \gamma_{a,ij}E_{ij} \qquad \begin{array}{l} a = 1,\ldots,N^2 \\ i,j = 1,\ldots,N \end{array} \tag{1.3}$$

The structure constants in the T_a-basis,

$$[T_a,T_b] = f^c_{ab}T_c \qquad a,b,c = 1,\ldots,N^2 \tag{1.4}$$

will then be given by

$$f^c_{ab} = \gamma_{a,ij}\gamma_{b,jk}(\Gamma^{-1})_{ik,c} - (a \leftrightarrow b). \tag{1.5}$$

Note that the N-dependence of $\gamma_{\ldots}$ and $f^{\cdot}_{\cdot\cdot}$ are suppressed. As $N \to \infty$ it may now very well happen that (1.5) converges (for each $a,b,c \in \mathbb{N}^3$) to some finite values $f^{\infty c}_{ab}$, while the numbers $\gamma_{a,ij}$ do *not* converge (or do not define a basis transformation between infinite dimensional vector spaces; unless stated otherwise all appearing Lie algebras are defined to consist of only finite(!) linear combinations of the given basis elements). Under some mild conditions on the $f^{\infty c}_{ab}$ (to ensure the validity of the Jacobi identity) the limit values of (1.5) may be used to define an infinite dimensional Lie algebra L:

$$[T_a,T_b] = f^c_{ab}T_c \qquad a,b,c = 1,\ldots,\infty \tag{1.6}$$

which may justly be called a $gl(N \to \infty)$ limit, but which need not at all be isomorphic to the $gl(\infty)$ defined above.

Let me give an example:

$$\begin{aligned} \gamma_{\boldsymbol{m},ij} &= \frac{iN}{4\pi M}\,\omega^{\frac{1}{2}m_1m_2+(i-1)m_1}\,\delta_{i+m_2,j(modN)} \\ \boldsymbol{m} = (m_1,m_2);\quad & m_1,m_2 = -\frac{N-1}{2},\ldots,+\frac{N-1}{2} \\ \omega = e^{4\pi i M/N};\quad & M,N \in \mathbb{N} \qquad N \text{ odd.} \end{aligned} \tag{1.7}$$

Then Eq. (1.5) leads to structure constants

$$f^{\boldsymbol{k}}_{\boldsymbol{m},\boldsymbol{n}} = \frac{N}{2\pi M}\sin\left(2\pi\frac{M}{N}(\boldsymbol{m}\times\boldsymbol{n})\right)\delta_{\boldsymbol{m}+\boldsymbol{n},\boldsymbol{k}} \tag{1.8}$$

As $N \to \infty$, (1.8) clearly converges, while (1.7) does not. Taking $N \to \infty$ in Eq. (1.8) one gets the infinite dimensional Lie algebra L^0,

$$\begin{aligned} [T_{\boldsymbol{m}}, T_{\boldsymbol{n}}] &= (\boldsymbol{m}\times\boldsymbol{n})T_{\boldsymbol{m}+\boldsymbol{n}} \\ \boldsymbol{m}\times\boldsymbol{n} = m_1n_2 - m_2n_1;&\quad \boldsymbol{m},\boldsymbol{n}\in\mathbb{Z}^2 \end{aligned} \tag{1.9}$$

as a $gl(N \to \infty)$ limit. As is well known, Eq. (1.9) defines a certain subalgebra of the complexification of $diff_a T^2 \oplus \mathbb{C}\cdot 1$ (the Lie algebra of infinitesimal area preserving diffeomorphisms of the torus, $\oplus U(1)$). Later we shall prove that L^0 is not isomophic to $gl(\infty)$. But before doing that, let us see what other possibilities Eq. (1.8) encodes: if one chooses M (as a function of N) such that

$$\lim_{N\to\infty}\frac{M}{N} = \Lambda \qquad \Lambda \in [0,1],\quad \Lambda \text{ irrational} \tag{1.10}$$

one gets infinitely many Lie algebras L^Λ

$$[T_{\boldsymbol{m}}, T_{\boldsymbol{n}}] = \frac{1}{2\pi\Lambda}\sin(2\pi\Lambda(\boldsymbol{m}\times\boldsymbol{n}))\,T_{\boldsymbol{m}+\boldsymbol{n}} \tag{1.11}$$

as limits $gl(N \to \infty)$. At first one could think that the L^Λ might all be the same; for $\Lambda \in (0,1/4)$, however, one can prove [9] that all the L^Λ are pairwise non-isomorphic (the reduction from (0,1) to (0,1/4) is caused by the fact that the sine function has 4 geometrically "identical" pieces within one period). The idea of any such non-isomorphy proof (between L and L', say) is to assume the existence of a Lie isomorphism, i.e., an invertible linear map Φ that respects the Lie structure, and to deduce a contradiction by using the commutation relations in L and L'.

In the case of $gl(\infty)$ and L^0 this looks as follows [9]:

$$\Phi_{ij} = \Phi(E_{ij}) = \Sigma C^{ij}_{\boldsymbol{m}} T_{\boldsymbol{m}} \tag{1.12}$$

with E_{ij} as in Eq. (1.2), and $T_{\boldsymbol{m}}$ as in Eq. (1.9). As E_{ij} commutes with E_{kl} unless $j = k$ or $l = i$, each fixed E_{ij} commutes with almost all E_{kk}, hence every Φ_{ij} with almost all Φ_{kk}. As the Φ_{kk} ($k \in \mathbb{Z}$) are mutually commuting (and linear dependence is transitive), one deduces that the $\hat{m}_{ij}$ (defined as the respective indices of the highest terms with respect to the lexicographic ordering of $\mathbb{Z}^2$

appearing in Eq. (1.12)) are all linear dependent, i.e. proportional to some fixed $\boldsymbol{r} \in \mathbb{Z}^2\backslash\{0\}$. This is the first step. The second step consists in taking multiple commutators of

$$T_{\boldsymbol{m}} = \Sigma d_{ij}^{\boldsymbol{m}} \Phi_{ij} \tag{1.13}$$

with suitable Φ_{ij}'s, in order to reduce the number of terms in Eq. (1.13) to one. Choosing $i \neq j \neq k \neq l$ (for given $\boldsymbol{m}$) such that $d_{li} \neq 0$, while $d_{jn} = 0$ for arbitrary n, one has indeed

$$[\,[T_{\boldsymbol{m}}, \Phi_{ij}], \Phi_{kl}\,] = -d_{li}^{\boldsymbol{m}} \Phi_{kj} \neq 0. \tag{1.14}$$

For $\boldsymbol{m} \not\propto \boldsymbol{r}$ this equation provides a contradiction, as $\hat{m}_{\text{r.h.s.}} = \hat{m}_{kj} \propto \boldsymbol{r}$ (according to the first step), but $\hat{m}_{\text{l.h.s.}} = \boldsymbol{m} + \hat{m}_{ij} + \hat{m}_{kl}$ cannot be proportional to $\boldsymbol{r}$ (as $\hat{m}_{ij} \propto \boldsymbol{r} \propto \hat{m}_{kl}$, but $\boldsymbol{m} \not\propto \boldsymbol{r}$).

2. Some New Algebras

As an intermediate step, let me generalize Eq. (1.9) and define a class of Lie algebras $L_{\boldsymbol{e}}$ (defined on a complex vector space with basis elements $T_{\boldsymbol{m}}$, with $\boldsymbol{m} \in \mathbb{Z}^2$) by the commutation relations

$$[T_{\boldsymbol{m}}, T_{\boldsymbol{n}}] = ((\boldsymbol{m} \times \boldsymbol{n}) + (\boldsymbol{m} - \boldsymbol{n}) \times \boldsymbol{e})\, T_{\boldsymbol{m}+\boldsymbol{n}}. \tag{2.1}$$

If $\boldsymbol{e} \in \mathbb{R}^2$, $\boldsymbol{e}$ may be restricted to lie in the fundamental region of $SL(2, \mathbb{Z})$ [13], as $L_{\boldsymbol{e}}$ and $L_{M\boldsymbol{e}}$ are clearly isomorphic if $M \in SL(2, \mathbb{Z})$.

Further note the subalgebras $L_{\boldsymbol{e}}^{+}(m_2 > -1)$ and the somewhat special (sub)algebra $L_{(0,1)}^{*}(m_2 > -2)$.

$L_{(0,1)}^{+}$ (or rather: a central extension of it) has actually been considered already in the literature [10]: it was proven to be a $N \to \infty$ limit of the well known chiral operator algebra W_N [14].

Rather than speculating about this interesting connection between (1+1)- and (2+1)-dimensions, let me introduce a class of associative algebras, denoted by $A(\Gamma, \rho; E)$, and defined by the following two relations:

$$T_m \cdot T_n = \rho^{(m,n)} T_{m+n} \tag{2.2}$$

$$T_m \cdot f = (\Gamma^m f) T_m \tag{2.3}$$

where f denotes an arbitrary element of some associative and commutative algebra E (let us think, e.g., of a space of functions $f(x)$, $x \in \mathbb{R}^n$), Γ denotes an algebra automorphism of E, and the T_m ($m \in \mathbb{Z}$) denote additional (and in general non-commuting) generators, whose multiplication law Eq. (2.2) contains "density functions" $\rho^{(m,n)} \in E$.

Due to Eqs. (2.2) and (2.3), the $\rho^{(m,n)}$ have to satisfy

$$\rho^{(m,n)} \cdot \rho^{(m+n,k)} = \rho^{(m,n+k)} \cdot (\Gamma^m \rho^{(n,k)}) \tag{2.4}$$

in order to make $A(\Gamma, \rho; E)$, i.e. the algebra of objects $T_m(f) = f \cdot T_m$ (and linear combinations thereof) associative. Somewhat more compactly, the multiplication in A can be written as

$$T_m(f) \cdot T_n(g) = T_{m+n}(h), \qquad h = \rho^{(m,n)} f \Gamma^m g. \tag{2.5}$$

Let me give a non-trivial solution to the consistency conditions (2.4):

$$\rho^{(m,n)} = \begin{cases} \prod_{r=m+n+1}^{m} (\Gamma^r \rho) & \text{for } {}^{m\geq 0\ n\leq 0}_{m+n\geq 0} \\ \prod_{r=1}^{m} (\Gamma^r \rho) & \text{for } {}^{m\geq 0\ n\leq 0}_{m+n\leq 0} \\ \prod_{r=m+1}^{0} (\Gamma^r \rho) & \text{for } {}^{m\leq 0\ n\geq 0}_{m+n\geq 0} \\ \prod_{r=m+1}^{m+n} (\Gamma^r \rho) & \text{for } {}^{m\leq 0\ n\geq 0}_{m+n\leq 0} \\ 1 & \text{otherwise} \end{cases} \tag{2.6}$$

where Γ is arbitrary, and ρ an arbitrary element of E.

A special case of Eq. (2.5), with E = space of functions of one real variable x, and *

$$\rho = \rho(x) = x^2 + x + \lambda, \qquad \lambda \in \mathbb{C}, \qquad \Gamma = e^{-\partial/\partial x} \tag{2.7}$$

is realized by representations U_λ of the universal enveloping algebra of $sl(2,\mathbb{C})$ (for a discussion of $[U_\lambda, U_\lambda]$ see [5]).

Let T_+, T_- and T_3 denote the generators of $sl(2,\mathbb{C})$, with commutation relations

$$[T_+, T_-] = -2T_3, \qquad [T_3, T_\pm] = \pm T_\pm. \tag{2.8}$$

Irreducible representations of the Lie algebra $sl(2,\mathbb{C})$ are then characterized by $\lambda \in \mathbb{C}$, where

$$T_+T_- - T_3^2 + T_3 = \lambda \cdot \mathbb{1}. \tag{2.9}$$

As (2.8) implies [15]

$$T_\pm^m f(T_3) = f(T_3 \mp m\mathbb{1}) T_\pm^m, \tag{2.10}$$

one can easily check the validity of Eq. (2.5) for

$$T_m(f) := f(T_3) \cdot T_\pm^{|m|} \quad (f \cdot m <, >, = 0), \tag{2.11}$$

where $\rho^{(m,n)}$ is given by Eq. (2.6), and ρ and Γ are as in Eq. (2.7).

* I would like to thank P. Schaller for reminding me of the usefulness of $e^{\partial/\partial x}$.

Returning to the general case, let us look at the Lie algebra structure induced by Eq. (2.5). After a slight redefinition, $X_m(f) := T_m(\Gamma^m f)$ for $m < 0$ (and $T_m(f)$ for $m \geq 0$) one finds (Lie algebras induced by such relations were first introduced in [16], and given the name "Continuum Analogues of Contragredient Lie Algebras", see also [10]; I would like to thank I. Bakas and A.M. Vershik for explaining their work to me)

$$\begin{aligned}
[X_0(f), X_{\pm 1}(g)] &= \pm X_{\pm}(gKf) \\
[X_0(f), X_0(g)] &= 0 \\
[X_{+1}(f), X_{-1}(g)] &= X_0(S(f.g)),
\end{aligned} \tag{2.12}$$

with $K = \rho^{(0,1)} - \rho^{(1,0)}\Gamma$, and $S = \rho^{(1,-1)} - \rho^{(-1,1)}\Gamma^{-1}$. At least for non-singular $\rho^{(m,n)}$, Eq. (2.4) implies $\rho^{(m,0)} = 1 = \rho^{(0,m)}$ and $\rho^{(m,-m)} = \Gamma^m \rho^{(-m,m)}$, which reduces K and S to

$$K = 1 - \Gamma, \qquad S = (1 - \Gamma^{-1})(\rho^{(1,-1)}\cdot\ ). \tag{2.13}$$

The factor $\rho^{(1,-1)}$ inside the action of S is important. In the case of L_λ (the Lie algebra induced by U_λ) Eqs. (2.12) and (2.13) allow a nice demonstation [17] of the finite dimensional representations of Eq. (2.8) (respectively L_λ), occuring when $\lambda = 1/4 - n^2 (n \in \mathbb{N})$. In that case $\rho^{(1,-1)} = x^2 - x + \lambda$ has two integer-spaced(!) zeros $x_\pm$ on the real axis, which allows one to consistently divide out an ideal I of finite codimension.

Density factors $\rho(\cdot)$ may also be used to generalize the Poisson algebra $diff_A T^2$ in yet another way: let $L_{\rho(t)}$ be the Lie algebra defined by (m,n $\in \mathbb{Z}$):

$$\begin{aligned}
[T_m(f), T_n(g)] &= T_{m+n}(h) \\
h &= \frac{1}{\rho_{m+n}} \left\{ m\rho_m \dot{f}(\rho_n \cdot g) - n\rho_n \dot{g}(\rho_m \cdot f) \right\}
\end{aligned} \tag{2.14}$$

where f, $g \in E$ and ρ_m are functions of some real variable t, say $\cdot = \partial/\partial t$. A realization of Eq. (2.14) is given by

$$T_m(f) = -ie^{im\varphi} \rho_m(t) f(t) \tag{2.15}$$

and [,]= usual Poisson bracket with respect to φ and t. Of course, ρ_m and E have to be chosen such that Eq. (2.14) is well defined, i.e., $h \in E$. An interesting example is

$$\rho_m(t) = (1 - t^2)^{|m|/2} \qquad t \in [-1, +1], \tag{2.16}$$

for which Eq. (2.14) becomes a Poisson algebra of functions on S^2. If the algebra E is the space of polynomials in t (which is equal to $-cos(\varphi)$, such that the expression in Eq. (2.14) is made a "dense" subalgebra of the complexification of $diff_A S^2$, and in the basis $T_{mk} = T_m(t^k) = -ie^{im\varphi} \cdot (1 - t^2)^{|m|/2} \cdot t^k$, $k \in \mathbb{N}$, the structure constants resulting from Eq. (2.14) can easily be calculated (note the similarity with Eq. (2.1)), we get:

$$\begin{aligned}[T_{mk}, T_{nl}] &= (ml - nk)T_{m+n,k+l-1} + \Sigma_{j=1}^{r} f_{kl}^{j}(m,n) T_{m+n,k+l-1+2j} \\ r &= \tfrac{1}{2}(|m| + |n| - |m+n|) \\ f_{kl}^{j} &= (ml - nk)\binom{r}{j}(-1)^{j} - (n|m| - |n|m)(-)j\binom{r-1}{j-1}.\end{aligned} \tag{2.17}$$

Presumably this is the simplest expression for the structure constants of $diff_A S^2$.

Acknowledgement

I would like to thank Martin Bordemann, Peter Schaller and Martin Schlichenmaier for many fruitful discussions.

Note Added: Shortly after my talk at the "Frühjahrsschule" I received a preprint [18], in which the class of Lie algebras L_e (Eq. (2.1)) has also been considered.

References

1. For an up to date introduction, see e.g. M. Jimbo: Int. J. Mod. Phys. A4 **15**, 3759 (1989)
2. Compare the contribution by K.H. Rehren (and references given there)
3. E.S. Fradkin, M.A. Vasiliev: Ann. Phys. **177**, 63 (1987)
4. E. Bergshoeff, M.P. Blencowe, K.S. Stelle: "Area Preserving Diffeomorphisms and Higher Spin Algebras" Comm. Math. Phys. (in press)
5. M. Bordemann, J. Hoppe, P. Schaller: Phys. Lett. B **232**, 199 (1989)
6. D.B. Fairlie, P. Fletcher, C.B. Zachos: Phys. Lett. B **218**, 203 (1989)
7. J. Hoppe: Int. J. Mod. Phys. A **4**, 5235 (1989)
8. E.G. Floratos: Phys. Lett. B **228**, 335 (1989); **232**, 467 (1989); **233**, 395 (1989)
9. J. Hoppe, P. Schaller: "Infinitely many versions of SU(∞)" Phys. Lett. B **237**, 407 (1990)
10. I. Bakas: "Area Preserving Diffeomorphisms and Higher Spin Fields in Two Dimensions ", talk given at the *Trieste Conference on Supermembranes and Physics in (2+1)-Dimensions* (1989) to appear in the proceedings, eds. M. Duff, C. Pope, E. Sezgin (World Scientific, Singapore); "The Structure of the W_∞ Algebra", Univ. of Maryland preprint 90-085 (November 1989)
11. J. Hoppe: MIT Ph.D. Thesis (1982)
12. M. Bordemann, J. Hoppe, P. Schaller, M. Schlichenmaier: "gl(∞) and Geometric Quantization", Karlsruhe preprint KA-THEP-05-1990 (February 1990)
13. See, e.g., M. Schlichenmaier: "An Introduction to Riemann Surfaces, Algebraic Curves and Moduli Spaces" (p. 71), SLN in Physics **322**, (1989)
14. A.B. Zamolodchikov: Teor. Mat. Fiz. **65**, 1205 (1985)
15. P. Dürr: private communication
16. M.V. Saveliev, A.M. Vershik: Comm. Math. Phys. **126**, (1989); "New Examples of Continuum Graded Lie Algebras", Phys. Lett. B
17. P. Schaller, private communication
18. R.M. Kashaev, M.V. Saveliev, S.A. Savelieva, A.M. Vershik: "On Nonlinear Equations Associated with Lie Algebras of Diffeomorphism Groups of Two-Dimensional Manifolds", Institute for High Energy Physics preprint 90-1 Protvino (1990)

Anomalies in Quantum Field Theory

Allen C. Hirshfeld

Institut für Physik der Universität
P.O. Box 500 500, D(W)-4600 Dortmund 50
Federal Republic of Germany

Abstract

A review is presented of the anomaly problem in quantum field theory. The problems encountered in evaluating the $\pi^0 \to 2\gamma$ decay are described, as well as the point-splitting method for dealing with these problems and deriving the chiral anomaly. We next show how the Wess-Zumino consistency conditions allow the understanding of these results in the context of the cohomology of Lie algebras. Then some concepts from the theory of characteristic classes are discussed, including the Chern character, in order to derive the descent equations which are the basis for the algebraic calculation of anomalies. Finally, the relation of the anomaly problem to the question of Schwinger terms in current algebra is clarified.

1. Introduction

The relationship between physics and mathematics is intimate but not always straightforward. Some areas of physics have yielded to precise mathematical analysis. In particular, it has by now been widely recognized that classical gauge field theory can be usefully described in the language of modern differential geometry [1]. This description has produced various results which would hardly have been achieved otherwise, for example the complete classification of self-dual solutions of the Yang-Mills equations [2]. In contrast, quantum field theory in four spacetime dimensions has up to now resisted all attempts at a rigorous mathematical formulation.

It is thus of particular interest that a significant connection has been discovered between certain problems in quantum field theory and some deep results of modern algebraic topology. This is all the more remarkable in that we are here talking about problems of eminent physical importance. Besides a host of more speculative applications, the chiral anomaly was and remains the key to our understanding of the $\pi^0 \to 2\gamma$ decay process. It not only explains the existence of this decay mode, despite contrary expectations based on traditional arguments of current algebra, it also yields a correct numerical value for the decay rate, upon inclusion of a factor of three due to the $SU(3)$-color degrees of freedom. Indeed, it was in this connection that Gell-Mann first introduced the SU(3) gauge group [3], which is the basis for the standard theory of the strong interactions, quantum chromodynamics (QCD).

The field theoretic treatment of anomalies is rather intricate, being intimately related to questions of regularization and renormalization of the terms in the perturbation expansion. The mathematical approach avoids these dynamical details and focuses attention on the symmetry aspects of the problem. This is a strategy which has frequently been used to advantage in elementary particle physics. In contrast to the most common physical situations we have in this case a non-trivial realization of the symmetry, connected to a non-vanishing cohomology class, i.e. to a non-trivial topology. Phenomena related to non-trivial topological effects are relatively new in physics, and are often the occasion of some initial confusion. The Bohm-Aharonov effect in quantum mechanics [4] and Dirac's magnetic monopole [5] are cases in point. However, from the mathematician's point of view the treatment of such situations is completely straightforward. At this point it behooves the physicist to learn enough of the underlying mathematics to understand the resolution of the physical problem.

Besides the gain in understanding, the differential geometric approach has proved to be an efficient tool for calculating anomalies in rather general settings, where the perturbative calculations would be at best extremely unwieldy [6].

In this article we first review in some detail the problem of the chiral anomaly in quantum field theory and its relevance for the $\pi^0 \to 2\gamma$ decay. We emphasize in particular in what way the result of these calculations is unexpected from the point of view of a naive application of symmetry arguments in the framework of the *canonical formalism.* We illustrate one of the field-theoretic methods for overcoming these difficulties, involving Schwinger's gauge invariant *point-splitting procedure.* We then indicate how, by the use of the *Wess-Zumino consistency conditions*, the problem may be related to a question in the *cohomology of Lie algebras.* This requires using the differential geometric formulation of the gauge theory in terms of connections on a *principal fibre bundle*, and an understanding of the rôle the *BRS transformations* play in incorporating the restrictions due to gauge invariance and the accompanying unphysical degrees of freedom. Having at this stage reformulated the problem in a mathematical language, we go on to show how such problems are treated in the relevant mathematical framework. The concept of the *Chern character*, whose origin is in the theory of characteristic classes, is introduced. The basic *transgression formula* of Chern and Weil concerning the independence of the Chern class on the connection used in its construction is derived. An expansion of this formula according to degree yields the *descent equations*, which are the basic tool in this approach for the calculation of cohomology classes. The term involving the *chiral anomaly* is calculated explicitly. Finally, we discuss a question of more recent origin, namely the verification of Faddeev's conjecture relating another of the terms in the descent equations with the *Schwinger terms* of current algebra.

2. Symmetries in Field Theory

Anomalies arise when the symmetries which a theory possesses at the classical level do not survive quantization. In order to provide a framework for our later discussions we review in this section the implementation of symmetries and their consequences. We begin with classical field theories.

2.1 The Canonical Formalism

A classical field theory is characterized by a Langrangian density $\mathcal{L}(\phi)$. Here ϕ stands for the set of fields involved in the theory $\{\phi_i\}$, $i = 1,\ldots,N$. For example, the ϕ_i may be the components of a higher-spin field. A variation of the fields $\delta\phi(x)$ is a symmetry transformation if

$$\delta\mathcal{L} = \mathcal{L}(\phi + \delta\phi) - \mathcal{L}(\phi) = 0 \tag{2.1}$$

(a total divergence is of course also allowed, but this is usually relevant when one is dealing with spacetime symmetries, which are not our immediate concern here). If $\mathcal{L}$ depends only on ϕ and $\partial^\mu\phi$, but not on higher derivatives, then

$$\begin{aligned}\delta\mathcal{L} &= \frac{\delta\mathcal{L}}{\delta\phi}\delta\phi + \frac{\delta\mathcal{L}}{\delta(\partial^\mu\phi)}\delta(\partial^\mu\phi) = \partial^\mu\left[\frac{\delta\mathcal{L}}{\delta(\partial^\mu\phi)}\right]\delta\phi + \frac{\delta\mathcal{L}}{\delta(\partial^\mu\phi)}\partial^\mu\delta\phi \\ &= \partial^\mu\left[\frac{\delta\mathcal{L}}{\delta(\partial^\mu\phi)}\delta\phi\right] = \partial^\mu(\pi_\mu\delta\phi) = 0,\end{aligned} \tag{2.2}$$

where π_μ is the canonical momenta

$$\pi_\mu = \frac{\delta\mathcal{L}}{\delta(\partial^\mu\phi)}, \tag{2.3}$$

and we have used the Euler-Lagrange equations:

$$\partial^\mu\left[\frac{\delta\mathcal{L}}{\delta(\partial^\mu\phi)}\right] - \frac{\delta\mathcal{L}}{\delta\phi} = 0. \tag{2.4}$$

This means that

$$J_\mu = -\pi_\mu\delta\phi \tag{2.5}$$

is the *conserved current* associated with the symmetry. This is essentially the content of *Noether's theorem.*

For homogeneous linear transformations the field variations take the form

$$\delta\phi = i\varepsilon^a T_a \phi \tag{2.6}$$

where the T_a are a set of $N \times N$ Hermitian matrices acting on the ϕ's, which satisfy the commutation relations

$$[T_a, T_b] = i f^c_{ab} T_c, \tag{2.7}$$

where the f^c_{ab} are the structure constants of the *symmetry group*. In the case of *global* symmetries the infinitesimal parameters ε^a are independent of x, and

$$\delta(\partial^\mu \phi) = i\varepsilon^a T_a (\partial^\mu \phi), \tag{2.8}$$

so that $\partial^\mu \phi$ transforms under the symmetry like ϕ itself. We say that $\partial^\mu \phi$ transforms *covariantly*. In this case we may factor out the infinitesimal parameter and write the current (2.5) in a particularly simple way:

$$J^a_\mu = -i\pi_\mu T^a \phi. \tag{2.9}$$

If the current is conserved, the charges

$$Q^a = \int d^3x\, J^a_0(t, \bar{x}) \tag{2.10}$$

are time independent.

We now go over to quantum field theory. The ϕ's become quantum field operators satisfying the equal-time commutation relations

$$\begin{aligned} [\phi(x), \phi(y)]^{ET}_\pm &= [\pi_0(x), \pi_0(y)]^{ET}_\pm = 0, \\ [\phi(x), \pi_0(y)]^{ET}_\pm &= i\delta^{(3)}(\bar{x} - \bar{y}), \end{aligned} \tag{2.11}$$

where ($\pm$) refers to the cases of boson and fermion fields, respectively. The current commutators are

$$[J^a_0(x), J^b_0(y)] = (-i)^2 [\pi_0(x) T^a \phi(x), \pi_0(y) T^b \phi(y)]. \tag{2.12}$$

Expand this, using the rule

$$[AB, C] = A[B, C]_\pm \mp [A, C]_\pm B. \tag{2.13}$$

The result is

$$\begin{aligned} [J^a_0(x), J^b_0(y)]^{ET} &= -\pi_0(x) T^a [\phi(x), \pi_0(y)]^{ET}_\pm T^b \phi(y) \\ &\quad \pm \pi_0(y) T^b T^a [\pi_0(x), \phi(y)]^{ET}_\pm \phi(x) \\ &= -i\delta^{(3)}(\bar{x} - \bar{y}) \pi_0(x) [T^a, T^b] \phi(y) \\ &= (-i) i f^c_{ab} \pi_0(x) T_c \phi(y) \delta^{(3)}(\bar{x} - \bar{y}) \\ &= i f^c_{ab} J^c_0(x) \delta^{(3)}(\bar{x} - \bar{y}). \end{aligned} \tag{2.14}$$

Note that this is expected to hold whether the current is conserved or not. From the above we get

$$[Q_a, Q_b] = \int d^3x \int d^3y \, [J_0^a(x), J_0^b(y)]^{ET}$$
$$= i f_{ab}^c \int d^3x \, J_0^c(x) = i f_{ab}^c Q_c. \tag{2.15}$$

The charges are generators of the symmetry, in the sense that

$$[Q^a, \phi(x)] = \int d^3y \, [J_0^a(y), \phi(x)]^{ET} = -i \int d^3y \, [\pi_0(y) T^a \phi(y), \phi(x)]^{ET}$$
$$= \pm i \int d^3y \, [\pi_0(y), \phi(x)]_{\pm}^{ET} T^a \phi(y) = -T^a \phi(x), \tag{2.16}$$

or

$$\delta\phi(x) = -i\varepsilon_a [Q^a, \phi(x)]. \tag{2.17}$$

Note that the Q^a are operators in Hilbert space, whereas the T^a operate on the indices of the fields ϕ.

We shall see in the following that in specific models the expectations expressed in this section, based on the canonical formulation of quantum field theory, are not always fulfilled. When the Noether current is not conserved we speak of an *anomaly*, when the equal-time current commutators (2.14) are violated we ascribe the failure to the presence of *Schwinger terms.*

2.2 Fermions and Chiral Symmetry

The Lagrangian for a massless free fermion field is

$$\mathcal{L} = i\bar{\psi}\gamma^\mu \partial_\mu \psi. \tag{2.18}$$

The canonical momenta are

$$\pi_\mu = i\bar{\psi}\gamma_\mu. \tag{2.19}$$

When the fermions carry a representation of the symmetry group with generators T^a the symmetry transformation is

$$\delta\psi = i\varepsilon_a T^a \psi$$
$$\delta\bar{\psi} = -i\bar{\psi} T^a \varepsilon_a, \tag{2.20}$$

so that

$$\delta\mathcal{L} = i(\delta\bar{\psi})\gamma^\mu \partial_\mu \psi + i\bar{\psi}\gamma^\mu \partial_\mu (\delta\psi) = 0. \tag{2.21}$$

The conserved current is

$$J_\mu^a = \bar{\psi}\gamma_\mu T^a \psi. \tag{2.22}$$

This Lagrangian actually possesses a larger symmetry: define the left-handed and right-handed fields by

$$\psi_L = \tfrac{1}{2}(1+\gamma_5)\psi, \quad \psi_R = \tfrac{1}{2}(1-\gamma_5)\psi, \tag{2.23}$$

and write the Lagrangian as

$$\mathcal{L} = i\bar{\psi}_L\gamma^\mu\partial_\mu\psi_L + i\bar{\psi}_R\gamma^\mu\partial_\mu\psi_R. \tag{2.24}$$

We still have a symmetry transformation if we transform the right-handed and the left-handed fields separately:

$$\delta\psi_L = i\varepsilon_L^a T_a\psi_L, \quad \delta\psi_R = i\varepsilon_R^a T_a\psi_R. \tag{2.25}$$

This symmetry transformation may be rewritten in terms of the infinitesimal parameters

$$\varepsilon^a = \tfrac{1}{2}(\varepsilon_R^a + \varepsilon_L^a) \quad \varepsilon_5^a = \tfrac{1}{2}(\varepsilon_R^a - \varepsilon_L^a) \tag{2.26}$$

in the form

$$\delta\psi = i(\varepsilon^a - \gamma_5\varepsilon_5^a)T_a\psi. \tag{2.27}$$

For $\varepsilon_5^a = 0$ this is just the normal symmetry transformation we had at the beginning, for $\varepsilon^a = 0$ it is called a *chiral transformation.*

A well-known model involving chiral symmetry is the σ *model* of Gell-Mann and Levy [7]. It involves an initially massless fermion isospin doublet ψ, the isospin-one pion π^a and an isoscalar σ. The Lagrangian is

$$\begin{aligned}\mathcal{L} = {} & i\bar{\psi}\gamma^\mu\partial_\mu\psi + \tfrac{1}{2}\partial^\mu\pi^a\partial_\mu\pi_a + \tfrac{1}{2}\partial^\mu\sigma\partial_\mu\sigma \\ & + ig\pi^a\bar{\psi}\gamma_5\tau_a\psi - g\sigma\bar{\psi}\psi - \tfrac{\lambda}{4}\Big[(\sigma^2+\pi_a^2) - f_\pi^2\Big]^2.\end{aligned} \tag{2.28}$$

Here τ_a $(a = 1,2,3)$ are the Pauli matrices. The fermion interacts with the pion and sigma fields via the Yukawa couplings, the last term generates a spontaneous symmetry breaking. λ is a dimensionless constant, f_π has the dimensions of a mass and is identified with the pion decay constant. The Lagrangian possesses a chiral symmetry when the fermion fields transform as above, and the pion and sigma fields transform as

$$\begin{aligned}\delta\sigma &= \varepsilon_5^a\pi_a, \\ \delta\pi_a &= -\varepsilon_{abc}\varepsilon^b\pi^c - \varepsilon_5^a\sigma.\end{aligned} \tag{2.29}$$

This can be understood as the transformation law of a vector in a four-dimensional Euclidean space, and reflects the fact that the symmetry of the Lagrangian can also be interpreted as $SO(4) \simeq SU(2)\otimes SU(2)$, where "$\simeq$" indicates a local Lie-group isomorphism. Because of the spontaneous symmetry breaking the sigma and pion fields can acquire non-vanishing vacuum expectation values, without loss of generality we may choose:

$$<\sigma> = f_\pi, \quad <\pi_a> = 0. \tag{2.30}$$

In terms of the shifted field $\sigma' = \sigma - f_\pi$, which *does* have a vanishing vacuum expectation value, the Lagrangian is

$$\begin{aligned}\mathcal{L} = & \, i\bar{\psi}\gamma^\mu\partial_\mu\psi + \tfrac{1}{2}\partial^\mu\pi_a\partial_\mu\pi_a + \tfrac{1}{2}\partial^\mu\sigma'\partial_\mu\sigma' \\ & + ig\pi_a\bar{\psi}\gamma_5\tau_a\psi - g\sigma'\bar{\psi}\psi - gf_\pi\bar{\psi}\psi \\ & - \tfrac{\lambda}{4}(\sigma'^2 + \pi_a^2 + 2f_\pi\sigma')^2. \end{aligned} \tag{2.31}$$

This describes nucleons with mass gf_π coupled to the scalar field σ' and massless pseudoscalar pions π_a. With f_π as measured in the decay $\pi \to \mu\nu_\mu$ the *Goldberger-Treiman relation*

$$m_N = g_{\pi NN} f_\pi \tag{2.32}$$

is empirically satisfied. Here $g = g_{\pi NN}$ is the pion-nucleon coupling constant and m_N the nucleon mass. The pions are here acting as Goldstone bosons; physical pions are of course not really massless, but they are indeed much lighter than the other hadrons.

Associated with the chiral symmetry of the Lagrangian (2.28) is the *axial vector current*

$$J^\mu_{5a} = (\partial^\mu\pi_a)\sigma - (\partial^\mu\sigma)\pi_a + \tfrac{1}{2}\bar{\psi}\gamma^\mu\gamma_5\tau_a\psi. \tag{2.33}$$

In terms if the shifted fields this has a piece proportional to $\partial^\mu\pi_a$:

$$J^\mu_{5a} = f_\pi\partial^\mu\pi_a + \ldots \tag{2.34}$$

The remaining terms are bilinear in the fields. This piece however gives the current a non-vanishing vacuum expectation value between the vacuum and the one-pion state:

$$<0|J^\mu_{5a}(0)|\pi_b> = if_\pi q^\mu\delta_{ab}, \tag{2.35}$$

where q^μ is the pion momentum. Such a non-vanishing vacuum expectation value is characteristic for spontaneous symmetry breaking. It exhibits the meaning of f_π as the pion decay constant controlling the decay $\pi \to \mu\nu$, since the final state is in the hadronic vacuum. Taking the divergence, and using the normalization condition for the pion field, $<0|\pi_a|\pi_b> = \delta_{ab}$, shows that the *PCAC relation* (partially conserved axial current)

$$\partial_\mu J^\mu_{5a}(x) = f_\pi m_\pi^2\pi_a(x) \tag{2.36}$$

is fulfilled in this model. We shall return to the PCAC relation in the following.

2.3 Gauge Symmetry

If in the symmetry transformation (2.20) the infinitesimal parameter ε_a has a spacetime dependence, $\varepsilon_a = \varepsilon_a(x)$, then the Lagrangian for free massless fermions,

$$\mathcal{L}_0 = i\bar{\psi}\gamma^\mu \partial_\mu \psi, \tag{2.37}$$

is no longer invariant with repect to the transformation

$$\delta\psi = i\varepsilon_a(x)T^a\psi. \tag{2.38}$$

In order to regain symmetry we must replace the ordinary derivative with the covariant derivative

$$D_\mu = \partial_\mu - iA_\mu^a T_a, \tag{2.39}$$

where A_μ^a is the gauge field, which transforms according to

$$\delta A_\mu^a = D_{\mu b}^a \varepsilon^b. \tag{2.40}$$

Here $D_{\mu b}^a$ is just the component expression for D_μ:

$$D_{\mu b}^a = \delta_b^a \partial_\mu + f_{cb}^a A_\mu^c. \tag{2.41}$$

This results from Eq. (2.39) when we remember that the gauge fields, and hence also the parameters ε^a, transform according to the adjoint representation of the symmetry group, and that in this represention the matrix elements of the generators are related to the structure constants of the group:

$$[T_a]_b^c = i f_{ab}^c. \tag{2.42}$$

We use in the following a compact notation in which

$$\begin{aligned} A_\mu &= -iA_\mu^a T_a \\ \varepsilon &= -i\varepsilon^a T_a, \end{aligned} \tag{2.43}$$

and we then have

$$\begin{aligned} D_\mu &= \partial_\mu + A_\mu, \\ \delta\psi &= -\varepsilon\psi, \quad \delta\bar{\psi} = \bar{\psi}\varepsilon, \\ \delta A_\mu &= D_\mu \varepsilon = \partial_\mu \varepsilon + [A_\mu, \varepsilon]. \end{aligned} \tag{2.44}$$

Using this compact notation it is very easy to verify that D_μ transforms covariantly:

$$\delta(D_\mu \psi) = -\varepsilon D_\mu \psi, \tag{2.45}$$

and hence that the Lagrangian

$$\mathcal{L} = i\bar{\psi}\gamma^\mu D_\mu \psi = \mathcal{L}_0 + \bar{\psi}\gamma^\mu T_a \psi A_\mu^a \tag{2.46}$$

is again invariant. The equations of motion are

$$D_\mu \psi = 0, \tag{2.47}$$

and the currents $J^\mu = -iJ_a^\mu T^a$ are now *covariantly conserved:*

$$D_\mu J^\mu = 0. \tag{2.48}$$

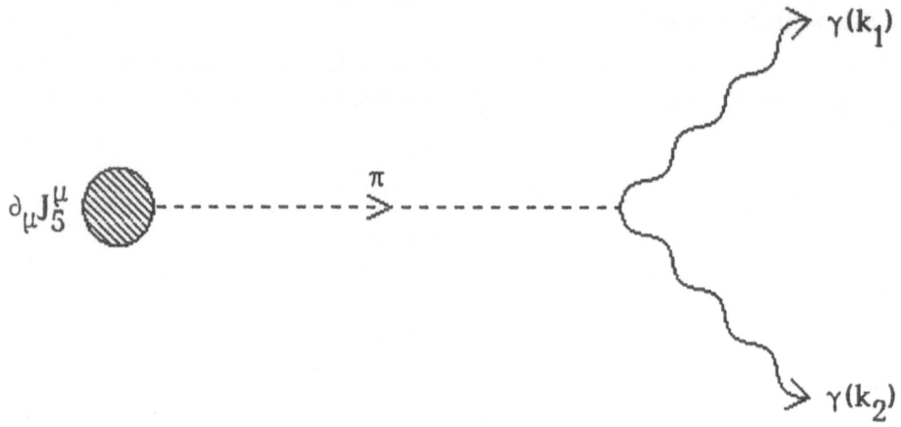

Fig. 1. The diagrammatic representation of the amplitude $<2\gamma|\partial_\mu J_5^\mu|0>$

3. $\pi^0 \to 2\gamma$ Decay

3.1 The PCAC Hypothesis

We have discussed in the last section the PCAC relation

$$\partial_\mu J_{5a}^\mu(x) = f_\pi q^2 \pi_a(x) \tag{3.1}$$

in the framework of the σ model. It is believed to have more general validity. The PCAC *hypothesis* is the assumption, based on extensive phenomenological experience, that the divergence of the axial current is dominated by the pion pole also off-mass-shell, at least in the region $0 \leq q^2 \leq m_\pi^2$ [8]. It is further expected that the "chiral limit" involving massless pions and fermions, i.e. $q^2 \to 0$, provides a good approximation to the real world. In the chiral limit the axial current is conserved. The physical fermion masses are supposed to result from spontaneous symmetry breaking involving the pion as a Goldstone boson, as in the σ model. However, these ideas lead to a difficulty involving the $\pi^0 \to 2\gamma$ decay, first noted by Veltman and Sutherland [9].

For the π^0 decay $a = 3$ is the relevant component of the axial current; in the remainder of this section we shall denote this component simply by J_5^μ. According to the PCAC hypothesis, the amplitude $< 2\gamma|\partial_\mu J_5^\mu|0 >$ would be controlled by the diagram shown in Fig. 1. This corresponds to the relation

$$<2\gamma|\partial_\mu J_5^\mu|0>=<2\gamma|\pi(q)> \frac{1}{q^2 - m_\pi^2} <\pi(q)|\partial_\mu J_5^\mu|0> . \tag{3.2}$$

We denote by $M(q^2)$ the amplitude $< 2\gamma|\pi(q) >$. The decay rate of the physical pion is then determined by

$$M(q^2 = m_\pi^2) = \lim_{q^2 \to m_\pi^2} \frac{q^2 - m_\pi^2}{f_\pi m_\pi^2} < 2\gamma|\partial_\mu J_5^\mu|0 > . \tag{3.3}$$

With (3.1) this becomes

$$M(q^2 = m_\pi^2) = \lim_{q^2 \to m_\pi^2} (q^2 - m_\pi^2) < 2\gamma|\pi|0 > . \tag{3.4}$$

This is just the LSZ (Lehmann-Symanzik-Zimmermann) reduction formula [10]. Accepting all this, the PCAC hypothesis would lead us to expect the amplitude to be well approximated by

$$M(q^2 = 0) = -\frac{1}{f_\pi} < 2\gamma|\partial_\mu J_5^\mu|0 > . \tag{3.5}$$

The conservation of the axial current in the chiral limit would then imply a vanishing decay rate for the process $\pi^0 \to 2\gamma$, the decay is, however, observed. It turns out that the problem does not involve the PCAC hypothesis, but rather the supposed vanishing of the divergence of the axial current in the chiral limit. As we shall see in the next subsection, this divergence does not vanish, contrary to classical expectations.

3.2 The Anomalous Divergence of the Axial Current

The fermionic part of the axial current is

$$J_{5a}^\mu(x) = \tfrac{1}{2}\bar{\psi}(x)\gamma_5\gamma^\mu\tau_a\psi(x). \tag{3.6}$$

Consider at first the contribution of a single fermion, say with $\tau_3 = +1$ and charge e. In quantum field theory the meaning of an expression like this is problematical, because the product of field operators at the same point is ill-defined. We might try to regulate the expression by "point-splitting", i.e., defining the current through the limit

$$J_5^\mu(x) = \lim_{\varepsilon \to 0} \tfrac{1}{2}\bar{\psi}(x + \tfrac{\varepsilon}{2})\gamma_5\gamma^\mu\psi(x - \tfrac{\varepsilon}{2}).$$

But this expression, in contrast to the first, is not gauge invariant. Such a regularization would therefore break the gauge invariance of the theory. In the *gauge invariant point-splitting* method suggested by Schwinger [11] the current is defined by the gauge invariant expression

$$J_5^\mu(x) = \lim_{\varepsilon \to 0} \tfrac{1}{2}\bar{\psi}(x + \tfrac{\varepsilon}{2})\gamma_5\gamma^\mu\psi(x - \tfrac{\varepsilon}{2})\exp\left(ie \int_{x-\varepsilon/2}^{x+\varepsilon/2} A_\mu(x)dx^\mu\right) . \tag{3.7}$$

This expression is easily seen to be invariant with respect to the gauge transformation $\psi(x) \to e^{ie\Lambda(x)}\psi(x)$; $\bar{\psi}(x) \to \bar{\psi}(x)e^{-ie\Lambda(x)}$; $A_\mu(x) \to A_\mu(x) + \partial_\mu\Lambda(x)$.

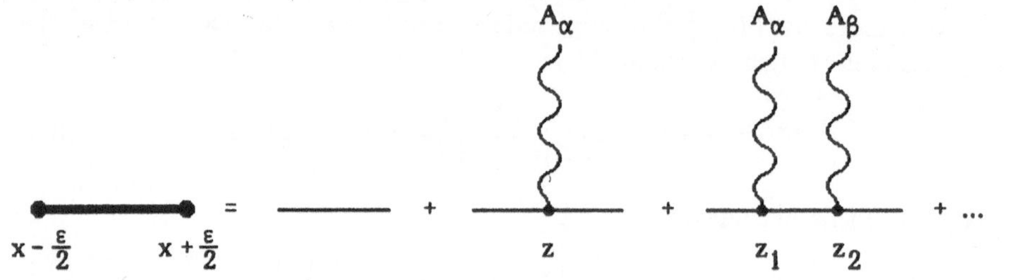

Fig. 2. Diagrammatic expansion of the Green's function

The equation of motion for $\psi(x)$ is

$$\partial_\mu \psi(x) = ieA_\mu(x)\psi(x), \tag{3.8}$$

so the divergence of the current may be written as

$$\begin{aligned}\partial_\mu J_5^\mu(x) = \lim_{\varepsilon\to 0}\frac{ie}{2}\Big[&\bar{\psi}\gamma_5\gamma^\mu\psi A_\mu(x-\tfrac{\varepsilon}{2})\\ &-\bar{\psi}\gamma_5\gamma^\mu\psi A_\mu(x+\tfrac{\varepsilon}{2})\\ &+\bar{\psi}\gamma_5\gamma^\mu\psi\partial_\mu\textstyle\int_{x-\varepsilon/2}^{x+\varepsilon/2}A_\nu dx^\nu\Big]\exp\left(ie\textstyle\int A_\nu dx^\nu\right).\end{aligned} \tag{3.9}$$

Retaining only the lowest order in ε yields

$$\partial_\mu J_5^\mu(x) = ieJ_5^\mu(x)\varepsilon^\nu(\partial_\mu A_\nu - \partial_\nu A_\mu) = ieJ_5^\mu(x)\varepsilon^\nu F_{\mu\nu}. \tag{3.10}$$

For the vacuum expectation value we find

$$<\partial_\mu J_5^\mu(x)>= ie <J_5^\mu(x)> \varepsilon^\nu F_{\mu\nu}. \tag{3.11}$$

This would vanish as $\varepsilon \to 0$ if $< J_5^\mu >$ were regular in this limit. However, we now demonstrate that it is singular.

We start the calculation by writing

$$\begin{aligned}<J_5^\mu(x)> &= \tfrac{1}{2} < \bar{\psi}_\alpha(x+\tfrac{\varepsilon}{2})[\gamma_5\gamma^\mu]_{\alpha\beta}\psi_\beta(x-\tfrac{\varepsilon}{2}) > \exp\left(ie\textstyle\int A_\nu dx^\nu\right)\\ &= \tfrac{1}{2}[\gamma_5\gamma^\mu]_{\alpha\beta} <T\psi_\beta(x-\tfrac{\varepsilon}{2})\bar{\psi}_\alpha(x+\tfrac{\varepsilon}{2})> \exp\left(ie\textstyle\int A_\nu dx^\nu\right)\\ &= \tfrac{1}{2}\,\mathrm{tr}\,\gamma_5\gamma^\mu G(x-\tfrac{\varepsilon}{2}, x+\tfrac{\varepsilon}{2})\exp\left(ie\textstyle\int A_\nu dx^\nu\right),\end{aligned} \tag{3.12}$$

where $G(x,y)$ is the fermionic Green's function. It may be handled by standard perturbative techniques; the first few terms in the diagrammatic expansion are shown in Fig. 2.

The corresponding formula is

$$G(x-\tfrac{\varepsilon}{2},x+\tfrac{\varepsilon}{2})$$
$$=G_0(-\varepsilon)+ie\int d^4z\,G_0(x-\tfrac{\varepsilon}{2}-z)\gamma^\alpha G_0(z-x-\tfrac{\varepsilon}{2})A_\alpha(z)$$
$$+(ie)^2\int d^4z_1\int d^4z_2 G_0(x-\tfrac{\varepsilon}{2}-z_1)\gamma^\alpha G_0(z_1-z_2)\gamma^\beta G_0(z_2-x-\tfrac{\varepsilon}{2})A_\alpha A_\beta$$
$$+\ldots \tag{3.13}$$

The free Green's function $G_0(x)$ behaves like x^{-3} as $x \to 0$, so the successive terms in the expansion go like $x^{-3}, x^4/x^6 = x^{-2}, x^8/x^9 = x^{-1}$ etc. The term of interest in this expansion is the term linear in A. The first term is divergent by power counting, but vanishes by symmetric integration. The linear term gives a finite contribution. The higher terms vanish in the limit we are interested in. The contribution of the linear term is

$$ie\int d^4z\,G_0(x-\tfrac{\varepsilon}{2}-z)\gamma^\alpha G_0(z-x-\tfrac{\varepsilon}{2})A_\alpha(z)$$
$$=ie\int\frac{d^4p}{(2\pi)^4}\frac{d^4q}{(2\pi)^4}e^{i\varepsilon p}e^{-iqx}G_0(p+\tfrac{1}{2}q)\gamma^\alpha G_0(p-\tfrac{1}{2}q)A_\alpha(q), \tag{3.14}$$

where $A(q)$ is the Fourier transform of $A(z)$. Hence

$$\varepsilon^\nu < J_5^\mu >$$
$$=\tfrac{ie}{2}\,\mathrm{tr}\,\gamma_5\gamma^\mu\int\frac{d^4p}{(2\pi)^4}\frac{d^4q}{(2\pi)^4}\varepsilon^\nu e^{i\varepsilon p}e^{-iqx}G_0(p+\tfrac{1}{2}q)\gamma^\alpha G_0(p-\tfrac{1}{2}q)A_\alpha(q)$$
$$=\tfrac{ie}{2}\,\mathrm{tr}\,\gamma_5\gamma^\mu\int\frac{d^4p}{(2\pi)^4}\frac{d^4q}{(2\pi)^4}\left[\frac{1}{i}\frac{\partial}{\partial p_\nu}e^{i\varepsilon p}\right]e^{-iqx}[\ldots]A_\alpha(q)$$
$$=-\tfrac{e}{2}\,\mathrm{tr}\,\gamma_5\gamma^\mu\int\frac{d^4p}{(2\pi)^4}\frac{d^4q}{(2\pi)^4}e^{i\varepsilon p}e^{-iqx}A_\alpha(q)\frac{\partial}{\partial p_\nu}[\ldots]. \tag{3.15}$$

By Gauss' theorem in four dimensions

$$\int d^4p\,\frac{\partial}{\partial p_\nu}f(p)=2i\pi^2P^\nu P^2 f(P), \tag{3.16}$$

so the p-integration yields

$$\int d^4p\,\frac{\partial}{\partial p^\nu}[\ldots]$$
$$=2i\pi^2P^\nu P^2\frac{\gamma^\rho(P+\frac{1}{2}q)_\rho}{P^2}\gamma^\alpha\frac{\gamma^\tau(P-\frac{1}{2}q)_\tau}{P^2}$$
$$=2i\pi^2P^\nu P^{-2}\gamma^\rho\gamma^\alpha\gamma^\tau(P_\rho P_\tau+\tfrac{1}{2}q_\rho P_\tau-\tfrac{1}{2}q_\tau P_\rho-\tfrac{1}{4}q_\rho q_\tau). \tag{3.17}$$

Now include the factor $\gamma_5\gamma^\mu$ and evaluate the trace:

$$\mathrm{tr}\,\gamma_5\gamma^\mu\gamma^\rho\gamma^\alpha\gamma^\tau=-4i\varepsilon^{\mu\rho\alpha\tau}. \tag{3.18}$$

The above expression then becomes

$$8\pi^2 P^\nu P^{-2} \varepsilon^{\mu\rho\alpha\tau} \tfrac{1}{2}(q_\rho P_\tau - q_\tau P_\rho) = 8\pi^2 P^\nu P^{-2} \varepsilon^{\mu\rho\alpha\tau} q_\rho P_\tau. \tag{3.19}$$

By symmetric integration $P^\nu P_\tau \to \frac{1}{4}\delta^\nu_\tau P^2$, and so

$$\begin{aligned}
\varepsilon^\nu < J_5^\mu > &= -\frac{e}{2} \int \frac{d^4q}{(2\pi)^8} e^{-iqx} A_\alpha(q) \cdot 2\pi^2 \varepsilon^{\mu\rho\alpha\nu} q_\rho \\
&= -\pi^2 \, e \, \varepsilon^{\mu\rho\alpha\nu} \int \frac{d^4q}{(2\pi)^8} \left[\frac{1}{-i} \frac{\partial}{\partial x^\rho} e^{-iqx} \right] A_\alpha(q) \\
&= i\pi^2 e \; \varepsilon^{\mu\rho\alpha\nu} \frac{1}{(2\pi)^4} \frac{\partial}{\partial x^\rho} A_\alpha(x) \\
&= \frac{ie}{16\pi^2} \; \varepsilon^{\mu\nu\rho\alpha} \partial_\rho A_\alpha = \frac{ie}{32\pi^2} \; \varepsilon^{\mu\nu\rho\alpha} F_{\rho\alpha}.
\end{aligned} \tag{3.20}$$

Inserting this expression into Eq. (3.11), we obtain the final result

$$< \partial_\mu J_5^\mu(x) > = \frac{-e^2}{32\pi^2} \; \varepsilon^{\mu\nu\rho\alpha} F_{\mu\nu}(x) F_{\rho\alpha}(x). \tag{3.21}$$

This calculation of the chiral anomaly is only meant to be illustrative of field theoretic methods for obtaining this result. We have followed here the analysis of Jackiw in Ref. [12]. Other methods include Fujikawa's path integral procedure [13], and the analysis of short distance singularities employed by Leutwyler [14]. The original perturbative calculations in terms of Feynman diagrams were performed by Adler [15], and by Bell and Jackiw [16].

3.3 The Decay Rate and the Quark Model

The calculation in the previous section suggests that in the chiral limit

$$\partial_\mu J_5^\mu = \frac{-e^2}{8\pi^2} \; \varepsilon^{\mu\nu\alpha\beta} \partial_\mu A_\nu \partial_\alpha A_\beta. \tag{3.22}$$

This yields the matrix element

$$< \gamma(k_1)\gamma(k_2) | \partial_\mu J_5^\mu | 0 > = \frac{e^2}{4\pi^2} \; \varepsilon^{\mu\nu\alpha\beta} \; k_{1\mu} k_{2\alpha} \epsilon_\nu(k_1) \epsilon_\beta(k_2), \tag{3.23}$$

where k_1, k_2 are the momenta of the photons in the final state and $\epsilon(k_1), \epsilon(k_2)$ their polarization vectors. The factor of two acquired in the passage from (3.22) to (3.23) is due to the Bose statistics of the photons. We may now use Eq. (3.5) to calculate the decay amplitude in the chiral limit:

$$\begin{aligned}
M(q^2 = 0) &= \frac{-e^2}{4\pi^2 f_\pi} \; \varepsilon^{\mu\nu\alpha\beta} \; k_{1\mu} k_{2\alpha} \epsilon_\nu(k_1) \epsilon_\beta(k_2) \\
&= \frac{-\alpha}{\pi f_\pi} \; \varepsilon^{\mu\nu\alpha\beta} \; k_{1\mu} k_{2\alpha} \epsilon_\nu(k_1) \epsilon_\beta(k_2).
\end{aligned} \tag{3.24}$$

Note that the *form* of the amplitude is uniquely determined by Lorentz invariance, Bose statistics, parity and gauge invariance.

To calculate the decay *rate* we need to square the amplitude:

$$\begin{aligned} |M|^2 &= \frac{\alpha^2}{\pi^2 f_\pi^2}\, \varepsilon^{\mu\nu\alpha\beta} k_{1\mu} k_{2\alpha} \varepsilon^{\rho\tau\gamma\delta} k_{1\rho} k_{2\gamma} g_{\nu\tau} g_{\beta\delta} \\ &= \frac{\alpha^2}{\pi^2 f_\pi^2}(-2)(g^{\mu\rho} g^{\alpha\gamma} - g^{\mu\alpha} g^{\rho\gamma}) k_{1\mu} k_{2\alpha} k_{1\rho} k_{2\gamma} \\ &= \frac{2\alpha^2}{\pi^2 f_\pi^2}\,(k_1 \cdot k_2)^2. \end{aligned} \tag{3.25}$$

Here $g_{\mu\nu}$ is the Minkowski metric tensor, and we have summed over the polarizations using $\epsilon_\mu(k_1)\epsilon_\nu(k_1) \to g_{\mu\nu}$. Now use the conservation of energy-momentum:

$$(k_1 + k_2)^2 = 2k_1 \cdot k_2 = m_\pi^2. \tag{3.26}$$

Hence

$$|M|^2 = \frac{\alpha^2 m_\pi^4}{2\pi^2 f_\pi^2}. \tag{3.27}$$

The decay rate for a process involving two identical bosons in the final state is

$$\Gamma(\pi^0 \to 2\gamma) = \frac{1}{2}\,\frac{|M|^2}{16\pi m_\pi} = \frac{\alpha^2 m_\pi^3}{64\pi^3 f_\pi^2} = 7.63\,\mathrm{eV}. \tag{3.28}$$

The experimental result is

$$\Gamma_{\text{exp}} = (7.37 \pm 1.5)\,\mathrm{eV}. \tag{3.29}$$

In the quark model the fermion current coupled to the electromagnetic field is

$$J_5^\mu = \tfrac{1}{2}\textstyle\sum_i \bar{\psi}_i \gamma_5 \gamma^\mu \tau_3^i \psi_i, \tag{3.30}$$

so compared to our previous calculation the amplitude M should be multiplied by a factor $X = \sum_i \tau_3^i Q_i^2$, where Q_i are the quark charges. The quarks (u,d,s) have the quantum numbers $Q_i = (\frac{2}{3}, -\frac{1}{3}, -\frac{1}{3})$ and $\tau_3 = (1, -1, 0)$, hence

$$X = (2/3)^2 - (1/3)^2 = 1/3. \tag{3.31}$$

This would destroy the compatibility we had already found between the theoretical and experimental values of the decay rate. This was Gell-Mann's original argument for introducing *quark colour* [3]; the resulting factor of three restores the previous agreement. For a modern view of elementary particle theory, including a discussion of the relevance of chiral Lagrangians, involving the pion as a massless Goldstone boson, as effective low-momentum field theories, see Georgi [17].

While the $\pi^0 \to 2\gamma$ decay is the most direct physical application of anomalies in field theories, there are many others. A particularly important application involves the anomaly cancellation in the standard model of quarks and leptons, achieved by demanding an equal number of quark and lepton families [18]. This has been thought to be a necessary condition for the renormalizability of the theory, and even for its consistency. However, recent studies of the chiral Schwinger model [19] have cast doubt on the necessity of this condition. For this view of the relationship between anomalies and renormalizability, see [20]. Then there is 't Hooft's resolution of the U(1) problem involving instantons [21], and the anomaly-matching condition for composite models [22]. Physical applications of anomalies in odd spacetime dimensions, including cosmic strings and domain walls, are discussed by Reuter in [23].

4. The Wess-Zumino Consistency Conditions

A quantum field theory is characterized by the generating functional

$$Z[A] = e^{iW[A]} = < Te^{i\int J^\mu_a(x) A^a_\mu(x) dx} > . \tag{4.1}$$

The quantum counterpart of the classical current is the expectation value

$$< J^\mu_a(x) > = \frac{1}{Z[A]}\frac{1}{i}\frac{\delta Z[A]}{\delta A^a_\mu(x)} = \frac{\delta W[A]}{\delta A^a_\mu(x)}, \tag{4.2}$$

which corresponds to the classical relation (2.46)

$$\mathcal{L} = \mathcal{L}_0 + J^\mu_a A^a_\mu, \tag{4.3}$$

and the notion of $W[A]$ as the "effective action".

The covariant divergence of the current may be written as

$$D_{\mu ab} < J^{\mu b}(x) > = \Big[\partial_\mu \frac{\delta}{\delta A^a_\mu(x)} + f^b_{ac} A^c_\mu(x) \frac{\delta}{\delta A^b_\mu(x)}\Big] W[A] = -X_a(x) W[A], \tag{4.4}$$

where

$$X_a(x) = -\partial_\mu \frac{\delta}{\delta A^a_\mu(x)} - f^c_{ab} A^b_\mu(x) \frac{\delta}{\delta A^c_\mu(x)}. \tag{4.5}$$

The *anomaly* is (minus) the value of the current divergence:

$$X_a(x) W[A] = G_a[A](x). \tag{4.6}$$

The operators $X_a(x)$ generate gauge transformations, because

$$\begin{aligned}
&\int dz\, \varepsilon^b(z) X_b(z) A^a_\mu(x) \\
&= -\int dz\, \varepsilon^b(z)\Big[\partial_\nu \frac{\delta}{\delta A^b_\nu(z)} + f^d_{bc} A^c_\nu(z) \frac{\delta}{\delta A^d_\nu(z)}\Big] A^a_\mu(x) \\
&= -\int dz\, \varepsilon^b(z)\Big[\partial_\mu \delta(x-z)\delta^a_b + f^a_{bc} A^c_\mu(z)\delta(z-x)\Big] \\
&= \partial_\mu \varepsilon^a(x) + f^a_{cb} A^c_\mu(x)\varepsilon^b(x) = D_\mu \varepsilon^a(x) = \delta_\varepsilon A^a_\mu(x).
\end{aligned} \tag{4.7}$$

Hence the vanishing of the anomaly is equivalent to the gauge invariance of the effective action:

$$\begin{aligned}
\int dz\, \varepsilon^a(z) X_a(z) W[A] &= -\int dz\, \varepsilon^a(z)[\delta^c_a \partial_\mu + f^c_{ab} A^b_\mu(z)] \frac{\delta W[A]}{\delta A^c_\mu(z)} \\
&= \int dz\, D_\mu \varepsilon^c(z) \frac{\delta W[A]}{\delta A^c_\mu(z)} \\
&= \int dz\, \delta_\varepsilon A^c_\mu(z) \frac{\delta W[A]}{\delta A^c_\mu(z)} = \delta_\varepsilon W[A].
\end{aligned} \tag{4.8}$$

The operators $X_a(x)$ form a representation of the Lie algebra:

$$[X_a(x), X_b(y)] = f^c_{ab} X_c(x)\delta(x-y). \tag{4.9}$$

This can be checked by a direct calculation using the definition (4.5). Operating with the above equation on $W[A]$ yields:

$$X_a(x)X_b(y)W[A] - X_b(y)X_a(x)W[A] = f^c_{ab} X_c(x) W[A]\delta(x-y), \qquad (4.10),$$

or,using the definition of the anomaly, Eq. (4.6), the *Wess-Zumino consistency condition* for the anomaly [24]:

$$X_a(x)G_b[A](y) - X_b(y)G_a[A](x) = f^c_{ab} G_c[A](x)\delta(x-y). \tag{4.11}$$

A trivial solution of the consistency condition is given by an expression of the form

$$\hat{G}_a(x) = X_a(x)\hat{W}[A], \tag{4.12}$$

with $\hat{W}[A]$ an arbitrary local functional of A. It can be shown that the replacement $W[A] \to W[A] + \hat{W}[A]$ merely corresponds to a finite renormalization of the effective action, and hence has no physical consequences. When $W[A] \to W[A] + \hat{W}[A]$ then of course $G_a \to G_a + \hat{G}_a$.

Already at this point we can begin to see that the Wess-Zumino consistency condition is something like the condition for the vanishing of the exterior derivative of a differential form:

$$d\omega([X_1, X_2]) = X_1(\omega(X_2)) - X_2(\omega(X_1)) - \omega([X_1, X_2]), \tag{4.13}$$

where ω is a one-form and X_1, X_2 are vector fields. A differential form whose exterior derivative vanishes is called *closed.* Since cohomology classes are just equivalence classes of closed forms we see that the Wess-Zumino condition turns the anomaly calculation into a problem in cohomology theory. The analogy can be made precise through the use of the BRS formalism, which is the subject of the next section. The original observation concerning the relevance of the BRS formalism and cohomology considerations for anomalies was made by Bonora and Cotta-Ramusino [25]. The presentation here follows that of Wess [26].

5. The BRS Formalism

Mathematically, a gauge field corresponds to a connection on a principal fibre bundle (see, e.g., the article by U. Kasper in this volume). The symmetry group of the theory corresponds to the structure group of the principal bundle. Let us suppose the group space of this group to be parametrized by some variables $(\lambda^1, \ldots, \lambda^N)$, then an element of the group (which is a *local* symmetry group) may be written as:

$$g(x;\lambda) = e^{i\lambda^a(x)T_a}, \tag{5.1}$$

where the T_a $(A = 1, \ldots, N)$ are the group generators.

We shall consider the connections, and any other relevant differential forms, to be defined on the total space parametrized by the variables $(x;\lambda)$. Then we are led to consider, besides the exterior derivative on the base space,

$$d = dx^\mu \frac{\partial}{\partial x^\mu}, \tag{5.2}$$

also the exterior derivative on the group space:

$$s = d\lambda^a \frac{\partial}{\partial \lambda^a} \tag{5.3}$$

The differentials on the base space (x) and the group space (λ) satisfy the usual anticommutation relations:

$$\begin{aligned} dx^\mu dx^\nu &= -dx^\nu dx^\mu, \\ dx^\mu d\lambda^a &= -d\lambda^a dx^\mu, \\ d\lambda^a d\lambda^b &= -d\lambda^b d\lambda^a. \end{aligned} \tag{5.4}$$

From here we find that s, like d, is nilpotent:

$$s^2 = 0, \tag{5.5}$$

and that s anticommutes with d:

$$sd + ds = 0. \tag{5.6}$$

We now introduce the *total exterior derivative* $\Delta = d + s$. Like d and s, it is nilpotent:

$$\Delta^2 = (d+s)^2 = d^2 + s^2 + ds + sd = 0. \tag{5.7}$$

The Lie algebra of the group is spanned by the left-invariant one-forms

$$u = g^{-1} sg. \tag{5.8}$$

These one-forms are seen to satisfy the *Maurer-Cartan equation,*

$$su = -g^{-1}(sg)g^{-1}(sg) = -u^2, \tag{5.9}$$

which may be interpreted to mean that the Riemann curvature of the Lie group vanishes.

Through its action on the group, the operator s also acts on the gauge potentials, if we define these potentials in such a way as to make their dependence on the group parameters explicit:

$$\tilde{A}(x;\lambda) = g^{-1}(x;\lambda)A(x)g(x;\lambda) + g^{-1}(x;\lambda)dg(x;\lambda). \tag{5.10}$$

Here $A(x) = \tilde{A}(x;0)$ are the usual gauge potentials defined on the base space with which the physicist usually works: $A(x) = -iA^a_\mu(x)T_a dx^\mu$. The action of s on $\tilde{A}$ is now:

$$\begin{aligned} s\tilde{A} &= -g^{-1}(sg)g^{-1}Ag - g^{-1}A(sg) - g^{-1}(sg)g^{-1}dg - g^{-1}d(sg) \\ &= -g^{-1}(sg)\tilde{A} - g^{-1}Agg^{-1}(sg) - g^{-1}d(gg^{-1}sg) \\ &= -g^{-1}(sg)\tilde{A} - g^{-1}Agg^{-1}(sg) - g^{-1}dgg^{-1}(sg) - d(g^{-1}sg) \\ &= -u\tilde{A} - \tilde{A}u - du = -du - [\tilde{A}, u] = -Du. \end{aligned} \tag{5.11}$$

Here and in the following the notation $[\,,\,]$ will be used to indicate a *graded* commutator: it is the usual commutator when at least one of the quantities in the brackets is even, it is the anticommutator when both are odd. Eq. (5.11) has exactly the form expected of a gauge transformation, which is indeed plausible if we think of s as moving $\tilde{A}$ along the fibre; such motion corresponds to the unphysical degrees of freedom which characterize a gauge transformation.

We begin to perceive at this stage that s corresponds to the familiar BRS operator used by physicists [27]. To make this relationship explicit we introduce the notation

$$u = -iC^a(x)T_a, \tag{5.12}$$

where the $C^a(x)$ are the Faddeev-Popov ghost fields [28], which anticommute because they are one-forms in group space; remember the definition of u in Eq. (5.8). In this notation the relation $su = -u^2$, Eq. (5.9), becomes

$$\begin{aligned} s(-iC^aT_a) &= -(-iC^bT_b)(-iC^cT_c) = C^bC^cT_bT_c \\ &= \tfrac{1}{2}C^bC^c[T_b, T_c] = \tfrac{i}{2}f^a_{bc}C^bC^cT_a, \end{aligned} \tag{5.13}$$

or

$$sC^a = -\tfrac{1}{2} f^a_{bc} C^b C^c, \tag{5.14}$$

which is the usual form for the BRS transformation of the ghost field. The BRS transformation of the gauge field follows from Eq. (5.11), it is

$$s(-iA^a_\mu(x) T_a dx^\mu) = -dx^\mu D_\mu(-iC^a(x)T_a), \tag{5.15}$$

or

$$sA^a_\mu(x) = D_\mu C^a(x), \tag{5.16}$$

where we have picked up a minus sign by commuting the factor dx^μ past $C^a(x)$, remember Eq. (5.4). We now use the property of the operators $X_a(x)$ as generators of gauge transformations, Eq. (4.7), to write this relation as

$$sA^a_\mu(x) = \int dz\, C^b(z) X_b(z) A^a_\mu(x). \tag{5.17}$$

The Wess-Zumino condition can now be written in a simple and informative form:

$$\begin{aligned} s\int dx\, C^a(x)G_a(x) &= \int dx \left[-\tfrac{1}{2} f^a_{bc} C^b(x) C^c(x) G_a(x) - C^a(x) sG_a(x)\right] \\ &= \int dx\, dy\, C^b(x)C^c(y)\left[-\tfrac{1}{2} f^a_{bc} G_a(x)\delta(x-y) - X_c(y)G_b(x)\right] \\ &= \tfrac{1}{2}\int dx\, dy\, C^b(x)C^c(y)\left[-f^a_{bc}G_a(x)\delta(x-y) - X_c(y)G_b(x) + X_b(x)G_c(y)\right] \\ &= 0. \end{aligned} \tag{5.18}$$

Finally, we introduce the notation

$$\omega^1_4 = C^a(x) G_a(x) d^4x, \tag{5.19}$$

which signifies that ω is a one-form in group space (this index corresponds to the *ghost number*) and a form of degree 4 in x-space. The consistency condition is now written as

$$s\omega^1_4 = -d\omega^2_3, \tag{5.20}$$

where an arbitrary form ω^2_3 is allowed by Stokes' law. This is the promised formulation of the Wess-Zumino condition in terms of an exterior derivative. It is also the form in which it appears in the descent equations, which are the mathematical equations which determine the cohomology classes of a Lie algebra, discussed in the following section.

6. Anomalies and Differential Geometry

In this section we shall explain the mathematical framework for treating anomalies, and illustrate how they are calculated in this method. We shall see that the Schwinger terms as well can be calculated in this way.

6.1 Chern Characters and Descent Equations

Let A be a connection on a 2n-dimensional manifold. Its associated curvature is

$$F = dA + A^2. \tag{6.1}$$

A transforms under an element g of the gauge group according to

$$\tilde{A} = g^{-1}Ag + g^{-1}dg \tag{6.2}$$

and F according to

$$\tilde{F} = d\tilde{A} + \tilde{A}^2 = g^{-1}Fg. \tag{6.3}$$

The Bianchi identity is

$$DF = dF + [A, F] = 0. \tag{6.4}$$

The *Chern character* of order n [29] is defined by

$$Ch_n(A) = \mathrm{tr}\, F^n. \tag{6.5}$$

It is an invariant polynomial:

$$\begin{aligned} Ch_n(\tilde{A}) &= \mathrm{tr}\, \tilde{F}^n = \mathrm{tr}\,(g^{-1}Fg)^n \\ &= \mathrm{tr}\,(F^n) = Ch_n(A). \end{aligned} \tag{6.6}$$

It is closed because of the Bianchi identity:

$$dCh_n(A) = DCh_n(A) = \mathrm{tr}\,(DF \cdot F^{n-1} + F \cdot DF \cdot F^{n-2} + \ldots) = 0. \tag{6.7}$$

Finally, it has the remarkable property that it is independent of the connection, up to a total divergence: If A and B are two different connections then

$$Ch_n(A) - Ch_n(B) = d\omega. \tag{6.8}$$

Proof : Introduce the interpolating connection $A_t = B + t(A - B)$ with $A_1 = A$ and $A_0 = B$. Since $(A - B)$ transforms as a tensor, A_t transforms as a connection. The associated curvature is

$$F_t = dA_t + A_t^2 = F_B + t(F_A - F_B) + (t^2 - t)(A - B)^2. \tag{6.9}$$

We easily compute

$$\begin{aligned}\frac{dF_t}{dt} &= d\left(\frac{dA_t}{dt}\right) + \left(\frac{dA_t}{dt}\right)A_t + A_t\left(\frac{dA_t}{dt}\right)\\ &= d(A-B) + (A-B)A_t + A_t(A-B)\\ &= d(A-B) + [A_t, A-B] = D_t(A-B).\end{aligned} \tag{6.10}$$

We can now evaluate the difference

$$\begin{aligned}Ch_n(A) - Ch_n(B) &= \int_0^1 dt\, \frac{d}{dt} Ch_n(A_t) = \int_0^1 dt\, \frac{d}{dt} \mathrm{tr}\, F_t^n\\ &= n\int_0^1 dt\, \mathrm{tr}\, [D_t(A-B)]F_t^{n-1}\\ &= n\int_0^1 dt\, \mathrm{tr}\, D_t[(A-B)F_t^{n-1}]\\ &= n\, d\int_0^1 dt\, \mathrm{tr}\, (A-B)F_t^{n-1},\end{aligned} \tag{6.11}$$

where we have used the Bianchi identity

$$D_t F_t = 0, \tag{6.12}$$

and the fact that $A-B$ and F_t are tensors, so that $\mathrm{tr}\,(A-B)F_t^{n-1}$ is a scalar.

We assume for simplicity that the underlying fibre bundle on which we are working is trivial. There then exists a frame in which the connection B vanishes. We shall work in the following in this frame; our results are nevertheless guaranteed to be valid in any frame, because of the invariance of the Chern characters. Thus

$$Ch_n(A) = n\, d\int_0^1 dt\, \mathrm{tr}\, AF_t^{n-1} = d\omega_{2n-1}^0(A, F). \tag{6.13}$$

Here A_t and F_t take the simplified forms $A_t = tA$ and $F_t = tdA + t^2A^2$. The indices associated with the form ω indicate that it is a $(2n-1)$-form in x-space and a zero-form in group space (compare Eq. (5.19)).

Instead of working with differential forms on the base manifold, as is the physicist's custom, we may work with corresponding differential forms on the total space of the fibre bundle. We should then use, instead of the exterior derivative on the base manifold d, the total exterior derivative Δ and the connection

$$\hat{A} = g^{-1}Ag + g^{-1}\Delta g = \tilde{A} + u. \tag{6.14}$$

Here u is the one-form defined in Eq. (5.8).

We now have a remarkable result for the curvature form $\hat{F}$:

$$\begin{aligned}\hat{F} &= \Delta\hat{A} + \hat{A}^2 = (d+s)(\tilde{A}+u) + (\tilde{A}+u)^2 \\ &= d\tilde{A} + \tilde{A}^2 = \tilde{F}. \end{aligned} \tag{6.15}$$

Stora calls this formula the *Russian formula* [30]. In [31] we likened it to the equivalence principle in general relativity, insofar as it exhibits the independence of physical results encoded in the curvature tensor on unphysical degrees of freedom (here the Faddeev-Popov ghosts, see Eq. (6.14)).

We now use the transformation property of $\tilde{F}$:

$$s\tilde{F} = [\tilde{F}, u] \tag{6.16}$$

to show that $\hat{F}$ also satisfies a Bianchi identity:

$$\begin{aligned}\hat{D}\hat{F} &= \Delta\hat{F} + [\hat{A}, \hat{F}] = (d+s)\tilde{F} + [\tilde{A}+u, \tilde{F}] \\ &= d\tilde{F} + [\tilde{F}, u] + [\tilde{A}, \tilde{F}] + [u, \tilde{F}] = D\tilde{F} = 0. \end{aligned} \tag{6.17}$$

Since the neccessary formal conditions are satisfied, our previous calculation of $Ch_n(A)$ carries over to the present case unchanged, and we find

$$Ch_n(\hat{A}) = \Delta\omega^0_{2n-1}(\hat{A}, \hat{F}), \tag{6.18}$$

with

$$\omega^0_{2n-1}(\hat{A}, \hat{F}) = n\int_0^1 dt\, \mathrm{tr}\, \hat{A}\hat{F}_t^{n-1}. \tag{6.19}$$

This result may be further simplified, since

$$Ch_n(\hat{A}) = \mathrm{tr}\, \hat{F}^n = \mathrm{tr}\, \tilde{F}^n = \mathrm{tr}\, F^n = Ch_n(A), \tag{6.20}$$

so we have

$$d\omega^0_{2n-1}(A, F) = (d+s)\omega^0_{2n-1}(\tilde{A}+u, \tilde{F}). \tag{6.21}$$

An important result follows from expanding this equation in powers of u, according to

$$\omega^0_{2n-1}(\tilde{A}+u, \tilde{F}) = \omega^0_{2n-1}(\tilde{A}, \tilde{F}) + \omega^1_{2n-2}(\tilde{A}, \tilde{F}) + \ldots + \omega^{2n-1}_0(\tilde{A}, \tilde{F}). \tag{6.22}$$

Comparing forms of the same degree yields the *descent equations*:

$$\begin{aligned} s\omega^0_{2n-1}(\tilde{A}, \tilde{F}) &= -d\omega^1_{2n-2}(\tilde{A}, \tilde{F}), \\ s\omega^1_{2n-2}(\tilde{A}, \tilde{F}) &= -d\omega^2_{2n-3}(\tilde{A}, \tilde{F}), \\ s\omega^2_{2n-3}(\tilde{A}, \tilde{F}) &= -d\omega^3_{2n-4}(\tilde{A}, \tilde{F}), \\ &\vdots \\ s\omega^{2n-1}_0(\tilde{A}, \tilde{F}) &= 0. \end{aligned} \tag{6.23}$$

For $n = 3$ the second line above is the Wess-Zumino consistency condition (compare Eq. (5.20)):

$$s\omega^1_4 = -d\omega^2_3. \tag{6.24}$$

6.2 Cohomology Calculation of the Anomaly

In this subsection we show how the ω_4^1 term, corresponding to the anomaly, may be calculated.

We start from the form

$$\tilde{F}_t = td\tilde{A} + t^2\tilde{A}^2 = t\tilde{F} + (t^2 - t)\tilde{A}^2. \tag{6.25}$$

In the same way we have

$$\begin{aligned}\hat{F}_t &= t\tilde{F} + (t^2 - t)(\tilde{A} + u)^2 \\ &= \tilde{F}_t + (t^2 - t)(u^2 + [\tilde{A}, u]).\end{aligned} \tag{6.26}$$

With these expressions we expand the form

$$\omega_{2n-1}^0(A + u, F) = n \int_0^1 dt \operatorname{tr}(\tilde{A} + u)\hat{F}_t^{n-1}. \tag{6.27}$$

For $n = 3$ this is

$$\begin{aligned}\omega_5^0(A + u, F) &= 3 \int_0^1 dt \operatorname{tr}(\tilde{A} + u)\hat{F}_t^2 \\ &= 3 \int_0^1 dt \operatorname{tr}(\tilde{A} + u)\{\tilde{F}_t + (t^2 - t)(u^2 + [\tilde{A}, u])\}^2.\end{aligned} \tag{6.28}$$

The term linear in u is

$$\begin{aligned}\omega_4^1 =& 3 \int_0^1 dt \operatorname{tr}\{\tilde{A}(t^2 - t)(\tilde{F}_t[\tilde{A}, u] + [\tilde{A}, u]\tilde{F}_t) + u\tilde{F}_t^2\} \\ &= \operatorname{tr} u(\tilde{A}d\tilde{A} + \tfrac{1}{2}\tilde{A}^3).\end{aligned} \tag{6.29}$$

Besides Eq. (6.26) we have only used Eq. (6.25).

The non-Abelian anomaly has been calculated by field theoretic methods [32] to be

$$G_a(x) = \frac{1}{24\pi^2}\varepsilon^{\mu\nu\alpha\beta} \operatorname{tr}[T_a\partial_\mu(A_\nu\partial_\alpha A_\beta + \tfrac{1}{2}A_\nu A_\alpha A_\beta)]. \tag{6.30}$$

Up to a proportionality constant, this is seen to be just the component form of Eq. (6.29). The cohomology calculation can give us only the form of the anomaly, the absolute value can be computed, except field theoretically, only by using the index theorem (see Römer's article in this volume). In the Abelian case the form of the anomaly obviously reduces to that discussed in Sect. 3. The differential geometric approach has been used by Zumino, Wu and Zee to calculate the chiral anomalies in arbitrary even and odd dimensions [33].

7. Schwinger Terms

We first define the Schwinger terms. We then work out in some detail a consistency condition on the Schwinger terms implied by the Jacobi identity. This turns out to coincide with one of the descent equations. This suggests a relationship between anomalies and Schwinger terms, first suggested by Faddeev [34]. We indicate how this relationship may be worked out explicitly. Further information on this subject is contained in a recent CERN preprint of Abud, Ader and Gieres [35].

7.1 A Consistency Condition for the Schwinger Terms

The operators $X_a(x)$ of Eq. (4.5) generate, according to Eq. (4.7), gauge transformations of the fields $A^a_\mu(x)$. They therefore also generate gauge transformations of functionals of the gauge fields. In general, however, our system consists of gauge fields *and* matter fields. It is therefore of interest to consider the generators of gauge transformations on the vector space of functionals of the gauge fields and the matter fields. We shall denote these generators by $M_a(x)$. They may be written as

$$M_a(x) = iX_a(x) + J^0_a(X), \tag{7.1}$$

where the $J^0_a(x)$ are the generators of gauge transformations on functionals of the matter fields $\psi(x)$. The generators $iX_a(x)$ satisfy (compare Eq. (4.9))

$$[\lambda_1 \circ iX, \lambda_2 \circ iX] = i(\lambda_1 \times \lambda_2) \circ iX, \tag{7.2}$$

where

$$\lambda_1 \circ X = \int dx\, \lambda_1^a(x) X_a(x) \tag{7.3}$$

and

$$(\lambda_1 \times \lambda_2)^a = f^a_{bc}\, \lambda_1^b\, \lambda_2^c. \tag{7.4}$$

The generators $M_a(x)$ do not in general build a closed algebra under commutation, their commutation relations turn out to be of the form

$$[\lambda_1 \circ M, \lambda_2 \circ M] = i(\lambda_1 \times \lambda_2) \circ M + \lambda_1 \circ Z[A] \circ \lambda_2, \tag{7.5}$$

with

$$\lambda_1 \circ Z[A] \circ \lambda_2 = \int d^3x\, d^3y\, \lambda_1^a(x)\, Z_{ab}[A](x,y)\, \lambda_2^b(y). \tag{7.6}$$

The term $Z[A]$ is called a *Schwinger term*, since terms of this kind were first discussed by Schwinger in the context of current algebra [36].

The Schwinger term Z_{ab} is antisymmetric by definition:

$$Z_{ab}(x,y) = -Z_{ba}(y,x). \tag{7.7}$$

A consistency condition on the Schwinger term follows from the Jacobi identity for the generators M_a:

$$\sum_{\substack{\lambda_1\lambda_2\lambda_3\\ \text{cyclic}}} \Big[[\lambda_1 \circ M, \lambda_2 \circ M], \lambda_3 \circ M\Big]$$
$$= \sum [i(\lambda_1 \times \lambda_2) \circ M + \lambda_1 \circ Z \circ \lambda_2, \lambda_3 \circ M]$$
$$= \sum \Big\{ -((\lambda_1 \times \lambda_2) \times \lambda_3) \circ M + i(\lambda_1 \times \lambda_2) \circ Z \circ \lambda_3 + [\lambda_1 \circ Z \circ \lambda_2, \lambda_3 \circ M] \Big\} = 0. \tag{7.8}$$

The first term sums to zero by itself:

$$\sum_{\lambda_1\lambda_2\lambda_3} f^c_{mn} f^m_{ab} \lambda^a_1 \lambda^b_2 \lambda^n_3 M_c = (f^c_{mn} f^m_{ab} + f^c_{ma} f^m_{bn} + f^c_{mb} f^m_{na}) \lambda^a_1 \lambda^b_2 \lambda^n_3 M_c = 0. \tag{7.9}$$

We are then left with

$$\sum_{\lambda_1\lambda_2\lambda_3} \Big\{ i(\lambda_1 \times \lambda_2) \circ Z \circ \lambda_3 - \lambda_1 \circ [\lambda_3 \circ M, Z] \circ \lambda_2 \Big\} = 0. \tag{7.10}$$

Since $Z[A]$ depends only on A, and not on the matter fields,

$$\Big[\lambda_3 \circ M, Z[A]\Big] = i \Big[\lambda^a_3 \circ X_a, Z[A]\Big] = i \int dx\, \lambda^a_3(x) X_a(x)\, Z[A] = isZ[A]. \tag{7.11}$$

Hence Eq. (7.10) may be written as

$$\sum_{\lambda_1\lambda_2\lambda_3} \Big\{ (\lambda_1 \times \lambda_2) \circ Z \circ \lambda_3 - \lambda_1 \circ \lambda_3 \circ X\, Z[A] \circ \lambda_2 \Big\} = 0. \tag{7.12}$$

In terms of the ghost fields the consistency condition for $Z[A]$ is

$$s(C^a Z_{ab} C^b) = -f^a_{mn} C^m C^n Z_{ab} C^b - C^a C^c X_c Z_{ab} C^b = 0, \tag{7.13}$$

where we have used Eq. (7.11). To see the equivalence of this with the previous form, just note that the cyclic sum is equivalent to antisymmetrizing with respect to the C- fields:

$$C^m C^n C^b f^a_{mn} Z_{ab} = \tfrac{1}{3} C^m C^n C^b (f^a_{mn} Z_{ab} + f^a_{nb} Z_{am} + f^a_{bm} Z_{an}), \tag{7.14}$$

and

$$C^a C^c C^b X_c Z_{ab} = \tfrac{1}{3} C^a C^c C^b (X_c Z_{ab} + X_a Z_{bc} + X_b Z_{ca}). \tag{7.15}$$

Finally, we write the condition (7.12) in a more explicit form:

$$s \int d^3x\, d^3y\, C^a(x)\, Z_{ab}(x,y)\, C^b(y) = 0. \tag{7.16}$$

The integrations are over 3-space since we are considering equal-time commutators. If we define

$$\omega_3^2 = C^a(x)\, Z_{ab}(x,y)\, C^b(y)\, d^3y \tag{7.17}$$

the condition takes the form

$$s\omega_3^2 = -d\omega_2^3, \tag{7.18}$$

which is nothing but the third of the descent equations (6.23), if we identify the expression in Eq. (7.17) with the corresponding form in the descent equations.

Like the anomaly, the Schwinger term is determined only up to terms of the form $s(\ldots)$. Adding such a term just amounts to a redefinition of the generators. That is, if we go over to the generators $M_a' = M_a + S_a[A]$, then

$$[\lambda_1 \circ M', \lambda_2 \circ M'] = (\lambda_1 \times \lambda_2) \circ M' + \lambda_1 \circ Z' \circ \lambda_2. \tag{7.19}$$

On the other hand, we have

$$\begin{aligned}
&[\lambda_1 \circ (M+S), \lambda_2 \circ (M+S)] \\
&= [\lambda_1 \circ M, \lambda_2 \circ M] + [\lambda_1 \circ M, \lambda_2 \circ S] + [\lambda_1 \circ S, \lambda_2 \circ M] \\
&= (\lambda_1 \times \lambda_2) \circ M + \lambda_1 \circ Z \circ \lambda_2 + [\lambda_1 \circ X, \lambda_2 \circ S] - [\lambda_2 \circ X, \lambda_1 \circ S].
\end{aligned} \tag{7.20}$$

Here we have used the fact that $S_a[A]$ depends only on the gauge fields, and is independent of the matter fields. We are assuming that the theory has been quantized in a gauge in which $A_0^a(x) = 0$ (the temporal gauge), and the spatial components of the vector potential A_i^a commute. Comparing Eq. (7.19) to Eq. (7.20) yields

$$\lambda_1 \circ Z' \circ \lambda_2 = \lambda_1 \circ Z \circ \lambda_2 - (\lambda_1 \times \lambda_2) \circ S + [\lambda_1 \circ X, \lambda_2 \circ S] - [\lambda_2 \circ X, \lambda_1 \circ S]. \tag{7.21}$$

In the ghost field notation this reads

$$C \circ Z' \circ C = C \circ Z \circ C + s(2\, C \circ S), \tag{7.22}$$

because the last term is, written more explicitly,

$$\begin{aligned}
s(2\, C \circ S) &= -f^a_{mn}\, C^m\, C^n\, S_a + 2\, C^a\, C^b\, X_b\, S_a \\
&= -(C \times C)^a \circ S_a + C^a\, C^b\, (X_b S_a - X_a S_b),
\end{aligned} \tag{7.23}$$

and this agrees with the expression in Eq. (7.21). The relation in Eq. (7.22) is what we wanted to establish. It means that the expressions Z and Z' are in the same cohomology class.

7.2 Anomalies and Schwinger Terms

It is clear from the discussion of the last section that anomalies, related to the non-vanishing divergence of some classically conserved current, and Schwinger terms, which arise in the breakdown of canonical equal-time commutation relations, are intimately related. For both occur in the descent equations, the different members of which arise just from the decomposition of a single relationship, namely Eq. (6.21). In this subsection we shall indicate how the relation between anomalies and Schwinger terms may be worked out directly.

We have identified the anomaly with the form in the descent equations ω_4^1. From the considerations of the previous subsection we expect the Schwinger terms to be related to the form ω_3^2. These forms are related by the third of the descent equations:

$$s\omega_4^1 = -d\omega_3^2 \tag{7.24}$$

We thus expect the gauge variation of the anomaly to be related to the spatial derivative of the Schwinger term. Recall the definition of the anomaly, Eq. (4.6):

$$X_a(x)W[A] = G_a(x). \tag{7.25}$$

When we calculate the gauge variation of this term we encounter derivatives of the form

$$\begin{aligned} \frac{\partial}{\partial A_\nu^b(y)} G_a(x) &= -\partial_\mu \frac{\partial}{\partial A_\mu^a(x)} \frac{\partial}{\partial A_\nu^b(y)} W[A] + \dots \\ &= -\partial_\mu < T\left(J_\mu^a(x) J_\nu^b(y)\right) > + \dots \\ &= -\partial_\mu [\theta(x^0 - y^0) J_\mu^a(x) J_\nu^b(y) + \theta(y^0 - x^0) J_\nu^b(y) J_\mu^a(x)] + \dots \\ &= \delta(x^0 - y^0)[J_0^a(x), J_\nu^b(y)] + \dots \end{aligned} \tag{7.26}$$

In this equation the symbol T indicates the usual time-ordering. The original Schwinger terms [36] occur in the equal-time commutators

$$\delta(x^0 - y^0)\,[J_0^a(x), J_0^b(y)] = f_{ab}^c\, J_c^0(x)\,\delta(x-y) + c_{ab}\,\delta(x-y) + S_{ab}\,\delta'(x-y). \tag{7.27}$$

Carrying out the differentiation and identifying the Schwinger terms in this way leads, in the case of a two-dimensional field theory, to the result

$$[M_a(x), M_b(y)] = i f_{ab}^c\, M_c(x)\,\delta(x-y) + \frac{1}{2\pi}\,\delta_{ab}\,\partial_x \delta(x-y). \tag{7.28}$$

These are the structure relations of an affine Kac-Moody algebra. The Schwinger term corresponds in this case to a central extension of the algebra [37].

In the four-dimensional case, the cohomological method leads to the result

$$\omega_3^2 = du\, du\, A. \tag{7.29}$$

This may be derived from the descent equations in the same way as we have illustrated in detail for the anomaly term ω_4^1. The result of a field theoretic calculation [38] is

$$\begin{aligned}[M_a(x), M_b(y)]\,\delta(x^0 - y^0) &= i f_{ab}^c\, M_c(x)\,\delta^{(4)}(x-y) \\ &+ \frac{1}{48\pi^2}\varepsilon^{ijk}\,\mathrm{tr}\Big\{[T_a, T_b]\,(\partial_i A_j A_k + A_i \partial_j A_k + A_i A_j A_k) \\ &+ T_a\,\partial_i(A_j T_b A_k)\Big\}\,\delta^{(4)}(x-y). \end{aligned} \tag{7.30}$$

The coordinate-independent form of the correction term in Eq. (7.30) is

$$\omega' = u^2(dA\,A + A\,dA + A^3) + ud(AuA). \tag{7.31}$$

The equivalence of the expressions in Eq. (7.29) and Eq. (7.31) is established by showing that they differ only by terms corresponding to a gauge transformation and a total derivative. This is indeed this case; we have

$$-2\,du\,du\,A = \omega' - s(u\,A^3 + u\,A\,dA + u\,dA\,A) - d(u\,du\,A + du\,u\,A - u\,A\,u\,A). \tag{7.32}$$

Remember that the cohomology calculation yields the appropriate form only up to a proportionality constant.

Appendix: Homological Algebra

In the algebraic approach the prerequisites for defining the relevant cohomologies are surprisingly simple. We start from scratch.

A1 Graded Differential Algebras

An *algebra* A over $\mathbb{R}$ is a real vector space together with a real bilinear map (a *product*) $A \times A \to A$. A *derivation* in an algebra A is a linear map $\theta : A \to A$ satisfying

$$\theta(xy) = \theta(x)y + x\theta(y), \qquad x, y \in A. \tag{A.1}$$

A *graded algebra* A over $\mathbb{R}$ is a graded vector space $A = \sum_{p \geq 0} A^p$, together with an algebra structure, such that

$$A^p \cdot A^q \subset A^{p+q}. \tag{A.2}$$

The elements of A^p are called *homogeneous elements of A of degree p*. A linear mapping L of A in itself is said to be *homogeneous of degree r* if $L(A^p) \subset A^{p+r}$ for all p. If

$$xy = (-1)^{pq} yx, \qquad x \in A^p,\ y \in A^q, \tag{A.3}$$

then A is called *anticommutative*. An *antiderivation* in a graded algebra A is a linear map $\alpha : A \to A$, homogeneous of *odd* degree, such that

$$\alpha(xy) = \alpha(x)y + (-1)^p x\alpha(y), \qquad x \in A^p,\ y \in A. \tag{A.4}$$

A *differential space* is a vector space X together with a linear map $\delta : X \to X$ satisfying $\delta^2 = 0$. δ is called the *differential operator* in X. The elements of the subspaces

$$Z(X) = \ker \delta \quad \text{and} \quad B(X) = \operatorname{Im} \delta \tag{A.5}$$

are called *cocycles* and *coboundaries*, respectively. The space

$$H(X) = Z(X)/B(X) \tag{A.6}$$

is called the *cohomology space* of X.

A graded space $X = \sum_{p \geq 0} X^p$ together with a differential operator δ homogeneous of degree $+1$ is called a *graded differential space*. In such a case the cocycle, coboundary and cohomology spaces are graded:

$$Z^p(X) = Z(X) \cap X^p, \qquad B^p(X) = B(X) \cap X^p \tag{A.7}$$

and

$$H^p(X) = Z^p(X)/B^p(X). \tag{A.8}$$

A *graded differential algebra* A is a graded algebra together with an antiderivation δ, homogeneous of degree one, such that $\delta^2 = 0$. In this case $Z(A)$ is a graded subalgebra and $B(A)$ is a graded ideal in $Z(A)$. Thus $H(A)$ becomes a graded algebra. It is called the *cohomology algebra* of A. If A is anticommutative, then so is $H(A)$.

The first example of a graded differential algebra which one usually learns is the space of *differential forms*; the associated cohomology is the *de Rham* cohomology, which we now describe. Let M be an n-manifold. The differential forms on M constitute a graded algebra, $A(M) = \sum_p A^p(M)$, with multiplication given by $(\Phi \wedge \Psi)(x) = \Phi(x) \wedge \Psi(x)$. The exterior derivative d is an antiderivation on $A(M)$, it acts on functions to yield the gradient, its action on one-forms is given by

$$d\omega(X,Y) = X(\omega(Y)) - Y(\omega(X)) - \omega([X,Y]). \qquad (A.9)$$

$(A(M), d)$ is a graded differential algebra; its cohomology is denoted by $H(M) = \sum_{p=0}^{n} H^p(M)$ and is called the *de Rham cohomology algebra* of M.

A2 Cohomology of Lie Algebras

Let $\mathcal{G}$ be a finite-dimensional Lie algebra. Its dual space is denoted by $\mathcal{G}^*$. The elements of $\mathcal{G}^*$ are the one-forms, the space of two-forms is $\Lambda^2\mathcal{G}^*$, etc. Define $s : \mathcal{G}^* \to \Lambda^2\mathcal{G}^*$ by

$$s\omega(X,Y) = -\omega([X,Y]). \qquad (A.10)$$

Note the similarity to Eq. (A.9): The Lie algebra may be considered to be generated by the left-invariant vector fields, and a left-invariant one-form, acting on a left-invariant vector field, yields a constant. Any vector field acting acting on a constant gives zero. Hence the first two terms on the r.h.s. of Eq. (A.9) vanish, and we are left with Eq. (A.10).

The operator s may now be extended to an antiderivation of the graded anticommutative algebra $\Lambda\mathcal{G}^* = \sum_{p=0}^{n} \Lambda^p\mathcal{G}^*$. Its action on the two-form $s\omega$ is then

$$\begin{aligned} s^2\omega(X,Y,Z) &= -\tfrac{1}{3}\Big(s\omega([X,Y],Z) + s\omega([Y,Z],X) + s\omega([Z,X],Y)\Big) \\ &= \tfrac{1}{3}\Big(\omega([[X,Y],Z]) + \omega([[Y,Z],X]) + \omega([[Z,X],Y])\Big) \\ &= \tfrac{1}{3}\omega([[X,Y],Z] + [[Y,Z],X] + [[Z,X],Y]) = 0, \qquad (A.11) \end{aligned}$$

where in the last line we have used the Jacobi identity. Since $s^2 = 0$ on the generators of the algebra, and s^2 is a derivation, $s^2 = 0$ is valid in general. In this way $(\Lambda\mathcal{G}^*, s)$ becomes a differential algebra, its cohomology is denoted by $H^*(\mathcal{G})$ and is called the *cohomology of the Lie Algebra* $\mathcal{G}$.

A3 Bigraded Differential Algebras

A *bigraded algebra* is an algebra A which admits a direct sum decomposition $A = \sum_{p,q} A^{p,q}$ and is such that the product satisfies

$$A^{p,q} \,.\, A^{p',q'} \subset A^{p+p',q+q'}. \qquad (A.12)$$

The elements of $A^{p,q}$ are *bihomogeneous of bidegree (p,q)*. A linear mapping of A in itself is *bihomogeneous of degree (r,s)* if $L(A^{p,q} \subset A^{p+r,q+s}$ for any (p,q). The elements of $A^k = \sum_{p+q=k} A^{p,q}$ are called *homogeneous of total degree k.* Thus a bigraded algebra is, in particular, a graded algebra for the grading corresponding to the total degree.

If a bigraded algebra A is equipped with two differentials $d^{1,0}$ and $d^{0,1}$ which are bihomogeneous of respective bidegrees $(1,0)$ and $(0,1)$, and if $d^{1,0}$ and $d^{0,1}$ anticommute:

$$d^{1,0}d^{0,1} + d^{0,1}d^{1,0} = 0, \qquad (A.13)$$

then A is a *bigraded differential algebra* with *total differential* $d = d^{1,0} + d^{0,1}$.

A graded differential algebra is a special case of a structure which is referred to in homological algebra as a *complex*; a bigraded differential algebra is a special case of a *double complex.*

We have seen in the main text that the mathematical structures discussed in this Appendix are relevant for the treatment of anomalies in quantum field theory. A detailed description of the application of these techniques to the anomaly problem is given by Dubois-Violette in [39]. We here merely quote the results. Let P denote the space of polynomial functionals of the gauge potentials. Let G be the group of gauge transformations, with Lie algebra $\mathcal{G}$. The infinitesimal right action of $\mathcal{G}$ on the gauge potentials induces a representation of $\mathcal{G}$ in P. Thus one may consider the complex $C^*(P;\mathcal{G})$ of cochains of $\mathcal{G}$ with values in P. Let δ denote the differential of $C^*(P;\mathcal{G})$. This complex must actually be restricted to the *local* complex $C^*_{loc}(P;\mathcal{G})$ involving local functionals of the gauge potentials. The cohomology of this differential space is $H^*_{loc}(P;\mathcal{G})$. The anomaly turns out to lie in $H^1_{loc}(P;\mathcal{G})$, the Schwinger term in $H^2_{loc}(P;\mathcal{G})$.

References

1. E. Binz, J. Sniatycki, H. Fischer: *Geometry of Classical Fields,* North Holland Mathematics Studies **154** (North Holland, Amsterdam 1988)
2. M.F. Atiyah, R.S. Ward: Comm. Math. Phys. **55**, 117 (1977)
3. M. Gell-Mann: Acta Phys. Austriaca Suppl. **IV**, 733 (1972)
4. T.T. Wu, C.N. Yang: Phys. Rev. **D12**, 3845 (1975)
5. P.A.M. Dirac: Proc. Roy. Soc. **A133**, 60 (1931)
6. L. Alvarez-Gaumé, E. Witten: Nucl. Phys. **B234**, 269 (1984)
7. M. Gell-Mann, M. Lévy: Nuov. Cim. **16**, 705 (1960)
8. S.B. Treiman: in *Lectures on Current Algebra and its Applications,* S.B. Treiman, R. Jackiw, D.J. Gross (Princeton University Press 1972)
9. D.G. Sutherland: Nucl. Phys. **B2**, 433 (1967); M. Veltman: Proc. Roy. Soc. **A301**, 107 (1967)
10. H. Lehmann, K. Symanzik, W. Zimmerman: Nouv. Cim. **1**, 205 (1951)
11. J. Schwinger: Phys. Rev. **82**, 664 (1951)
12. R. Jackiw: in *Lectures on Current Algebra and its Applications,* S.B. Treimann, R. Jackiw, D.J. Gross (Princeton University press 1972)
13. K. Fujikawa: Phys. Rev. Lett. **42**, 1195 (1979)
14. H. Leutwyler: Helvetica Physica Acta **59**, 201 (1986); reprinted in *The Fundamental Interaction,* eds. J. Debrus, A.C. Hirshfeld (Plenum New York, London 1988)
15. S. Adler: Phys. Rev. **177**, 2426 (1969)
16. J.S. Bell, R. Jackiw: Nuov. Cim. **60**, 47 (1969)
17. H. Georgi: *Weak Interactions and Modern Particle Theory* (Benjamin, Reading Massachusetts 1984)
18. C. Bouchiat, J. Iliopoulos, P. Meyer: Phys. Lett. **B38**, 519 (1972)
19. R. Jackiw, R. Rajaraman: Phys. Rev. Lett. **54** 1219 (1985)
20. N.K. Falck: in *The Fundamental Interaction,* eds. J. Debrus, A.C. Hirshfeld (Plenum, New York, London 1988)
21. G. 't Hooft: Phys. Rev. **D14**, 3432 (1976)
22. G. 't Hooft: in *Recent Developments in Gauge Theories,* eds. G. 't Hooft et. al. (Plenum, New York, London 1980)
23. M. Reuter: in *The Fundamental Interaction,* eds. J. Debrus, A.C. Hirshfeld (Plenum, New York, London 1988)
24. J. Wess, B. Zumino: Phys. Lett. **B37**, 95 (1971)
25. L. Bonora, P. Cotta-Ramusino: Comm. Math. Phys. **87**, 589 (1983)
26. J. Wess: Lectures at the 5th Adriatic Meeting on Particle Physics, Dubrovnik, Yugoslavia (1986)
27. C. Becchi, A. Rouet, R. Stora: Ann. Phys. **98**, 287 (1976)
28. L.D. Faddeev, V.N. Popov: Phys. Lett. **B25**, 29 (1967)
29. W. Greub, S. Halperin, R. Vanstone: *Connections, Curvature and Cohomology,* (Academic Press, New York 1973)
30. R. Stora: in *New Developments in Quantum Field Theory and Statistical Mechanics* eds. H. Levy, P. Mitter (Plenum, New York, London 1977)
31. A.C. Hirshfeld, H. Leschke: Phys. Lett. **101B**, 48 (1981)
32. W.A. Bardeen: Phys. Rev. **184** 1848 (1969)
33. B. Zumino, Y.-S. Wu, A. Zee: Nucl. Phys. **B239** 477 (1984); B. Zumino: in *Relativity, Groups and Topology II,* ed. R. Stora (North Holland, Amsterdam 1984)
34. L.D. Faddeev: Phys. Lett. **B145**, 81 (1984)
35. M. Abud, J.-P. Ader, F. Gieres: CERN preprint TH.5576 (1989)
36. J. Schwinger: Phys. Rev. Lett. **3**, 296 (1959)
37. J. Mickelsson: *Current Algebras* (Plenum, New York, London 1990)
38. T. Nishikawa, I. Tsutsui: Nucl. Phys. **B308**, 544 (1988)
39. M. Dubois-Viollette: in *Fields and Geometry,* ed. A. Jadczyk (World Scientific, Singapore 1986)

The Rôle of Stratification in Anomalies

N.A. Papadopoulos

Institut für Physik der Johannes-Gutenburg-Universität,
Staudinger Weg 7, Postfach 3980, D(W)-6500 Mainz,
Federal Republic of Germany

Abstract

The relation between symmetries and anomalies is discussed, with particular emphasis on the rôle of the group action. The action of the full gauge group on the space of gauge potentials is investigated in connection with chiral anomalies in gauge theories. This allows the treatment of all gauge anomalies within the framework of the Atiyah-Singer Index Theorems.

1. Introduction

Anomalies are closely connected with symmetries. They were found in certain diagrams in perturbation theory [1]. Although this is still the most useful method to detect them [2], in some cases at least, they are characterized by universal properties which are deeply related to the structure of gauge field theories and their quantization. New insights have emerged, mainly in the last decade, which reveal algebraic and geometric structures of which no one could have dreamed in the sixties [3 – 6].

Anomalies are connected with a change in realization of a symmetry as one goes from the classical to the quantum level. Here we shall restrict ourselves to chiral anomalies within a gauge theory. We shall take the G-theoretical point of view [7 – 9], which focuses attention on the group action aspects. The relevant group action is that of the gauge group $\mathcal{G}$. Since we are looking for anomalies stemming from the full group $\mathcal{G}$, it is understood that we have to go beyond perturbation theory methods. This is possible within the framework of the Atiyah and Singer Index Theorems [10]. They have first been applied by Atiyah and Singer to detect anomalies stemming from a subgroup $\mathcal{G}^*$, the freely acting part of $\mathcal{G}$ on the space of gauge potentials $\mathcal{A}$. For a non-abelian gauge theory this leads to the topological anomalies which appear as a twist in the determinant bundle.

Given the fact that not all anomalies are detected as topological anomalies, we may suspect that the non-freely acting part of $\mathcal{G}$, $\mathcal{G}/\mathcal{G}^*$, may also play an important rôle. In addition, we may ask if it is possible to determine all anomalies in an unified approach based on the use of Atiyah-Singer Index Theorems. This is indeed possible, as we shall discuss below, within the G-theoretical point of view [7], taking into account the group action of the full group $\mathcal{G}$ on the space of connections $\mathcal{A}$ [10]. This leads, as we shall see, to a stratification on $\mathcal{A}$.

It is remarkable that the above unified approach to all gauge anomalies has allowed to show recently that the standard model in elementary particle physics is indeed anomaly free with respect to the full gauge group [11], and not only with respect to the Lie algebra of $\mathcal{G}$, as is usually done in perturbation theory.

In the following, we shall first discuss some well-known facts about anomalies, with particular emphasis on the infinitesimal group action aspects which are represented by the Noether Theorem (Section 2). The freely acting part $\mathcal{G}^*$ of the gauge group is discussed in Section 3. In the case of a non-abelian Yang-Mills theory it leads, via the Atiyah-Singer approach, to the topological anomalies, detected as a twist in the determinant bundle. The action of the subgroup $\mathcal{G}^*$ on the space of connections $\mathcal{A}$ is simple in the sense that it leads to only one sort of orbit. The treatment of the full gauge group $\mathcal{G}$ is in general much more complicated, since the corresponding actions on $\mathcal{A}$ lead to various sorts of orbits. So in order to be able to discuss the anomalies stemming from $\mathcal{G}$, we first have to deal with the stratification of $\mathcal{G}$ on $\mathcal{A}$ (Section 4). The treatment of all chiral anomalies arising from the action of the full gauge group is discussed in Section 5, followed by some conclusions.

2. Noether Theorem and Anomalies

In the conventional approach to chiral anomalies in QED, the appearance of a γ_5-coupling in the triangle diagram leads to an unexpected anomalous non-conservation of the corresponding axial current. This has serious implications for the quantized theory, since it destroys its renormalizability when chiral fermions are present. The same thing happens, of course, not only in QED but in every gauge theory. It may be useful at this stage to review briefly what is really unexpected about these facts.

It is the Noether Theorem which leads to our initial expectations. The Noether Theorem in its general form yields the connection between an infinitesimal group action on the Lagrangian of the theory and the divergence of the corresponding current

$$\delta_\xi \mathcal{L} = \partial_\mu J^\mu(\xi). \tag{2.1}$$

Spacetime transformations are here omitted for simplicity. ξ represents a Lie algebra element of a compact Lie group G of internal symmetry transformations and $J^\mu(\xi)$ the corresponding current.

It is important to realize that the Noether Theorem is in the first place just a statement about the infinitesimal group action, and in the case of an invariant Lagrangian, ($\delta_\xi \mathcal{L} = 0$) leads to the usual form of current conservation, $\partial_\mu J^\mu(\xi) = 0$.

Up to now the discussion concerns the classical theory. Proceeding to the quantized level, we know that the theory has to be regularized, and counter terms added to the classical Lagrangian $\mathcal{L}$. So we may think that the quantized theory corresponds to an effective Lagrangian $\mathcal{L}^{\text{eff}}$. Now the question is: starting with an invariant Lagrangian $\mathcal{L}$ with $\delta_\xi \mathcal{L} = 0$, what is the corresponding action on the effective Lagrangian $\mathcal{L}^{\text{eff}}$?

If $\delta_\xi \mathcal{L}^{\text{eff}} \neq 0$, the quantized theory no longer has the symmetry of the classical theory, and we get anomalies. The unexpected fact is the transition from an invariant Lagrangian $\mathcal{L}$ to a noninvariant Lagrangian $\mathcal{L}^{\text{eff}}$. As we see from the above considerations, the question of anomalies, by use of the Noether theorem, is a question about infinitesimal group actions on the theory. In what follows we shall extend this point of view to finite transformations. In this sense, what we are looking for is nothing but a generalization of the Noether Theorem. Since we shall deal with chiral anomalies in a gauge theory, the group action we are interested in is that of the gauge transformations.

3. Group Action and Anomalies

3.1 Gauge Theory and Gauge Group

In this subsection we are going to use almost all the notions we have learned in the lectures on geometrical foundations (see also [12]). Our starting point is a gauge theory with a compact Lie group G as structure group. In addition to the gauge fields we shall assume the existence of fermionic fields on spacetime M. The most natural description for the gauge fields, as we have seen in the above mentioned lectures on geometry, is as a connection A on a principal G-bundle $P = P(M, G)$. The principal bundle P brings some advantages for the fermionic fields as well. The fermionic fields may be described as vector valued functions on P with the property of equivariance: $\psi(pg) = g^{-1}\psi(p)$, with $p \in P$ and $g \in G$. This is much simpler than the alternative description in terms of sections in an associated vector bundle. We do not have to assume that P is a trivial bundle ($P = M \times G$). For purposes of integration we may think of the spacetime M as a compact space.

We shall denote the space of connections by $\mathcal{A} = \{A\}$ and the space of fermionic fields by $H = \{\psi\}$. The gauge transformations $\mathcal{G} = \{U\}$ act on these fields in the usual fashion:

$$\begin{aligned}(\mathcal{A} \times H) \times \mathcal{G} &\to \mathcal{A} \times H \\ (A, \psi)U &\mapsto (AU, U^{-1}\psi).\end{aligned} \tag{3.1}$$

In local coordinates the gauge transformation is given as

$$\begin{aligned}AU &:= \tilde{U}^{-1}A\tilde{U} + \tilde{U}^{-1}d\tilde{U} \\ (U\psi)(x) &:= \tilde{U}(x)\psi(x) \quad \text{with } x \in M.\end{aligned} \tag{3.2}$$

Since the action of the gauge group $\mathcal{G}$ is particularly important for our considerations, we are going to discuss it in some detail. One of the advantages of using the principal bundle P is that the gauge transformations can be expressed as point transformations on P. This is a great advantage, since transformations on the finite-dimensional space P are much simpler than transformations on the infinite-dimensional space $\mathcal{A} \times H$.

The gauge group $\mathcal{G} = Aut_{\text{vert}}P$ is the set of vertical automorphisms of P. These are transformations on P with $U(pg) = U(p)g$.

The gauge transformations can also be characterized as the *equivariant* G-valued functions on P, $C_{\text{eq}}(P, G) = \{\bar{U}\}$, i.e., functions $\bar{U} : P \to G$ with $\bar{U}(pg) = g^{-1}\bar{U}(p)g$ for $g \in G$. For a trivial principal bundle $P = M \times G$ these functions are equivalent to functions $\tilde{U} : M \to G$; these are the usual gauge transformations in particle physics phenomenology. We have the isomorphism

$$\mathcal{G} \equiv C_{\text{eq}}(P, G). \tag{3.3}$$

The gauge group $\mathcal{G}$ acts in two levels: on the space P

$$\begin{aligned}P \times \mathcal{G} &\to P \\ (p, U) &\mapsto U(p),\end{aligned} \tag{3.4}$$

and on the space of connections $\mathcal{A} = \{A\}$ by the induced action

$$\begin{aligned}\mathcal{A} \times \mathcal{G} &\to \mathcal{A} \\ (A, U) &\mapsto U^*A.\end{aligned} \tag{3.5}$$

$\mathcal{G}$ acts similarly on the fermionic fields.

The classical action, corresponding to a Lagrangian $\mathcal{L}$, is given by

$$S(A, \psi, \bar{\psi}) = \int d^4x \, \mathcal{L}(A, \psi, \bar{\psi}). \tag{3.6}$$

The action S is invariant with respect to gauge transformations, we have

$$S(U^*A, U^*\psi, \overline{U^*\psi}) = S(A, \psi, \bar{\psi}). \tag{3.7}$$

The notations U^*A, $U^*\psi$ denote the induced action on fields on P. This is given explicitly in Eq. (3.2).

Proceeding to the quantum theory, we consider the generating functional Z, where the fermionic degrees of freedom have been integrated out:

$$Z(A) = \int \mathcal{D}\psi\mathcal{D}\bar{\psi}\ e^{iS(A,\psi,\bar{\psi})}. \tag{3.8}$$

We assume that the theory has been properly regularized, i.e., that some definite regularization procedure has been chosen. It may be useful in this connection to deal with an effective action $\Gamma(A)$, which can be obtained from $-i\log Z$ by means of a Legendre transformation. We may think of it as corresponding to an effective regularized Lagrangian $\mathcal{L}^{\text{eff}}$.

Anomalies may appear if the above regulularized theory is no longer gauge invariant, that is if $Z(AU) \neq Z(A)$.

3.2 Topological Anomalies

Since we are interested in the action of the gauge group $\mathcal{G}$ on $Z[A]$ (or equivalently on $\Gamma[A]$), we have first to understand the action of $\mathcal{G}$ on the space of connections $\mathcal{A}$. This is in general a nontrivial problem.

As a first step we consider a subgroup of $\mathcal{G}$, the *pointed group* $\mathcal{G}^*$. $\mathcal{G}^*$ is the stability group of a point $p \in P$:

$$\mathcal{G}^* = \mathcal{G}_{p_0} := \{U \in \mathcal{G}/U(p_0) = p_0\}. \tag{3.9}$$

$\mathcal{G}^*$ is a normal subgroup of $\mathcal{G}$, and we have the exact sequence

$$1 \quad \rightarrow \quad \mathcal{G}^* \quad \rightarrow \quad \mathcal{G} \quad \rightarrow \quad \mathcal{G}/\mathcal{G}^* \quad \rightarrow \quad 1 \tag{3.10}$$

with $\mathcal{G}/\mathcal{G}^* \cong G$.

At the level of fields $\{A\} = \mathcal{A}$, the subgroup $\mathcal{G}^*$ has the important property that it acts freely on $\mathcal{A}$. So we may consider $\mathcal{A}$ as a $\mathcal{G}^*$-principal bundle:

$$\mathcal{G}^* \quad \rightarrow \quad \mathcal{A} \quad \rightarrow \quad \mathcal{M}^* \quad \text{with } \mathcal{M}^* = \mathcal{A}/\mathcal{G}^*. \tag{3.11}$$

This procedure guarantees that the space of "gauge equivalent" ($\mathcal{G}^*$-equivalent) connections $\mathcal{M}^*$ is a smooth manifold. As we shall see below, this is not true in general for the space of the (really) gauge equivalent ($\mathcal{G}$ -equivalent) connections $\mathcal{A}/\mathcal{G}$.

Concerning the anomalies, the observation of Atiyah and Singer was the following [4]: In the background quantization procedure the functional integral over the fermionic degrees of freedom leads to a specific $\mathbb{C}$-line bundle over $\mathcal{M}^*$, which is called, for reasons we shall discuss below, the *determinant bundle*

$$DET^* = \mathcal{M}^* \tilde{\times} \mathbb{C}. \tag{3.12}$$

The symbol $\tilde{\times}$ indicates that the bundle DET^* may be twisted. The assertion of Atiyah and Singer is that if this determinant bundle really turns out to be twisted, we find an anomaly. We may call this anomaly a *topological anomaly* (sometimes it is called a nonabelian gauge anomaly since in the generic cases it can occur with a nonabelian structure group).

First we would like to give a rough idea of how the determinant bundle comes in. We work here in the Euclidean regime and consider the spacetime M as a compact Riemannian manifold. Starting with the generating functional we get a first hint from the Gaussian integration:

$$Z(A) = \int \mathcal{D}\psi\mathcal{D}\bar{\psi}\; e^{-S(A)}\; e^{-\int \bar{\psi}\not{\partial}_A \psi} = e^{-S(A)} \det \not{\partial}_A, \tag{3.13}$$

where $\not{\partial}_A$ is the Weyl-Dirac operator. A more detailed derivation involves going through the details of the Berezin integration, and separating out the zero modes of the Weyl-Dirac operator [11]. This leads to

$$\int \mathcal{D}\psi\mathcal{D}\bar{\psi}\; e^{-\bar{\psi}\not{\partial}_A \psi} \sim \prod \lambda_k \int d\bar{\psi}_1^0 \dots d\bar{\psi}_i^0 d\psi_1^0 \dots d\psi_j^0, \tag{3.14}$$

where λ_k are the nonzero eigenvalues, $\bar{\psi}_i^0$ are the basis elements of the kernel of $\not{\partial}_A$ and ψ_j^0 are the basis elements of the cokernel of $\not{\partial}_A$. In this way we obtain the index bundle and the determinat bundle, and the connecion between chiral anomalies and the Atiyah-Singer index theorem can be fixed.

In the case of a twisted determinant bundle we detect a gauge anomaly, since we cannot perform the functional integral over the space of gauge inequivalent connections.

A different way to present these results is to take the G-theory (or equivariant) point of view. As we shall see, in the G-theoretical approach anomalies are connected less with the breaking of a symmetry than with a change in the realization of the symmetry when we switch from the classical to the quantum mechanical formulation.

Starting with a classical action $S(A,\psi,\bar{\psi})$ which is *strictly* invariant under gauge transformations

$$S(Ag,\psi g,\bar{\psi} g) = S(A,\psi,\bar{\psi}) \quad \text{with } g \in \mathcal{G}, \tag{3.15}$$

in quantum theory we consider the generating functional

$$Z : \mathcal{A} \to \mathbb{C} \tag{3.16}$$

given by

$$Z(A) = \int \mathcal{D}\psi \mathcal{D}\bar{\psi}\; e^{-S(A)-\bar{\psi}\not{\partial}_A \psi}. \tag{3.17}$$

There are two possibilities for the symmetry property of Z. With $g \in \mathcal{G}$ we may have

(i) $Z(Ag) = Z(A)$, Z is strictly invariant, or
(ii) $Z(Ag) = \rho(A,g)^{-1} Z(A)$,
Z is equivariant or invariant (not strictly invariant).

Here ρ is an A-dependent "realization" on $\mathbb{C}$. The adjective "invariant" for Z is also justified in case (ii), since Z and its transform have the same graph.

So Z remains invariant after quantization. In case (i), in which Z is strictly $\mathcal{G}$-invariant, no gauge anomalies are present. In case (ii), in which Z is invariant but not *strictly* invariant, gauge anomalies are present. We may summarize this by saying that the theory contains anomalies if the passage from the classical to the quantum level is accompanied by a change in the realization of the symmetry.

This phrasing is very general, including both gauge and chiral anomalies. Since these anomalies are connected with the equivariance properties of the theory they may be called *equivariant anomalies.* Such anomalies are not necessarily topological, but include the topological anomalies. A prominent example of an equivariant anomaly which is not topological is the $U(1)$-anomaly.

Within the G-theory framework gauge anomalies appear in connection with the reduction procedure for the generating functional $Z(A)$. For this purpose we consider Z as a trivial section (in the space where its graph is living) in the determinant bundle $\mathcal{A} \times \mathbb{C}$:

$$Z : \mathcal{A} \to \mathcal{A} \times \mathbb{C} =: DET. \tag{3.18}$$

In order to get the connection with the work of Atiyah and Singer [4, 13], we consider the reduction relative to the pointed gauge group $\mathcal{G}^*$. Dividing $\mathcal{A}$ by $\mathcal{G}^*$ we obtain the reduced section Z^*:

$$Z^* : \mathcal{A}/\mathcal{G}^* \to \mathcal{A} \times \mathbb{C}/\mathcal{G}^* =: DET^*. \tag{3.19}$$

As discussed before, we have the two possibilities:

(i) Z is strictly invariant. Then $\mathcal{G}^*$ acts trivially on $\mathbb{C}$, and we obtain

$$\mathcal{A} \times \mathbb{C}/\mathcal{G}^* = \mathcal{A}/\mathcal{G}^* \times \mathbb{C}/\mathcal{G}^* = \mathcal{A}/\mathcal{G}^* \times \mathbb{C} = \mathcal{M}^* \times \mathbb{C}, \tag{3.20}$$

so that

$$Z^* : \mathcal{M}^* \to \mathcal{M}^* \times \mathbb{C}.$$

Since DET^* is trivial, the theory can be reduced on $\mathcal{M}$.

(ii) Z is equivariant but not strictly invariant. Then $\mathcal{G}^*$ *acts nontrivially* on $\mathbb{C}$ and we have

$$Z^* : \mathcal{M}^* \to \mathcal{M}^* \tilde{\times} \mathbb{C} = DET^*. \tag{3.21}$$

In this case DET^* is twisted and we cannot perform the functional integration $\int \mathcal{D}A\ Z(A)$. This is the topological anomaly ($\mathcal{G}^*$-anomaly). The reduction of the (trivial) section to a (trivial) section is not possible.

In this way we obtain a new interpretation of gauge anomalies as obstructions from the quantization to the reduction procedure. This G-theoretical point of view is more general than the original one introduced by Atiyah and Singer since it allows the treatment of all $\mathcal{G}$-anomalies.

4. Stratification on the Space of Connections

As we have seen in the previous section, the question of anomalies stemming from the $\mathcal{G}^*$-action may be treated in a very satisfactory geometrical way. But $\mathcal{G}^*$ is only one part of the gauge group, and even if there is only "very little" left of the full gauge group, i.e. the part $\mathcal{G}/\mathcal{G}^* \cong G$ isomorphic to the structure group, this part is physically very relevant, since important phenomena like the $U(1)$-anomaly correspond precisely to this part.

In order to go beyond the $\mathcal{G}^*$-subgroup, we have to consider the action of the full group $\mathcal{G}$ on the space of connections $\mathcal{A}$. We have already remarked the fact that the action of $\mathcal{G}$ on $\mathcal{A}$ is in general more complicated than the action of $\mathcal{G}^*$ on $\mathcal{A}$. It leads to the *stratification* of $\mathcal{A}$. This means that the space $\mathcal{A}$ is divided into disjoint parts, the *strata*. Every stratum itself contains again disjoint $\mathcal{G}$-orbits of a given fixed type. This type corresponds to the conjugacy class of the stability group and characterizes the elements of a stratum. So the stratum may be considered as a fibre bundle, the typical fibre being the $\mathcal{G}$-orbits of the particular fixed type, and we may call it the *orbit bundle*. Particular orbit bundles are well-known: a principal bundle $P(M, G)$ is a special orbit bundle with the group G as typical fibre. The same is true for the infinite dimensional principal bundle $\mathcal{A}(\mathcal{A}/\mathcal{G}^*, \mathcal{G}^*)$, with the subgroup $\mathcal{G}$ as a typical fibre.

There are two questions which occur naturally in connection with stratification: What is the general structure of an orbit bundle and how can we find the set of strata which belongs to a certain group action (the stratification problem). The first question can be treated in a quite general way and the answer is in a sense universal, since a generic orbit bundle is independent of the specific action. We shall therefore explain shortly its structure for the finite dimensional case. The second question (the stratification) is a serious problem, but for the case we are interested in in connection with anomalies, the $\mathcal{G}$-action on $\mathcal{A}$, it was essentially solved, in a sense we shall discuss below [9].

We now turn to the first question, the structure of an orbit bundle. We consider the action of a compact Lie group on a (connected) manifold X,

$$X \times G \to X. \tag{4.1}$$

Every point $x \in X$ is characterized by a subgroup of G, its stability group J_x is given by

$$J_x := \{ g \in G \mid xg = x \}. \tag{4.2}$$

The space X is an orbit bundle if every stability group J_x is conjugate to a fixed subgroup J of G. This means that for every $x \in X$ there exists a $g \in G$ such that

$$J_x = g^{-1} J g. \tag{4.3}$$

The subgroup J (for $J := J_{x_0}$, $x_0 \in X$), or more precisely the conjugacy class (J) of J, characterizes the type of the orbit bundle. It follows that all orbits of the G-action are isomorphic to the space $G/J := \{ Jp \mid p \in G\}$. Furthermore, there exists a subspace X^J of X with elements given by

$$X^J := \{ x \in X \mid J_x = J\}. \tag{4.4}$$

These are the points of X with exactly the same stability group. It can be shown that X^J is a principal bundle with structure group $N(J)/J$, with $N(J)$ the normalizer of J in G,

$$N(J) := \{ n \in G \mid nJ = Jn\} \tag{4.5}$$

and that the space $X \equiv X^{(J)}$ is a fibre bundle associated to the X^J fibre bundles with typical fibres given by the orbit G/J. So we have [14]:

$$X^{(J)} = X^J \times_{N(J)/J} G/J. \tag{4.6}$$

This explains the name of X as an orbit bundle of the G-action. Denoting the orbit space of $X^{(J)}$ by M_J ($M_J = X^{(J)}/G$), we may represent the orbit bundle $X^{(J)}$ also in the form

$$X^{(J)} = M_J \tilde{\times} G/J, \tag{4.7}$$

where the symbol $\tilde{\times}$ indicates that the fibre bundle $X^{(J)}$ is nontrivial.

We now turn to our actual problem. A stratum of the $\mathcal{G}$-action on $\mathcal{A}$ of the orbit type (J) is given, in a notation similar to that used in Eq. (4.6), by

$$\mathcal{A}^{(J)} = \mathcal{A}^J \times_{N(J)/J} \mathcal{G}/J. \tag{4.8}$$

This needs some further explanation. J is the stability group of the $\mathcal{G}$-action on $\mathcal{A}$, corresponding to a given element $A_0 \in \mathcal{A}$ ($J = J_{A_0}$). But what is the physical meaning of the stability group J in the present case? We note that the elements of $\mathcal{A}$ are not simply "points", as was the case in the previous example involving $X^{(J)}$. Here the elements A of $\mathcal{A}$ are gauge potentials with "inner structure". In order to clarify the situation, we consider a gauge transformation $j \in \mathcal{G}$ which leaves A_0 invariant (so that this j is a symmetry of A_0):

$$A_0 = j^{-1} A_0 j + j^{-1} dj. \tag{4.9}$$

From this equation we may expect that A_0 remains invariant if $j^{-1}dj = 0$ and $jA_0j^{-1} = A_0$ is valid. This gives two conditions for j: the analytic says that j should be essentially constant, the algebraic one that the Ad action of j on the A_0 values should be trivial. The last condition means that if A_0 takes values in the full Lie G, then $j(p_0)$[1] is an element of the centralizer of G (see below), which in this case is the center $C(G)$ of G. In general we expect that a given $A_0 \in \mathcal{A}$ must be a reducible connection [9, 15], which means that A_0 takes its values not in the full Lie G, but that there is a subgroup H_{A_0} of G and A_0 takes its values in Lie H_{A_0}. In this case $j(p_0)$ belongs to the centralizer of H_{A_0}, $j(p_0) \in \beta(J) = Z(H_{A_0})$. The choice of $p_0 \in P$ allows the isomorphism β, $J \cong \beta(J)$ with $J < \mathcal{G}$ and $\beta(J) < G$. In addition, we write $H_{A_0} = H_{A_0}(p_0)$. The centralizer of H in G is given by

$$Z(H) = Z_G(H) := \{\, z \in G \mid zh = hz \quad \forall h \in H\}. \tag{4.10}$$

H_{A_0} is the holonomy group of A_0. The above considerations are nothing but a plausibility argument for the connection between the stability group $J = J_{A_0}$ of A_0 (which is the maximal symmetry group of A_0) and the holonomy group H_{A_0} of A_0 [15].

So we may state again the magic theorem

$$J_A = Z(H_A). \tag{4.11}$$

This theorem is not directly useful, since in our case we know only the symmetry group J_A, not its holonomy group H_A. Nevertheless, this theorem indicates a deep connection between the (gauge) symmetry and the reducibility of a given connection. It can be used to give further useful corollaries and to solve the problem of stratification [9].

In order to proceed, we first need to look more closely at some of the above considerations. The holonomy group $H_A(p_0)$ is connected with the holonomy bundle $Q_A(p_0)$, $(Q_A(p_0) = P(M, H_A(p_0)) \leq P(M,G))$, and the elements of J_A are constants on $Q_A(p_0)$. So we have

$$J_A = \{\, j \in \mathcal{G} \mid \bar{j}\Big|_{Q_A(p_0)} = \text{constant} = \bar{j}(p_0) \in \beta(J_A) < G\}. \tag{4.12}$$

An important observation is that every $j \in J_A$ is constant on a "maximal" principal bundle $Q_J(p_0)$ which contains all $Q_A(p_0)$ with $A \in \mathcal{A}^j$. So we have $Q_J(p_0) = P(M, H_J)$ and $Q_A(p_0) \leq Q_J(p_0) \leq P(M,G)$. The maximal group H_J is the maximal subgroup in G with $H_A(p_0) \leq H_J(p_0)$ and $J = Z(H_J(p_0))$.

The miracle now is that knowing J, we cannot know H_A, but we know the H_J, since it is given by $H_J = Z(J_A)$. This means that if we only know J_A, we cannot derive the holonomy bundle of A, but we can derive the maximal bundle

---
1 We take $p_0 \in P$ and make use of the various isomorphisms of the group $\mathcal{G}$, as discussed in Section (3.1).

Q_J. This maximal bundle, a subbundle of $P(M,G)$, completely characterizes the stratum $\mathcal{A}^{(J)}$ [9]. To be more precise, Q_J corresponds directly to $\mathcal{A}^J$, the standard principal bundle of the stratum $\mathcal{A}^{(J)}$: All elements of $\mathcal{A}^J$ have exactly the same maximal group H_J and exactly the same maximal bundle Q_J. We must also remark that the elements of $\mathcal{A}^J$ have in general different holonomy groups.

It can furthermore be shown that in the same sense $\mathcal{A}^{(J)}$ may be considered as a $\mathcal{G}$-orbit of $\mathcal{A}^J$, and we may define (Q_J) to be the $\mathcal{G}\times G$-orbit of Q_J. In this way we obtain a direct correspondence between $\mathcal{A}^{(J)}$ and (Q_J). This implies a relation between the strata $\mathcal{A}$ and maximal bundles in $P(M,G)$, so we have the bijection [9]:

$$\{\mathcal{A}^{(J)}\} \cong \{(Q_J)\}. \tag{4.13}$$

We have thus succeeded in reducing the problem of stratification on an infinite dimensional space $\mathcal{A}$ to an equivalent algebraic and topological problem on a finite dimensional space $P(M,G)$. The algebraic part is related to the fact that we have to deal with all subgroups J of G with $Z(J) = H_J$ (and $ZZ(J) = J$), and the topological part to the imbedding of Q_J in P. It is interesting to note that this corresponds directly to the right hand side of Eq. (4.9).

As an example we consider the trivial principal bundle $P = S^2 \times SU(2)$. The maximal subgroups are Z_2 and $U(1)$. The possible maximal subbundles are P itself and Q_n for $n \in \mathbb{N}$ with

$$Q_0 = S^2 \times \mathbb{Z}_2, \quad Q_1 = S^1 \times U(1), \quad Q_2 = S^3 \tag{4.14}$$

and $Q_n = S^3/\mathbb{Z}_{n-1}$ for $n \geq 3$. So the set of orbit types is given by the set of the above maximal subbundles in P,

$$\{(Q_J)\} = \{\, P \text{ and } (Q_n) \text{ with } n \in \mathbb{N}\}. \tag{4.15}$$

The stability groups J_n are given by

$$J_n := \{\, j/j : Q_n \longrightarrow Z_{SU(2)}(H_n), \quad \text{constant}\}, \tag{4.16}$$

where $H_0 = \mathbb{Z}_2$, $H_n = U(1)$ for $n \geq 1$ and

$$Z_{SU(2)}(H_0) = SU(2), \quad Z_{SU(2)}(H_n) = U(1). \tag{4.17}$$

In addition we have, corresponding to P, the stability group

$$Z_{SU(2)}(SU(2)) = C(SU(2)) \cong \mathbb{Z}_2, \tag{4.18}$$

which corresponds to the main stratum $\mathcal{A}^{\mathbb{Z}_2} = \bar{\mathcal{A}}$ (see below). So we obtain from Eq. (4.15) the stratification of $\mathcal{A}$,

$$\mathcal{A} = \mathcal{A}^{(\mathbb{Z}_2)} \bigcup_{n\in\mathbb{N}} \mathcal{A}^{(J_n)}. \tag{4.19}$$

Here we would like to summarize what we have obtained till now. The action of the gauge group $\mathcal{G}$ on the space of connections leads in general to a nontrivial stratification on $\mathcal{A}$, given by the disjoint sum of strata:

$$\mathcal{A} = \mathcal{A}^{(J_0)} \cup \mathcal{A}^{(J_1)} \cup \ldots \mathcal{A}^{(J_k)} \ldots \quad . \tag{4.20}$$

As we have already seen, each stratum has the structure of an orbit bundle. The number of strata is, as anticipated, countable. There exists one special stratum (denoted by $\bar{\mathcal{A}} = \mathcal{A}^{(J_0)}$) which is dense in $\mathcal{A}$ [17]. The gauge potentials which belong to the main stratum have the smallest possible symmetry. Their stability group is isomorphic to the center of G ($J_0 = C(G)$).

Since we are interested in the space of gauge-inequivalent connections, we have to consider the orbit space of $\mathcal{A}$ given by $\mathcal{M} = \mathcal{A}/\mathcal{G}$. The problem now is that in general $\mathcal{M}$ is not a manifold, since the various $\mathcal{M}_k = \mathcal{A}^{(J_k)}/\mathcal{G}$ do not fit smoothly together. It is remarkable that here in physics we meet a space (like $\mathcal{M}$), which is in general not a smooth space. This was also emphasized recently in [18] and is the result of the gauge symmetry. So as we have seen, the stratification has generally the following form:

$$\begin{aligned} \mathcal{A} &= \mathcal{A}^{(J_0)} \cup \mathcal{A}^{(J_1)} \cup \ldots \mathcal{A}^{(J_k)} \ldots \\ \mathcal{M} &= \mathcal{M}_0 \cup \mathcal{M}_1 \cup \ldots \mathcal{M}_k \cup \ldots \end{aligned} \tag{4.20}$$

There is no doubt that the stratification is given here by the physical situation and so constitutes a serious complication.

5. Stratification and Anomalies

We consider a gauge theory with a compact Lie group G as structure group and we restrict ourselves to the sector given by the principal bundle $P(M,G)$. The gauge transformations $\mathcal{G}$, as we have seen, contain a freely acting part $\mathcal{G}^*$ on the space of connections $\mathcal{A}$ and a nonfreely acting part $\mathcal{G}/\mathcal{G}^* \cong G$.

Taking the G-theoretical point of view [9], the possible anomalies in the theory may be classified according to certain subgroups of the $\mathcal{G}$-action on $\mathcal{A}$. We remember that the anomalies possess a certain universal character and do not depend on the details of the particular Lagrangian we start with. It is also useful to remember the connection of this point of view with the usual perturbation theory approach. In perturbation theory we can test only a part of the possible gauge anomalies, namely those corresponding to the part of the group $\mathcal{G}$ near the identity. If we wish to examine all possible gauge anomalies we have to go beyond perturbation theory. As shown in [11], the Atiyah-Singer Index Theorems [10, 14] are the appropriate instruments to deal with this problem.

In Section 3 we have seen that in the case of a nonabelian structure group G, the freely acting part $\mathcal{G}^*$ of the gauge group leads to the topological anomalies. If we want to go beyond the $\mathcal{G}^*$-subgroup we are faced with the problem of stratification. As we have seen in the previous section, the stratification leads to certain subgroups of $\mathcal{G}$, the stability groups J_k, which are isomorphic to the subgroups $\beta(J_k)$ of G.

Now we are in a position to classify all anomalies according to the G-theoretical point of view. In addition to the $\mathcal{G}^*$-anomalies (topological anomalies) we have anomalies generated by the nonfreely acting part $\mathcal{G}/\mathcal{G}^* \cong G$. These are the various stabilizer (J_k) anomalies and correspondingly the remaining anomalies coming from the remaining part $G - J_k$. It was shown in [8] that if no $\mathcal{G}^*$-anomalies and no stabilizer anomalies are present, no remaining anomalies are present either. It follows that from the remaining part no new anomalies may appear.

On the other hand, anomalies stemming from the stabilizers J_k are well known to exist. The center anomalies, corresponding to the stabilizer $C(G)$ from the main orbit bundle $\bar{\mathcal{A}}$, are a special case of stabilizer anomalies.

In the case of $SU(2)$ we have a global anomaly, the well known *Witten anomaly* [19]. It was also discussed from the G-theoretical point of view in [8,20]. Its detection from that point of view is very simple and the connection with the Atiyah-Singer Index Theorems is direct.

After these discussions it seems plausible, given a gauge theory, to ask not only for the anomalies which can be detected by perturbation theory, but also for anomalies which are related to the full gauge group. This is particularly relevant for the standard model in elementary particle physics, which in the usual sense is, of course, anomaly-free. As was shown in [11], it is also anomaly-free with respect to the full gauge group.

Acknowledgements

I would like to thank A. Heil, A. Kersch, B. Reifenhäuser and F. Scheck for many discussions. I would also like to thank J. Debrus and A. Hirshfeld for their kind hospitality in Bad Honnef.

References

1. J. Steinberger: Phys. Rev. **76**, 1180 (1949)
2. S.L. Adler: Phys. Rev. **177**, 2426 (1969);
   J.S. Bell and R. Jackiw: Nuovo Cimento **A60**, 47 (1969);
   W.A. Bardeen: Phys. Rev. **184**, 1848 (1969)
3. H. Römer: In *Differential Geometric Methods in Theoretical Physics*, ed. by H.D. Doebner, Lecture Notes in Physics **139** (Springer, Berlin, Heidelberg 1978);
   R. Stora: In *Non-Perturbative Methods*, ed. by S. Narrison (World Scientific, Singapore 1985);
   B. Zumino: In *Relativity, Groups and Topology II*, Les Houches (North Holland, Amsterdam 1983);
   H. Leutwyler: In *The Fundamental Interaction*, ed. by J. Debrus and A.C. Hirshfeld (Plenum Press, New York, London 1988)
4. M.F. Atiyah and I.M. Singer: Proc. Nat. Acad. Sci. USA **81**, 2597 (1984);
   I.M. Singer: Société Mathématique de France, Astérisque **323** (1985)
5. L.D. Faddeev and S.L. Shatashvili: Theor. Math. Phys. **60**, 770 (1985);
   K. Fujikawa: Phys. Rev. **D21**, 2848 (1980);
   A.Yu. Morozov: Sov. Phys. Usp. Zg. **11** (1986);
   J. Sidenius: preprint NORDITA-85/33;
   L. Alvarez-Gaume: preprint HUTP-85/A092
6. D. Kastler and R. Stora: preprint CPT-86 (1985);
   M. Dubois-Violette: preprint Orsay LPTHE-86-12;
   L. Bonora, P. Cotta-Ramusino, M. Rinaldi and J. Stasheff: Comm. Math. Phys. **112**, 237 (1987); Comm. Math. Phys. **114**, 381 (1988)
7. A. Heil, N.A. Papadopoulos, B. Reifenhäuser and F. Scheck: Nucl. Phys. **B293**, 445 (1987);
   A. Heil, A. Kersch, N.A. Papadopoulos, B. Reifenhäuser, F. Scheck and H. Vogel: Jour. Geom. Phys. **6.2**, 237 (1989);
   F. Scheck: "Geometrical Approaches to Particle Physics", Mainz preprint (1990)
8. A. Heil, A. Kersch, N.A. Papadopoulos, B. Reifenhäuser and F. Scheck: Ann. Phys. (New York) **206**, 200 (1990)
9. A. Heil, A. Kersch, N.A. Papadopoulos, B. Reifenhäuser and F. Scheck: "Structure of the space of reducible Yang-Mills potentials", Mainz preprint (1990)
10. M.F. Atiyah and I.M. Singer: Ann. Math. **87**, 484 (1968); Ann. Math. **87**, 546 (1968); Ann. Math. **92**, 139 (1970)
11. A. Kersch: "Anomalien im Rahmen des Indextheorems", Diss. Mainz (1990)
12. E. Binz, J. Sniatycki and H. Fischer: *Geometry of Classical Fields*, North Holland Mathematics Studies 154 (North Holland, Amsterdam 1988)
13. M.F. Atiyah: Scuola Normale Superiore, Pisa (1979); Lecture Notes in Physics **208** (Springer, Berlin, Heidelberg 1984)
14. G.E. Bredon: *Introduction to Compact Transformation Groups* (Academic Press, New York 1972);
    K. Jänich, Lecture Notes in Mathematics **59** (Springer, Berlin, Heidelberg 1968)

Field Theoretical Applications of the Index Theorem

– A Pedagogical Introduction

H. Römer

Physikalisches Institut der Universität
Hermann-Herder-Straße 3, D(W)-7800 Freiburg
Federal Republic of Germany

Abstract

The following topics are covered in these lectures:

Introduction to the index theorem, anomalies and index theorems, vector bundles and characteristic classes, special cases and important applications, the family index theorem, determinant bundles and gauge anomalies. An outlook to further applications concludes the article.

1. Introduction to the Index Theorem

For a given linear operator $D : V \to W$ between hermitian vector spaces the following subspaces may be defined:

$$\begin{aligned} \text{The } \textit{kernel } \ker D &:= \{v \in V \mid Dv = 0 \in W\}, \\ \text{the } \textit{image } \operatorname{im} D &:= DV := \{Dv \in W \mid v \in V\} \quad \text{and} \\ \text{the } \textit{cokernel } \operatorname{coker} D &:= W/\operatorname{im} D \cong (\operatorname{im} D)^{\perp} = \ker D^{*}. \end{aligned} \tag{1.1}$$

As a first example, consider a linear operator D which maps a m-dimensional space V into a n-dimensional space W. In this case:

$$\begin{aligned} \dim(\ker D) &= m - \dim(\operatorname{im} D) \quad \text{and} \\ \dim(\operatorname{coker} D) &= n - \dim(\operatorname{im} D). \end{aligned} \tag{1.2}$$

By subtraction we get a quantity called the *index* of the operator D, which in this very special case is obviously independent of D:

$$\operatorname{index} D := \dim(\ker D) - \dim(\operatorname{coker} D) = m - n. \tag{1.3}$$

A different situation arises when linear operators mediate between infinite-dimensional spaces. Consider the Hilbert space

$$\mathcal{H} = [|0\rangle, |1\rangle, |2\rangle, \dots] \tag{1.4}$$

which is spanned by a countable infinite basis $\{|j\rangle\}$. Let us concentrate on the surjective displacement operator S with the property

$$\begin{aligned} S|0\rangle = 0, \qquad S|j\rangle &= |j-1\rangle \quad \text{for } j > 0 \\ \text{and similarly } S^{\dagger}|j\rangle &= |j+1\rangle \quad \text{for } j \geq 0. \end{aligned} \tag{1.5}$$

Here we get:

$$\operatorname{index} S = 1 \quad \text{and} \quad \operatorname{index} S^{\dagger} = -1. \tag{1.6}$$

The examples just shown have the common property that the dimensions of kernel and cokernel are finite, so that the index can be calculated simply as the difference between them. Such operators are called *Fredholm operators.* A further example for a Fredholm operator is the operator $1 + K$, where K is compact. The index of this operator vanishes: $\operatorname{index}(1 + K) = 0$. The most important property of the index of a Fredholm operator F is the fact that it can be shown to be invariant under smooth deformations of F.

In the following we will restrict our attention to differential operators that mediate between complex vector bundles E and F, both of which have as base space a compact orientable manifold M without boundary. Let $v \in \Gamma(E)$ be a section in E. Then the action of D on v is locally

$$(Dv)_i = \sum_{|\alpha| \leq m} a_{ij}^{\alpha} \frac{\partial^{|\alpha|}}{\partial x^{\alpha}} v_j(x). \tag{1.7}$$

Here m is the *order* of D and α a multi-index, i.e. $\alpha = (\alpha_1, \alpha_2, \dots, \alpha_n)$ with $|\alpha| = \alpha_1 + \alpha_2 + \dots + \alpha_n$ and

$$\frac{\partial^{|\alpha|}}{\partial x^{\alpha}} = \left(\frac{\partial}{\partial x_1}\right)^{\alpha_1} \left(\frac{\partial}{\partial x_2}\right)^{\alpha_2} \cdots \left(\frac{\partial}{\partial x_n}\right)^{\alpha_n}. \tag{1.8}$$

The *leading symbol* σ_D is defined as

$$(\sigma_D(x, \xi))_{ij} := \sum_{|\alpha| = m} a_{ij}^{\alpha} \xi^{\alpha}. \tag{1.9}$$

The operator D is called *elliptic* if this leading symbol is invertible for all (x, ξ) with $\xi \neq 0$. An example for an elliptic differential operator is the *Laplace operator*

$$\Delta := \sum_{i=1}^{n} \frac{\partial^2}{\partial x_i^2} \quad \text{with } \sigma_{\Delta}(x, \xi) = \xi^2. \tag{1.10}$$

In contrast to this, the *d'Alembert operator*

$$\Box := \frac{\partial^2}{\partial t^2} - \Delta \quad \text{with } \sigma_{\Box}(x, \xi) = \xi_0^2 - \sum \xi_i^2 \tag{1.11}$$

is *not* elliptic, because the leading symbol vanishes for $\xi_0^2 = \sum \xi_i^2$.

Also in more general cases it is possible to express the index of D in terms of deformation invariant quantities. A fairly general form of the index theorem is

$$\operatorname{index} D = (-1)^{n(n-1)} \frac{\operatorname{ch} E - \operatorname{ch} F}{\mathrm{e}(TM)} \operatorname{td}(T_{\mathbb{C}}M)[M]. \tag{1.12}$$

Here ch E is the *Chern character* of E, $T_{\mathbb{C}}M$ the complexified tangential bundle of M and td$(T_{\mathbb{C}}M)$ the *Todd class*, while e(TM) is the Euler class and $[M]$ the evaluation on M, a kind of integration over M. These concepts will be explained in later sections. Here we give a list of cases in which index theorems are applicable.

1.1 The Index Theorem for Chain Complexes

The *chain complex*

$$0 \rightarrow \Gamma(E_0) \xrightarrow{D_0} \Gamma(E_1) \xrightarrow{D_1} \ldots \xrightarrow{D_{n-1}} \Gamma(E_n) \rightarrow 0 \tag{1.13}$$

is called *elliptic*, if $D_{i+1}D_i = 0$ and the sequence of the corresponding leading symbols is exact. The difference $\operatorname{ch} E - \operatorname{ch} F$ of Chern characters is here substituted by the alternating sum $\sum(-1)^i \operatorname{ch} E_i$.

1.2 The Index Theorem with Boundary

The index theorem can be generalized to manifolds M with nonvanishing boundary $\partial M \neq \emptyset$. In this case the index theorem takes the form

$$\operatorname{index} D = \alpha[M] + \beta[\partial M] + \xi[\partial M], \tag{1.14}$$

where $\alpha[M]$ is a closed n-form on M, the same as in the index theorem without boundary integrated over M. $\beta[\partial M]$ is a local boundary contribution, an $(n-1)$-form integrated over ∂M, and $\xi[\partial M]$ is a non-local boundary term, closely related to the spectral asymmetry $\eta = \lim_{s\to 0} \sum_\lambda \operatorname{sign}(\lambda)|\lambda|^{-s}$ of the restricted operator $D|_{\partial M}$, where λ denotes the eigenvalues of $D|_{\partial M}$.

1.3 The G-Index Theorem

If E and F are G-vector bundles, where G is a Lie group which acts on sections s in E, respectively F, according to

$$(gs)(m) := gs(g^{-1}m), \quad g \in G, \tag{1.15}$$

we have the *equivariant* or *G-index theorem.*

If D commutes with G, its kernel and cokernel can be regarded as representation spaces for G, and we can define a *g-index*

$$\mathrm{ind}_g\, D = \mathrm{tr}\, g|_{\ker D} - \mathrm{tr}\, g|_{\mathrm{coker}\, D}, \quad g \in G. \tag{1.16}$$

If M^g is the set of fixed points under the action of $g \in G$ and N^g the normal bundle in the neighbourhood of these fixed points, the index theorem takes the following form:

$$\mathrm{ind}_g\, D = \frac{i^*(\mathrm{ch}_g\, E - \mathrm{ch}_g\, F)\, \mathrm{td}(T_{\mathbb{C}} M^g)}{\mathrm{e}(TM^g)\, \mathrm{ch}_g(\Lambda_{-1} N^g \otimes \mathbb{C})}[M^g]. \tag{1.17}$$

1.4 The Family Index Theorem

Finally, if $(D_y)_{y\in Y}$ is a family of differential operators we have the *family index theorem*, which will be explained in a later section.

2. Anomalies and Index Theorems

In this section we discuss some physical applications of these seemingly abstract relations.

2.1 The Heat Kernel Method

Let Δ be a nonnegative elliptic operator (e.g. DD^* or D^*D for an elliptic operator D) with eigenvalues λ and eigenfunctions ψ_λ, i.e. $\Delta\psi_\lambda = \lambda\psi_\lambda$. For a complete set of these functions we have:

$$\begin{aligned}(h\psi)(x) := (e^{-t\Delta}\psi)(x) &= \sum_\lambda a_\lambda (e^{-t\Delta}\psi_\lambda)(x) = \sum_\lambda a_\lambda e^{-t\lambda}\psi_\lambda(x) \\ &= \sum_\lambda e^{-t\lambda}\psi_\lambda(x) \int \psi^*_\lambda(y)\psi(y)\, d^4y \\ &= \int \sum_\lambda e^{-t\lambda}\psi_\lambda(x)\psi^*_\lambda(y)\psi(y)\, d^4y =: \int h(t,x,y)\psi(y)\, d^4y.\end{aligned} \tag{2.1}$$

The integral kernel $h(t,x,y)$ is called the *heat kernel* and satisfies the equations

$$\left(\frac{\partial}{\partial t} + \Delta\right) h(t,x,y) = 0 \quad \text{and} \quad h(0,x,y) = \delta(x-y), \tag{2.2}$$

which is a heat equation with singular starting function.

For $t > 0$ we can expand $h(t,x,x)$ asymptotically to get

$$h(t,x,x) \longrightarrow \sum_{r \geq -n/2} t^r \mu_r(x) \quad \text{for } t \longrightarrow +0. \tag{2.3}$$

The *Seeley algorithm* allows us to calculate the coefficients μ_r recursively as polynomials in the coefficients of Δ and their derivatives with respect to x. We define the trace of the heat kernel as

$$h(t) := \operatorname{tr} \int h(t,x,x)\, d^4x = \sum_\lambda e^{-\lambda t} \longrightarrow \sum_r t^r \operatorname{tr} \int \mu_r(x)\, d^4x. \tag{2.4}$$

The quantities $\operatorname{tr} \int \mu_r(x)\, d^4x$ have geometrical meaning, for example $r = -1$ gives the surface, $r = -1/2$ the circumference and $r = 0$ the Euler characteristic of the base manifold.

2.2 The Connection with the Index

There is an interesting connection between the index of a differential operator and the heat kernel of its square. Suppose $D : \Gamma(E) \to \Gamma(F)$ is a differential operator which maps sections of the vector bundle E into sections of the vector bundle F. We can construct two positive definite differential operators

$$\begin{aligned} \Delta_E &:= D^*D : \Gamma(E) \to \Gamma(E), \\ \Delta_F &:= DD^* : \Gamma(F) \to \Gamma(F). \end{aligned} \tag{2.5}$$

We shall now show that Δ_E and Δ_F have the same spectrum, and the eigenvalues have the same multiplicities, except for the zero modes. Suppose that $\lambda \neq 0$ and

$$\Delta_E \psi_\lambda = D^* D \psi_\lambda = \lambda \psi_\lambda. \tag{2.6}$$

Then we get

$$\Delta_F D \psi_\lambda = D \Delta_E \psi_\lambda = \lambda D \psi_\lambda. \tag{2.7}$$

So if ψ_λ is eigenfunction of Δ_E, $D\psi_\lambda$ is an eigenfunction of Δ_F with the same eigenvalue λ, or else $D\psi_\lambda = 0$. But this contradicts the assumption that λ be nonvanishing, because if ψ_λ is an element of the kernel of D, it is also an element of the kernel of Δ_E, and so $\lambda = 0$. Now the difference

$$\begin{aligned} h_E(t) - h_F(t) &= \sum_{\lambda_E} e^{-\lambda_E t} - \sum_{\lambda_F} e^{-\lambda_F t} = \sum_{\lambda_E = 0} 1 - \sum_{\lambda_F = 0} 1 = \\ &= \dim\ker \Delta_E - \dim\ker \Delta_F = \\ &= \dim\ker D - \dim\ker D^* = \dim\ker D - \dim\operatorname{coker} D = \\ &= \operatorname{index} D \end{aligned} \tag{2.8}$$

is independent of t, and so the left-hand-side gets contributions only from the term of the asymptotic series with $r = 0$. So we finally find

$$\operatorname{index} D = \operatorname{tr} \int (\mu_0^E(x) - \mu_0^F(x))\, d^4x. \tag{2.9}$$

2.3 The Effective Action

Let D be a kinetic operator, e.g. the Dirac operator or the Laplace operator. Then we can calculate the field-theoretical generating function $\mathcal{Z}$ by the path integral quantization method as

$$\begin{aligned} \mathcal{Z} = \exp(i\Gamma_{\text{eff}}) &= \int [d\psi] \exp\{(\frac{1}{\hbar})(-S_E[\psi] + \int K\psi \, d^4x)\} \\ &= \int [d\psi] \exp\{-\frac{1}{\hbar}\int(\psi D\psi - K\psi)\, d^4x\} \\ &\sim (\det D)^{\mp 1/2} \end{aligned} \tag{2.10}$$

("−" for bosons, "+" for fermions), where we get the action S_E by Wick rotation. Thus the effective action may be written as

$$\Gamma_{\text{eff}} = \frac{\mp 1}{2} \ln(\det D). \tag{2.11}$$

If this effective action changes under symmetry transformations of S_E we say that the theory has an *anomaly*. We distinguish between two kinds of anomalies:

1. The *generalized axial anomalies*, which appear in the context of chiral symmetry transformations $\psi \mapsto e^{i\alpha\gamma_5}\psi$.
2. The *gauge anomalies* for the Dirac operator $\not{D} = \gamma(\partial + A)$.
   They violate the Ward identities of gauge invariance and are thus fatal for the theory.
   These anomalies are treated by two different kinds of index theorems: *Local gauge anomalies* arise from gauge transformations which are continuously connected to the identity. They use the family index theorem. *Global gauge anomalies* are treated by the index theorem for operators acting on sections of a vector bundle whose base is a manifold with boundary.

2.4 Zeta Function Regularisation

We now concentrate on generalized axial anomalies and examine a specific example in order to illustrate how such calculations are actually done. Consider therefore the Lagrangian

$$\mathcal{L}(x) = \bar{\psi}(x)\, \not{D}\, \psi(x). \tag{2.12}$$

Here the operator $\not{D} := \Gamma_i \nabla_i$ acts on sections in a vector bundle E. They are subject to a gauge transformation $\psi \mapsto e^{i\alpha\Gamma}\psi$, with $\Gamma^2 = 1$ and $[\Gamma_i, \Gamma]_+ = 0$. The invariance of $\mathcal{L}$ under this transformation leads to the *classical Noether current*

$$J_i(x) = \bar{\psi}(x)\Gamma_i\Gamma\psi(x). \tag{2.13}$$

The generating functional of the theory involves the gauge potential $A_i(x)$:

$$\begin{aligned} \mathcal{Z} &= \int [d\bar{\psi} d\psi] \exp\{(\frac{1}{\hbar})(-S_E[\psi] + \textstyle\int A_i(x) J_i(x)\, d^4x)\} \\ &= \int [d\bar{\psi} d\psi] \exp\{ -\frac{1}{\hbar} \textstyle\int \bar{\psi}(x)(\not{D} - A_i(x)\Gamma_i \Gamma)\psi(x)\, d^4x\} \\ &= \det(\not{D} - A_i(x)\Gamma_i\Gamma). \end{aligned} \tag{2.14}$$

We calculate the quantum Noether current by varying with respect to $A_i(x)$:

$$\begin{aligned} \langle J_i \rangle(x) :&= \frac{1}{\mathcal{Z}} \frac{\delta}{\delta A_i(x)} \mathcal{Z}|_{A_i=0} = \frac{\delta}{\delta A_i(x)} \ln \mathcal{Z}|_{A_i=0} \\ &= \frac{\delta}{\delta A_i(x)} \ln \det(\not{D} - A_i(x)\Gamma_i\Gamma)|_{A_i=0} \\ &= \frac{\delta}{\delta A_i(x)} \operatorname{tr} \ln(\not{D} - A_i(x)\Gamma_i\Gamma)|_{A_i=0}; \end{aligned}$$

$$\langle J_i \rangle(x) = -\operatorname{tr}(\Gamma_i \Gamma \not{D}^{-1}). \tag{2.15}$$

Its divergence can be calculated (see Appendix 1) according to:

$$\nabla_i \langle J_i \rangle(x) = 2 \operatorname{tr}\Big(\Gamma \sum_\lambda \psi_\lambda(x) \bar{\psi}_\lambda(x)\Big). \tag{2.16}$$

Considering ψ_λ as eigenfunctions of $\not{D}^2$, we now use zeta function regularisation and obtain

$$\nabla_i \langle J_i \rangle(x) = 2 \lim_{s \to 0} \operatorname{tr}\Big(\Gamma \sum_\lambda \lambda^{-2s} \psi_\lambda(x) \bar{\psi}_\lambda(x)\Big) = 2 \lim_{s \to 0} \operatorname{tr}(\Gamma \zeta(s, x, x)). \tag{2.17}$$

Here the zeta function is defined by

$$\begin{aligned} \zeta(s, x, y) :&= \frac{1}{\Gamma(s)} \int_0^\infty dt\, t^{s-1} h(t, x, y) \\ &= \frac{1}{\Gamma(s)} \int_0^\infty dt\, t^{s-1} \sum_\lambda e^{-\lambda t} \psi_\lambda(x) \bar{\psi}_\lambda(y) \\ &= \sum_\lambda \lambda^{-s} \psi_\lambda(x) \bar{\psi}_\lambda(y). \end{aligned} \tag{2.18}$$

If we replace $h(s, x, x)$ by its asymptotic form we get the surprising result $\lim_{s \to 0} \zeta(s, x, x) = \mu_0(x)$, and so

$$\nabla_i \langle J_i \rangle(x) = 2 \operatorname{tr}(\Gamma \mu_0(x)). \tag{2.19}$$

We can divide the sets $\Gamma(E)$ into two subspaces: The functions that are even under the action of Γ, called $\Gamma_+(E)$, and those that are odd, $\Gamma_-(E)$. $\not{D}$ maps these two subspaces into each other,

$$\not{D} : \Gamma_+(E) \to \Gamma_-(E) \quad \text{respectively} \quad \not{D} : \Gamma_-(E) \to \Gamma_+(E). \tag{2.20}$$

In the same manner we can decompose $h(x)$, and also $\mu_0(x)$, into two parts,

$$\begin{aligned} \mu_0(x) &= \mu_+(x) + \mu_-(x) \quad \text{with} \\ \Gamma\mu_+(x) &= \mu_+(x) \quad \text{and} \quad \Gamma\mu_-(x) = -\mu_-(x). \end{aligned} \tag{2.21}$$

The divergence of the current is then:

$$\nabla_i \langle J_i \rangle = 2\,\mathrm{tr}(\mu_+(x) - \mu_-(x)). \tag{2.22}$$

The integration of this expression gives the index of $\not{D} : \Gamma_+ E \to \Gamma_- E$.

3. Vector Bundles and Characteristic Classes

We wish to classify principal bundles ξ with reference to their non-triviality. A principal bundle is described by a *total space* P_ξ, a *base space* B_ξ and a *projection* $\pi_\xi : P_\xi \to B_\xi$. Its *typical fibre* G_ξ is isomorphic to its *structure group*.

Our first concern will be to transfer the structure of the principal bundle from one base space to another. To this end we need the following

3.1 Definition:

Let $f : B' \to B_\xi$ be a mapping between two base spaces and $\xi = (P_\xi, \pi_\xi, B_\xi)$ be a principal bundle over B_ξ, then the *induced bundle* $\xi' = (P_{\xi'}, \pi_{\xi'}, B_{\xi'})$ with base space $B_{\xi'} = B'$ involves:

The total space $P_{\xi'}$, given by

$$P_{\xi'} = \{(b', p) \in B' \times P_{\xi'} \mid f(b') = \pi_\xi(p)\}, \tag{3.1}$$

i.e. the fibres are transported from base points of the first bundle to those of the second,

the projection $\pi_{\xi'}$, given by

$$\pi_{\xi'}(b', p) = b' \qquad \text{with} \quad f(b') = \pi_\xi(p), \tag{3.2}$$

and the structure functions, that describe the change from one chart to another and satisfy $g'_{ij} = g_{ij} \circ f$.

We shall write

$$\xi' = f^* \xi. \tag{3.3}$$

3.2 Proposition:

If f and f' are *homotopic* base space mappings, $f \simeq f'$, the induced bundles are isomorphic:

$$f^*\xi \cong f'^*\xi. \tag{3.4}$$

The inverse holds only for the universal bundle:

3.3 Definition and Proposition:

For "every" structure group G there is a *universal principal bundle* ξ_G with the following properties:

(i.) The total space P_G is contractable,
(ii.) Every principal bundle ξ has the form $\xi = f^*\xi_G$, with a suitable choice of structure group G, and
(iii.) $\xi \cong \xi'$ if and only if $f \simeq f'$.

Fibre bundles with arbitrary fibre F, for instance vector spaces, can be obtained from principal bundles by a process called *association*:

Let $\pi : P \to B$ be a G-principal bundle and $\rho : G \to \mathrm{Aut}\, F$ a representation of the group G, then we define a fibre bundle $\hat{\pi} : \hat{P} \to B$ with structure group G, base B and fibre F in the following way: $\hat{P} := P \times_G F$ is the set of equivalence classes in $P \times F$ under the action $(p, f)g = (pg, \rho(g^{-1})f)$. The projection is given by $\hat{\pi}(p, f)G := \pi(f)$, and the structure functions are $\hat{g}_{ij} := \rho(g_{ij})$. By association from universal principal bundles one obtains universal bundles with arbitrary fibre F.

Example:

Consider an n-dimensional manifold M. The tangent bundle TM can be induced from a universal bundle, for M can be imbedded into $\mathbb{R}^k$ with a suitably large value of k. The tangent planes are mapped onto n-planes in $\mathbb{R}^k$. The set of n-planes in $\mathbb{R}^k$ is the Grassmannian manifold G(n,k), the natural n-vector bundle over $G(n, k)$ is denoted by $E(n, k)$. If $f : M \to \mathbb{R}^k$ is the embedding then $TM = f^*E(n, k)$.

We now come to the main subject of this section, the characteristic classes. As the name signifies, these quantities can be used to characterize a bundle:

3.4 Definition:

A mapping $\chi : \xi \to \chi(\xi) \in H^*(B_\xi)$ is called a *characteristic class*, if it is natural with respect to the induction,

$$\chi(f^*\xi) = f^*\chi(\xi). \tag{3.5}$$

We can therefore construct such a characteristic class for the universal bundle and transfer it to ξ by means of Eq. (3.5). Principal bundles, hence, can be identified with cohomology classes of the base B_G of P_G.

Now we want to look at a special kind of characteristic classes, which is in a certain sense generic. It should be noticed here that in the following we shall characterize the bundle ξ by its total space E:

3.5 Definition:

The *Chern class* $\mathrm{c}_i(E)$ of a complex vector bundle (E, π, B) is characterized by the following properties:

(i.) $\mathrm{c}_i(E) \in H^{2i}(B)$,
where $H^{2i}(B)$ is the $2i$-th cohomology class of the base space B,

(ii.) $\mathrm{c}_0(E) = 1$, $\mathrm{c}_i(E) = 0$ for $i > \dim E$.
For the plane bundle over the complex two-dimensional space, $E(1,2)$, we have that
1. $\mathrm{c}_1(E(1,2))$ is the generator of the cohomology of $\mathbb{C}P^1 \cong S^2$ and
2. $\mathrm{c}_i(E(1,2)) = 0$ for $i > 1$.

(iii.) If we define the *total Chern class* as $\mathrm{c}(E) := \sum \mathrm{c}_i(E)$, we have the rule $\mathrm{c}(E_1 \oplus E_2) = \mathrm{c}(E_1)\,\mathrm{c}(E_2)$, where the product is defined in the cohomology $H^*(B)$ of the base space.

For a line bundle (i.e. for $\dim E_i = 1$) the only nonvanishing and nontrivial Chern class is $x_i := \mathrm{c}_1(E_i)$. So we have:

$$\mathrm{c}(\oplus E_i) = \prod \mathrm{c}(E_i) = \prod (\mathrm{c}_0(E_i) + \mathrm{c}_1(E_i)) = \prod (1 + x_i). \tag{3.6}$$

This property is important because the *splitting principle* says that each bundle is in a certain sense isomorphic to a sum of line bundles. More precisely, for every vector bundle E we can introduce a splittable vector bundle f^*E, such that f^* gives an injective map of H^*B.

A further property of line bundles is

$$\mathrm{c}_1(E_1 \otimes E_2) = \mathrm{c}_1(E_1) + \mathrm{c}_1(E_2). \tag{3.7}$$

Applying this to the trivial line bundle $E \otimes E^*$ we get:

$$0 = \mathrm{c}_1(E \otimes E^*) = \mathrm{c}_1(E) + \mathrm{c}_1(E^*) \quad \Rightarrow \quad \mathrm{c}_1(E^*) = -\,\mathrm{c}_1(E). \tag{3.8}$$

We now present some further definitions:

3.6 Definition:

The *Todd class* is given by

$$\mathrm{td}(E) = \prod \frac{x_i}{1 - e^{-x_i}}. \tag{3.9}$$

Hence $\mathrm{td}(E \oplus F) = \mathrm{td}\,E \cdot \mathrm{td}\,F$.

3.7 Definition and Proposition:

The *Chern character* is defined as

$$\mathrm{ch}(E) = \sum e^{x_i}. \tag{3.10}$$

For this characteristic class we have the properties

(i.) $\mathrm{ch}(E_1 \oplus E_2) = \mathrm{ch}(E_1) + \mathrm{ch}(E_2)$ and
(ii.) $\mathrm{ch}(E_1 \otimes E_2) = \mathrm{ch}(E_1) \cdot \mathrm{ch}(E_2)$.

ch is a homomorphism from a semi-ring of complex vector spaces into a cohomology ring.

3.8 Definition:

We complexify the tangent bundle TM of M to get the bundle $T_{\mathbb{C}}M$. In the formal splitting of $T_{\mathbb{C}}M$ the line bundles E_i occur in complex conjugate pairs E_i, $E_i{}^*$. Therefore, the quantities $x_i = c_1(E_i)$ are pairwise equal up to a sign. With a suitable choice of these signs the *Euler class* is given by

$$\mathrm{e}(TM) = \prod_{i=1}^{n/2} x_i \tag{3.11}$$

and the *Euler characteristic* of the manifold M,

$$\chi := \int_M \mathrm{e}. \tag{3.12}$$

We now turn to the question of the relevance of characteristic classes in physics. Given a connection one-form $A := dx^\mu a_\mu$ on a vector bundle $\xi = (E, \pi, B)$, we can form the covariant derivative $\nabla := d + A$ with $d := dx^\mu \partial/\partial x^\mu =: dx^\mu \partial_\mu$ and a curvature two-form $F := dx^\mu dx^\nu F_{\mu\nu}$ with

$$F_{\mu\nu} := \partial_\mu A_\nu - \partial_\nu A_\mu + [A_\mu, A_\nu]_-, \quad \text{i.e.} \quad F = [\nabla, \nabla]_-. \tag{3.13}$$

Given an invariant polynomial P, invariant on the Lie algebra, we can define a characteristic class by

$$\chi_P(\xi) := P(F). \tag{3.14}$$

In fact, this expression is closed, $dP(F) = 0$, for a different connection A', $P(F) - P(F')$ is exact, i.e., the cohomology class of $P(F)$ does not depend on the connection, and $\chi_P(f^*\xi) = f^*\chi_P(\xi)$. In particular the Chern class can be constructed in this way:

3.9 Proposition:

Let E be a complex vector bundle and F the curvature of any connection on E, then

$$\sum_r t^r \, c_r(E) = \det\left(1 + \frac{it}{2\pi} F\right). \tag{3.15}$$

To prove this proposition it is necessary to demonstrate all the properties which are listed in Def. 3.5. This will be left to the reader. We shall instead look here at a few special points:

3.10 Corollary:

The first two Chern classes are given by

$$\begin{aligned} c_1(E) &= \operatorname{tr}\left(\frac{iF}{2\pi}\right) \quad \text{and} \\ c_2(E) &= \frac{1}{2}\left(\left(\operatorname{tr}\left(\frac{iF}{2\pi}\right)\right)^2 - \operatorname{tr}\left(\left(\frac{iF}{2\pi}\right)\right)^2\right). \end{aligned} \tag{3.16}$$

3.11 Corollary:

The induced Chern character has the form

$$\operatorname{ch}(E) = \operatorname{tr}\left(\exp\left(\frac{iF}{2\pi}\right)\right). \tag{3.17}$$

We have now collected all the concepts necessary to understand the content of index theorem:

$$\operatorname{index} D = (-1)^{n(n+1)/2} \frac{\operatorname{ch}(E) - \operatorname{ch}(F)}{e(TM)} \operatorname{td}(T_{\mathbb{C}}M)[M], \tag{3.18}$$

for the operator $D : \Gamma(E) \rightarrow \Gamma(F)$. The right-hand-side of Eq. (3.18) contains a characteristic class, which is a closed differential form, nonhomogeneous in its degree, which in turn is a polynomial in the Riemannian curvature R on M with respect to any Riemannian metric on M, and in the Yang-Mills curvature with respect to any connection on E and F. This polynomial is calculable by the formal splitting methods described above. Its contibution of degree n, in the situation of Sect. 2, is proportional to a generalized axial anomaly. Integrating this part of degree n over the compact manifold M gives index D.

Even for non-compact manifolds or manifolds with boundary, the n-form on the right-hand-side is the same, because the zeta function algorithm is local.

4. Special Cases and Important Applications

We want to consider the set $\Omega^* M$ of all differential forms on M. There we can define a scalar product

$$(\alpha, \beta) := \int_M \alpha \wedge *\beta, \tag{4.1}$$

where $*$ is known as the *Hodge star operator*, which maps a p-form onto the so-called dual $(n-p)$-form.

4.1 First Example:

The exterior derivative d, as well as its adjoint d^*, are mappings from $\Omega^* M$ into itself. The operator $D = d + d^*$ maps even forms into odd forms and vice versa. We are interested in the index of this operator.

$\ker(d + d^*) = \ker d \cap \ker d^*$ consists of harmonic forms, which by definition are both closed and coclosed. Because of the Hodge decomposition theorem $\Omega^* M = \ker d^* \oplus \operatorname{im} d$, we have

$$\ker d = \ker(d + d^*) + \operatorname{im} d. \tag{4.2}$$

In other words, every cohomology class has precisely one harmonic representative, and the r-th Betti number is just the number of harmonic r-forms. This means that the index of the operator $D = d + d^*|_{\text{even}}$ is just the alternating sum of the Betti numbers b_r. A direct evaluation of the right-hand-side of the index theorem gives the Euler class $e(TM)$. Thus

$$\operatorname{index} D = \sum_r (-1)^r b_r = e(TM)[M] = \chi. \tag{4.3}$$

4.2 Second Example:

The Hodge star operator can be made involutive by supplementing it with a suitable phase:

$$\tau := \epsilon(p)*, \qquad \tau^2 = 1. \tag{4.4}$$

It is easily seen that for $n = 4l$, τ anticommutes with $d + d^*$, and we can consider the index of the operator $D = d + d^*$, restricted to the subspace of $\Omega^* M$, which is even under τ. Then, on the one hand, $\operatorname{index} D$ is the number of harmonic forms even under τ minus the number of τ-odd harmonic forms. This quanitity is the so-called *signature of* M and is, in fact, given by the initial index of the form

$$(\alpha, \beta) = \int_M \alpha \wedge *\beta \qquad \text{on} \quad H^{2l}(M) \tag{4.5}$$

(all contributions of degree $\neq 2l$ cancel).

The right-hand-side of the index theorem, evaluated by formal splitting methods, yields the *Hirzebruch's L-genus* of the manifold,

$$L = \prod_{r=1}^{2l} \frac{x_i}{\tanh x_i} = 1 - \frac{1}{3}\sum x_i^2 + \dots . \tag{4.6}$$

Thus

$$\begin{aligned} \operatorname{index} D &= L[M] \\ &= \frac{1}{48\pi^2}\epsilon^{ikrs} R_{ikuv} R_{rsuv} \qquad (\text{for } n = 4). \end{aligned} \tag{4.7}$$

4.3 Third Example:

For a complex manifold M the *Theorem of Riemann and Roch* supplies the index

$$\operatorname{ind} \bar{d} = \dim H^0(M, O(V)) - \dim H^1(M, O(V)) = c_1(V)[M] + 1 - g, \tag{4.8}$$

where g is the genus of the Riemannian plane and $\bar{d} = dz \wedge \partial/\partial\bar{z}$.

4.4 Forth Example:

Consider the Dirac operator $\not{D}$ on a twisted tensor bundle. We have

$$\operatorname{ind} \not{D} = \hat{A}(M)\, \mathrm{ch}(V)|_{\dim M}, \tag{4.9}$$

where

$$\hat{A}(M) := \prod \frac{-x_i/2}{\sinh(x_i/2)} = (1 + \frac{1}{24}\sum x_i^2 + \dots) \tag{4.10}$$

is called the *Â-genus* of M.

According to Sect. 2, the index of the Dirac operator, restricted to spinors of positive chirality, is directly related to the anomaly of the axial current $j_k = \bar{\psi}\gamma_5\gamma_k\psi$. But also the first and the second example derived above are special cases of the general situation described in Sect. 2. This is due to the identity

$$d\omega = \sum e^i \wedge \nabla_{e_i}\omega, \tag{4.11}$$

where e_i and e^i are vierbein- and dual vierbein-fields and ∇ a torsion free connection. Then

$$d^*\omega = \sum_i i_{e_i} \nabla_{e_i}\omega, \tag{4.12}$$

where i_{e_i} means insertion and

$$D\omega = (d + d^*)\omega = \sum_i \Gamma^i \nabla_i \omega = \sum_i (e^i \wedge (+i_{e_i}))\nabla_{e_i}\omega. \tag{4.13}$$

Taking $\Gamma\omega = (-1)^{\deg\omega}\omega$ for the first and $\Gamma\omega = \tau\omega$ for the second example, we rediscover the situation of a generalized axial current and find currents, whose anomalies are related to the Euler characteristic and to Hirzebruch's L-genus.

5. The Family Index Theorem, Determinant Bundles and Gauge Anomalies

In this section we attempt to describe the topological meaning of the anomalies in quantum field theory.

In the theory of the weak interaction we know that it is important to distinguish the space of fermions with positive chirality from the space of those with negative chirality. The Dirac operator

$$\not{D}_A : \Gamma(S^+) \to \Gamma(S^-) \tag{5.1}$$

mediates between these two spaces, i.e. between vector bundles of different dimension. By calculating the effective action corresponding to the Lagrangian $\mathcal{L} = \psi \not{D}_A \psi$ we arrive at the expression $\det \not{D}_A$. This formal expression is not yet well defined: $\not{D}_A$ is a linear operator between two different vector spaces, and the determinant of such operators is ambiguous. The determinant of $\not{D}_A^* \not{D}_A : \Gamma(S^+) \to \Gamma(S^+)$ is defined, which shows that the ambiguity of $\det \not{D}_A$ concerns only the phase of this quantity. A possible definition of $\det \not{D}_A$ would be the following:

Choose a fixed operator $P : \Gamma(S^-) \to \Gamma(S^+)$ and define

$$\det{}_P \not{D}_A := \det(P \not{D}_A). \tag{5.2}$$

Actually the regularization of $\det \not{D}_A$ corresponds to such a procedure. Although $\det_P \not{D}_A$ depends on the choice of P, variations of $\ln \det_P \not{D}_A$ with respect to the gauge potential A are independent of P.

The regularized quantity $\det \not{D}_A$ is now, contrary to formal expectations, not generally gauge invariant. Its variation under an infinitesimal gauge transformation ξ,

$$\delta_\xi \ln \det \not{D}_A = \int d^n x \; \xi(x)\alpha(A) = \langle \xi, \alpha[A] \rangle \tag{5.3}$$

is the continuous gauge anomaly of the gauge theory with chiral fermions. In a consistent theory it has to be cancelled out. A physical example is the gauge anomaly responsible for the decay of the π^0-meson and its concellation by fermionic contributions in the standard model. The anomaly $\alpha(A)$ is an n-form on the space time manifold M, and the integrated anomaly $\alpha[A]$ has to be considered as a closed 1-form on the gauge group $\mathcal{G}$.

The Wess-Zumino consistency condition

$$[\delta_\xi, \delta_\eta] \ln \det \not{D}_A = \delta_{[\xi,\eta]} \ln \det \not{D}_A \tag{5.4}$$

is just identical with the closedness of $\alpha[A]$: $\delta\alpha = 0$.

In a geometrical framework, det $\not{D}_A$ is not a globally defined function on the set $\mathcal{A}$ of gauge potentials, but rather a section of a determinant line bundle

$$\mathrm{DET}\,\not{D} = \mathcal{A} \times_{\mathcal{G}} \mathbb{C} \tag{5.5}$$

over the space $\mathcal{A}/\mathcal{G}$ of gauge equivalence classes of connections. The gauge variation of the transition functions of this bundle is just the gauge anomaly.

Let us now give a brief discussion of the topological properties of the determinant bundle and its relationship to the gauge anomaly. By $\mathcal{H}S^+$ and $\mathcal{H}S^-$ we denote the Hilbert bundles over $\mathcal{A}/\mathcal{G}$ of sections of S^+ and S^-. Then

$$\begin{aligned}\mathrm{DET}\,\not{D} &= \mathrm{Hom}(\Lambda^{\max}\mathcal{H}S^+, \Lambda^{\max}\mathcal{H}S^-)\\ &= \mathrm{Hom}(\Lambda^{\max}\ker\not{D}, \Lambda^{\max}\ker\not{D}^*),\end{aligned} \tag{5.6}$$

because everything except the contributions of the zero modes cancels out. Now a simple calculation gives

$$\mathrm{ch}_1(\mathrm{DET}\,\not{D}) = \mathrm{c}_1(\mathrm{DET}\,\not{D}) = \mathrm{ch}_1(\mathrm{IND}\,\not{D}), \tag{5.7}$$

where the so-called *index bundle*

$$\mathrm{IND}\,\not{D} := \ker\not{D} - \ker\not{D}^* \tag{5.8}$$

is a well defined virtual bundle: ker $\not{D}$ and ker $\not{D}^*$ are families of finite dimensional vector spaces, labelled by $\mathcal{A}/\mathcal{G}$. Their difference, however, is a virtual bundle because of the deformation invariance of the index. For calculating $\mathrm{ch}_1(\mathrm{IND}\,\not{D})$ it suffices to restrict the bundle to a two-sphere $Y \subset \mathcal{A}/\mathcal{G}$. Then the family index theorem gives

$$\mathrm{ch}(\mathrm{IND}\,\not{D}) = \int_M \hat{A}(M)\,\mathrm{ch}\,V \quad \in H^*(Y). \tag{5.9}$$

The integrand is a closed form on $M \times Y$, whose part of degree n in space-time M is separated and integrated over M to give a closed form on Y.

The exact sequence ($\mathcal{A}$ is contractable, thus $H^i(\mathcal{A}) = 0$)

$$0 = H^1(\mathcal{A}) \to H^1(\mathcal{G}) \xrightarrow{\delta} H^2(\mathcal{A}/\mathcal{G}) \to H^2(\mathcal{A}) = 0 \tag{5.10}$$

gives an isomorphism

$$\delta : \alpha \to \mathrm{ch}_1\,\mathrm{DET}\,\not{D} = \mathrm{ch}_1\,\mathrm{IND}\,\not{D}, \tag{5.11}$$

where α is the anomaly, considered as a 1-form over $\mathcal{G}$.

The density in Eq. (5.3) is an $(n+1)$-form on $M \times Y$ of degree n on M and degree 1 on $\mathcal{G}$. Its local expression can be obtained by transgression. If we decompose it into degrees on M and Y, we get

$$\omega := \hat{A}\ \mathrm{ch}\,V|_{n+2} = \omega_{n+2,0} + \omega_{n+1,1} + \dots . \tag{5.12}$$

Denoting exterior derivations in M and Y by d and δ and using

$$(d+\delta)^2 = d^2 = \delta^2 = 0, \qquad d\delta + \delta d = 0, \tag{5.13}$$

and closedness of ω gives

$$\begin{aligned} d\omega_{n+2,0} &= 0. \qquad \text{Hence, locally} \\ \omega_{n+2,0} &= d\omega_{n+1,0}, \qquad \text{and from } \delta\omega_{n+2,0} = 0 \\ \delta\omega_{n+1,0} &= -d\delta\omega_{n+1,0} = 0, \qquad \text{thus} \\ \delta\omega_{n+1,0} &= \omega_{n+1,1} = d\omega_{n,1}, \end{aligned} \tag{5.14}$$

where $d\omega_{n,1}$ is the anomaly density.
The consistency condition follows from

$$d\delta\omega_{n,1} = -\delta d\omega_{n,1} = -\delta^2\omega_{n+1,0}. \tag{5.15}$$

6. Outlook to Further Applications

In this final section we list some applications of the index theorem, which cannot be dealt with in detail here.

6.1 Higher Spins and Dimensions

The preceeding methods apply to arbitrary space-time dimensions and to spinor bundles of arbitrary spin. Rarita-Schwinger fields, which appear in supersymmetric theories, are of particular importance. They are treated as sections of the virtual bundles $S^{\pm}(M) \otimes T_{\mathbb{C}}M \ominus S^{\pm}(M)$.

6.2 Index Theorem for Manifolds with Boundary and G-Index Theorem

For manifolds with boundary there are boundary contributions to the index, whereas the anomalies are still given by the same expressions as for manifolds without boundary. Non-compact manifolds can be treated by shifting boundaries to infinity. Using the G-index theorem one can also evaluate boundary contributions to the index on asymptotically local euclidean gravitational instantons, which at infinity look like S^3/Γ, where Γ is some discrete group. The boundary contribution ξ in Eq. (1.14) is also needed to evaluate global gauge anomalies, i.e. non-invariance of the effective action under gauge transformations, which cannot be smoothly deformed into the unit transformation.

6.3 Anomalies in String Theories

String theories contain an infinity of spinor fields. The cancellation of gauge anomalies is delicate and an important consistency constraint. It turns out to be related to the modular invariance of the theory and to the G-index theorem of loop spaces, on which $G = U(1)$ acts in a natural way, and to the so-called *elliptic genera*.

Appendix

A.1 The Divergence of the Quantum Noether Current

In a complete system of eigenfunctions of $\not{D}$ with $\not{D}\psi_\lambda = \lambda\psi_\lambda$ and $\not{D}^2\psi_\lambda = \lambda^2\psi_\lambda$, the quantum Noether current of Eq. (2.15) is given by

$$\langle J_i\rangle(x) = -\operatorname{tr}\Big(\sum_\lambda \bar{\psi}_\lambda(x)\Gamma_i\Gamma\lambda^{-1}\psi_\lambda(x)\Big), \tag{A1.1}$$

where the trace is not strictly necessary, but is helpful in the following. If we calculate the divergence, we get:

$$\nabla_i\langle J_i\rangle(x) = -\operatorname{tr}\Big(\sum_\lambda \big((\nabla_i\bar{\psi}_\lambda(x))\Gamma_i\Gamma\lambda^{-1}\psi_\lambda(x) + \bar{\psi}_\lambda(x)\Gamma_i\Gamma\lambda^{-1}(\nabla_i\psi_\lambda(x))\big)\Big). \tag{A1.2}$$

Next we use the property $[\Gamma_i, \Gamma]_+ = 0$ and

$$\begin{aligned} &\Gamma_i\nabla_i\psi_\lambda(x) = \lambda\psi_\lambda(x) \quad\Leftrightarrow\quad \nabla_i\psi_\lambda^\dagger(x)\Gamma_i^\dagger = \lambda\psi_\lambda^\dagger(x) \quad\Leftrightarrow \\ &\nabla_i\psi_\lambda^\dagger(x)\Gamma_i^\dagger\Gamma_0 = \lambda\bar{\psi}_\lambda(x) \quad\Leftrightarrow\quad -\nabla_i\bar{\psi}_\lambda(x)\Gamma_i = \lambda\bar{\psi}_\lambda(x), \end{aligned} \tag{A1.3}$$

where we used $\Gamma_i^\dagger\Gamma_0 = -\Gamma_0\Gamma_i$. So we get

$$\nabla_i\langle J_i\rangle(x) = 2\operatorname{tr}\Big(\sum_\lambda \bar{\psi}_\lambda(x)\Gamma\psi_\lambda(x)\Big) = 2\operatorname{tr}\Big(\Gamma\sum_\lambda \psi_\lambda(x)\bar{\psi}_\lambda(x)\Big). \tag{A1.4}$$

Acknowledgement

This contribution is largely based on notes of my lectures in Bad Honnef, taken and worked out by Stefan Groote. I should like to thank him for his careful work. I also thank the organizers A. Hirshfeld and J. Debrus for their hospitality.

Comments on the References

The literature on applications of the index theorem is vast. We shall restrict ourselves to a small choice of representative references. More comprehensive lists can be found in the reviews quoted below. The references are structured as follows:

- Index Theorem and Generalizations .... [1 – 3]
- Reviews .... [4 – 6]
- Axial Anomalies .... [7 – 14]
- Boundary Contributions .... [15 – 16]
- Gauge Anomalies .... [17 – 21]
- String Anomalies and Elliptic Genera .... [22]

References

[1] M. F. Atiyah, I. M. Singer, Ann. of Math. **87**, 485, 546 (1968); **93** 1, 119,139 (1971)
[2] M. F. Atiyah, G. B. Segal, Ann. of Math. **87**, 531 (1968)
[3] M. F. Atiyah, V. J. Petodi, I. M. Singer, Math. Proc. Camb. Phil. Soc. **77**, 43 (1975); **78**, 405 (1975); **79**,71 (1976)
[4] H. Römer, *Proceedings of the International Conference on Differential Geometric Methods*, Clausthal 1978, H. D. Doebner (ed.), Springer Lecture Notes in Physics 139 (1981)
[5] T. Egushi, P. B. Gilkey, A. J. Hanson, Physics Reports **66**, Nr.6 (1980)
[6] L. Alvarez-Gaumé, *Fundamental Problems of Gauge Field Theory*, G. Velo, A. S. Wightman (eds.), NATO ASI Series, Series **B**: Physics, Vol.141, Plenum Press 1986
[7] J. Kiskis, Phys. Rev. **D15**, 2329 (1977)
[8] R. Jackiv, C. Rebbi, Phys. Rev. **D16**, 1052 (1977)
[9] N. K. Nielsen, H. Römer, B. Schroer, Phys. Lett. **70B**, 445 (1977)
[10] M. T. Grisaru, N. K. Nielsen, H. Römer, P. van Nieuwenhuizen, Nucl. Phys. **B140**, 477 (1978)
[11] H. Römer, *Proceedings of the XIth International Colloquium on Group Theoretical Methods*, Bebek, Istanbul, M. Serdaroğlu, E. Inönü (eds.), Springer Lecture Notes in Physics **180** (1983)
[12] H. Römer, Phys. Lett. **83B**, 172 (1979)
[13] H. Römer, Phys. Lett. **101B**, 55 (1981)
[14] P. van Nieuwenhuizen, H. Römer, Phys. Lett. **162B**, 290 (1985)
[15] A. J. Hanson, H. Römer, Phys. Lett. **80B**, 58 (1978)
[16] G. W. Gibbons, C. N. Pope, H. Römer, Nucl. Phys. **B157**, 377 (1979)
[17] R. Stora, *Cargèse Lecture Notes 1983*, G. 't Hooft (ed.), Plenum Press 1984
[18] B. Zumino, *Les Houches Lecture Notes 1983*, B. S. DeWitt, R. Stora (eds.), North Holland 1984
[19] B. Zumino, Y. S. Wu, A. Zee, Nucl. Phys. **B239**, 477 (1984)
[20] M. F. Atiyah, I. M. Singer, Proc. Nat. Acad. Sci. USA **81**, 2597 (1984)
[21] N. K. Nielsen, H. Römer, Phys. Lett. **154**, 141 (1985)
[22] E. Witten, Comm. Math. Phys. **109**, 525 (1987)

All Solutions of the Wess-Zumino Consistency Conditions

Friedemann Brandt[1], *Norbert Dragon*[1], *Maximilian Kreuzer*[2]

[1] Institut für Theoretische Physik der Universität,
Appelstraße 2, D(W)-3000 Hannover 1,
Federal Republic of Germany
[2] Institute for Theoretical Physics, Univ. of California,
Santa Barbara, CA 93106, USA

Abstract

For the case of a compact gauge group we list all solutions to the Wess-Zumino consistency equations which have to be satisfied by anomalies. We describe the main algebraic tools and theorems required for this complete classification. Our results answer the question whether in nonrenormalizable gauge theories there exist additional up-to-now unknown anomalies in the negative.

1. Introduction

The evaluation of loop diagrams in Quantum Field Theory leads to divergent integrals if one naively applies the Feynman rules. To define the divergent diagrams one needs a regularization, or refined Feynman rules, to cancel the divergencies in a consistent way. It may happen that no regularization respects all symmetries of the classical theory and that these symmetries cannot be restored by appropriate counterterms in the limit which removes the unphysical regularization. The renormalized quantum theory then possesses an anomaly.

Mathematically, an anomaly can be defined as a variation of the effective action with respect to the generators of the anomalous broken symmetry. Wess and Zumino [9] found that anomalies have to satisfy consistency conditions which follow from the Lie algebra of the gauge group. Therefore, if one knows all solutions to these consistency conditions one can check whether the corresponding possible anomalies really occur in perturbation theory.

The introduction of the nilpotent BRS-operator allows to formulate the consistency conditions as a cohomological problem, and to define the anomaly (in mathematical language) as a nontrivial 1-cocycle, i.e. (in our language) as a nontrivial solution to the Wess-Zumino consistency conditions with ghost number one.

But one can also study the consistency conditions for ghost numbers different from one. The solutions for ghost number zero determine all gauge invariant local actions and for ghost number two (and form-degree $(D-1)$ in D dimensions) they are related to Schwinger terms, which show up in anomalous equal time commutation relations of currents [6]. A physical interpretation for solutions with higher ghost numbers is not known up to now. But studying the consistency conditions one is led directly to solutions with higher ghost numbers (via the Descent Equations (5.1)).

For a compact gauge group we determined all solutions of the consistency conditions for the Yang-Mills [2, 3, 4] and the gravitational case [5], i.e. we computed all solutions in arbitrary space-time dimension with arbitrary ghost number and without any restrictions on their mass dimension or the order of derivatives of the fields appearing in the solutions. Therefore our investigation is not restricted to renormalizable theories. Our result is that there exist two kinds of solutions, both familiar to physicists, and no additional up-to-now unknown solutions: solutions whose densities are invariant (like invariant Lagrangians) and solutions whose integrands transform into a total derivative (like Chern-Simons actions and chiral anomalies, see Section 3) and which are always constructable from generalized Chern-Simons forms. We describe our proof of this result only for the Yang-Mills case. The gravitational case can be treated in an analogous way but needs some more work (and pages).

In the second section we define the problem and in the third we describe the result (especially for ghost numbers zero and one). The fourth section collects and comments four theorems needed for the proof which follows in section five. Finally, the sixth section gives an outview on further areas of research.

2. Defining the Problem: Wess-Zumino Consistency Condition, Algebra and Field Content

Consider a Lorentz invariant local functional a of the fields ϕ, which is a collective designation for the Yang-Mills gauge fields A_μ^I, matter fields ψ, ghosts C^I, antighosts $\bar{C}^I$ and auxiliary fields B^I, i.e., a can be written as a D-dimensional integral over the volume form $\mathcal{A}^G$ in D-dimensional space-time which is a polynomial in all fields and their derivatives (collectively denoted by $[\phi]$) up to an arbitrary (but finite) order:

$$a = \int \mathcal{A}^G([\phi]), \quad \phi = (A_\mu^I, \psi, C^I, \bar{C}^I, B^I), \quad [\phi] = (\phi, \partial_\mu \phi, \partial_\mu \partial_\nu \phi, \ldots). \quad (2.1)$$

The superscript of $\mathcal{A}$ indicates its ghost number G, i.e. its degree of homogeneity in the ghosts $[C^I]$ minus the degree of homogeneity in the antighosts $[\bar{C}^I]$.

Using the BRS-operator s (see Eqs. (2.3) to (2.6) for its definition) the WZ-consistency conditions read $sa = 0$. This equation for the functional sa has to be satisfied identically in the fields $[\phi]$. For the integrand $\mathcal{A}^G$ this means that $s\mathcal{A}^G$ has to be total a derivative,

$$s\mathcal{A}^G([\phi]) + d\mathcal{A}^{G+1}([\phi]) = 0. \quad (2.2)$$

We note that by solving Eq. (2.2) we in fact solve the equation $sa =$ (boundary terms), which is more general than $sa = 0$. Namely, in general Eq. (2.2) implies that sa is a sum of boundary-terms (integrals over $(D-1)$-dimensional boundaries whose number may be larger than one in topological nontrivial cases), and one may discuss whether sa vanishes or not after one has found all a transforming into such boundary terms.

Eq. (2.2) has to be satisfied identically in the fields $[\phi]$ irrespective of their x-dependence. Therefore we define all operations like s or d on the $[\phi]$, which are infinitely many independent variables restricted only by the obvious algebraic identities $\partial_\mu \partial_\nu \phi = \partial_\nu \partial_\mu \phi$, etc., originating from $[\partial_\mu, \partial_\nu] = 0$.

The BRS-operator s acts on the ϕ as

$$\begin{aligned} sA_\mu^I &= \partial_\mu C^I + f_{JK}{}^I C^J A_\mu^K, \qquad sC^I = \tfrac{1}{2} f_{JK}{}^I C^J C^K, \\ s\psi &= -C^I \delta_I \psi, \quad \delta_I \psi = -T_I \psi, \quad s\bar{C}^I = B^I, \quad sB^I = 0. \end{aligned} \quad (2.3)$$

The δ_I are the generators of the gauge group and the T_I are a matrix representation of the δ_I,

$$[\delta_I, \delta_J] = f_{IJ}{}^K \delta_K, \quad [T_I, T_J] = f_{IJ}{}^K T_K. \quad (2.4)$$

As is seen from Eq. (2.3) we assume the matter fields to transform linearly under the gauge group (in the presence of nonlinearly transforming matter fields there are no additional solutions but some solutions become trivial, [3]).

The action of s on derivatives of the ϕ follows from Eq. (2.3) by

$$[s, \partial_\mu] = 0. \tag{2.5}$$

On products of the $[\phi]$ and dx^μ s is defined by the product rule for graded differential operators $\mathcal{D}$,

$$\mathcal{D}(AB) = (\mathcal{D}A)B + (-)^{|A|\,|\mathcal{D}|}A(\mathcal{D}B), \quad A, B \in \{[\phi], dx^\mu\}. \tag{2.6a}$$

The grading $|\ |$ takes the value zero for A^I_μ, B^I, bosonic matter fields and their partial derivatives, and the value one for $C^I, \bar{C}^I, dx^\mu$, fermionic matter fields and the BRS-operator. The grading determines the commutation relations of the $[\phi]$ and dx^μ:

$$\begin{aligned} &|A^I_\mu| = |B^I| = |\partial_\mu| = 0, \qquad |\psi| = \begin{cases} 0 & \text{if } \psi \text{ is a boson} \\ 1 & \text{if } \psi \text{ is a fermion,} \end{cases} \\ &|C^I| = |\bar{C}^I| = |dx^\mu| = |s| = 1, \\ &AB = (-)^{|A|\,|B|}BA, \quad A, B \in \{[\phi], dx^\mu\}. \end{aligned} \tag{2.6}$$

The action of s on the sum of monomials of the $[\phi]$ and dx^μ is defined as the sum of its action on the monomials. Therefore on forms of the $[\phi]$ s acts as a graded first order differential operator.

The exterior derivative

$$d = dx^\mu \partial_\mu, \quad s\, dx^\mu = 0, \quad \partial_\mu\, dx^\nu = 0 \tag{2.7}$$

appearing in Eq. (2.2) is also a graded differential operator which is defined on the $[\phi]$ (and dx^μ), not on x^μ, i.e. ∂_μ is *not* defined as $\partial/\partial x^\mu$ but as a linear operator which maps e.g. the variable ψ to the variable $\partial_\mu \psi$ and as a linear first order differential operator on forms of the $[\phi]$. As a consequence, the cohomology[1] of d is *not* given by the familiar Poincaré Lemma (for differential forms $\omega_p(x)$ on contractible coordinate patches), but by a lemma which we call the *Algebraic Poincaré Lemma* (see Eq. (4.3)).

The BRS-transformation of the ghosts, Eq. (2.3), is chosen such that s is nilpotent. Because differentials dx^μ anticommute and derivatives commute, $[\partial_\nu, \partial_\mu] = 0$, d also is nilpotent, $d^2 = 0$. From Eqs. (2.5) to (2.7) s and d anticommute,

$$s^2 = d^2 = \{s, d\} = (s + d)^2 = 0. \tag{2.8}$$

As a consequence of Eq. (2.8) each form

$$\mathcal{A}^G = s\mathcal{B}^{G-1}([\phi]) + d\mathcal{B}^G([\phi]) + (\text{const})\, d^D x$$

---

[1] Consider a graded differential operator $\mathcal{D}$, which is nilpotent, $\mathcal{D}^2 = 0$, and well defined on variables ξ and certain functions $f(\xi)$. By the cohomology of $\mathcal{D}$ (in a space of functions $f(\xi)$) we denote the kernel of $\mathcal{D}$ divided by its image, i.e. the functions $f(\xi)$ which are closed with respect to $\mathcal{D}$ (i.e. which satisfy $\mathcal{D}f = 0$) modulo the functions $g(\xi)$ which are exact (i.e. $g = \mathcal{D}h$).

satisfies Eq. (2.2) (with $\mathcal{A}^{G+1} = s\mathcal{B}^G$), i.e. Eq. (2.2) defines an equivalence class of solutions, where two solutions are called equivalent if they differ only by such a trivial solution.

For ghost number zero trivial solutions change the physically irrelevant gauge fixing part of the action, for ghost number one trivial solutions correspond to removable, i.e. non-anomalous symmetry breaking. So the mathematical notion of equivalence corresponds to physical equivalence. Neglecting trivial solutions we demand

$$\mathcal{A}^G \neq s\mathcal{B}^{G-1}([\phi]) + d\mathcal{B}^G([\phi]) + (\text{const})\, d^D x. \tag{2.9}$$

3. Results

There are two kinds of solutions to Eqs. (2.2) and (2.9). Solutions of the first kind are called *Lagrangian* (or *trace-*) *solutions* $\mathcal{A}^G_{\text{trace}}$, those of the second kind *chiral solutions* $\mathcal{A}^G_{\text{chiral}}$. Both kinds are (up to trivial terms) independent of the antighosts $[\bar{C}^I]$, the auxiliary fields $[B^I]$ and of derivatives of the ghosts $[\partial_\mu C^I]$.

3.1 Lagrangian Solutions:

These solutions solve $s\mathcal{A}^G_{\text{trace}} = 0$, i.e. they can be taken such that the total derivative $d\mathcal{A}^{G+1}$ appearing in Eq. (2.2) vanishes. Lagrangian solutions are of the form $\mathcal{A}^G_{\text{trace}} = \mathcal{L}^G_{\text{inv}} d^D x$ where $\mathcal{L}^G_{\text{inv}}$ is a superfield in functions $\Theta_K(C^I)$ (see below) whose component fields are δ_I-invariant polynomials in the matter fields ψ, the field strengths $F^I_{\mu\nu}$,

$$F^I_{\mu\nu} = \partial_\mu A^I_\nu - \partial_\nu A^I_\mu - f_{JK}{}^I A^J_\mu A^K_\nu , \tag{3.1}$$

and symmetrized covariant derivatives of the matter fields and field strengths (denoted collectively by $\{\psi, F^I_{\mu\nu}\}$). The component fields have nonvanishing Euler derivatives with respect to ψ or A^I_μ, or are pure constants. So

$$\mathcal{A}^G_{\text{trace}} = \mathcal{L}^G_{\text{inv}}(\Theta_K, \{\psi, F^I_{\mu\nu}\}) d^D x. \tag{3.2}$$

The Θ_K are nontrivial s-invariant by themselves and span the Lie algebra cohomology, i.e. each solution $f(C^I)$ (depending only on the ghosts C^I) of $sf = 0$ is, up to trivial terms, a function of the Θ_K (see Eq. (4.16)). The Θ_K are in one-to-one correspondence to the independent Casimir operators (see Example 3 in Section 4) of the gauge group $\mathcal{G}$, their number equals the rank of $\mathcal{G}$.

Explicitly the Θ_K are given by

$$\Theta_K(C^I) = \tfrac{m!(m-1)!}{(2m-1)!}\,\mathrm{tr}(C^{2m-1}), \qquad C = C^I T_I, \qquad m = m_K, \qquad K = 1,\dots,\mathrm{rank}\,(\mathcal{G}), \tag{3.3}$$

where the T_I are an appropriate matrix representation of the δ_I (either the fundamental or the spinor representation, [7]) and m_K is the order of the Kth Casimir operator. Generators belonging to abelian factors are Casimir operators by themselves (because they commute with each generator δ_I) and therefore to each abelian factor of the gauge group there belongs a Θ_K with $m_K = 1$ which is given just by the corresponding abelian ghost itself, $\Theta_K = C^K$ ($sC = 0$ for abelian ghosts C).

For $G = 0$ Eq. (3.2) gives just those invariant actions whose Lagrangians themselves are invariant.

For $G = 1$ Eq. (3.2) is linear in the abelian ghosts (because they are the Θ_K with $G = 1$) and therefore anomalies corresponding to these solutions can appear only if the gauge group contains at least one abelian factor. The prominent example is the trace anomaly of dilatations.

3.2 Chiral Solutions:

These solutions do not depend on $[\psi]$, and satisfy Eq. (2.2) either with a nonvanishing $d\mathcal{A}^{G+1}$, or they are superfields in the Θ_K with component fields which are nonconstant functions of the f_K defined by Eq. (3.5) (and thus the Euler derivative of the component fields vanishes, which distinguishes these chiral solutions from the Lagrangian solutions). Chiral solutions depend (up to trivial terms) only on the gauge field one-form A, the field strength two-form F and the ghost matrices C,

$$C = C^I T_I, \quad A = dx^\mu A^I_\mu T_I, \quad F = \frac{1}{2} dx^\mu dx^\nu F^I_{\mu\nu} T_I. \tag{3.4a}$$

All chiral solutions $\mathcal{A}^G_{\mathrm{chiral}}$ are constructable from polynomials $P(\tilde{q}_K, f_K)$ of generalized Chern-Simons-forms $\tilde{q}_K$ and Chern-forms f_K (see Eqs. (3.5) and (3.8) below). These polynomials are in general linear combinations of terms with different ghost numbers and $\mathcal{A}^G_{\mathrm{chiral}}$ is the term with ghost number G,

$$\mathcal{A}^G_{\mathrm{chiral}} = [P(\tilde{q}_K, f_K)]_G. \tag{3.4b}$$

The polynomials $P(\tilde{q}_K, f_K)$ are restricted by conditions which ensure that they give nonequivalent solutions to Eqs. (2.2) and (2.9). Here we do not discuss these conditions for the general case, which can be found in [2, 3], but give two examples for chiral solutions. The first example introduces the $\tilde{q}_K$ and f_K and the second example discusses those parts of P, Eq. (3.4b), which are bilinear in the $\tilde{q}_K$ and f_K.

Example 1: Fundamental Chiral Solutions

The following construction of solutions to Eq. (2.2) is the standard one and can be found e.g. in [8]: To the Kth Casimir (with order m_K) there corresponds the $2m_K$-Chern-form f_K,

$$f_K = \mathrm{tr}(F^{m_K}). \tag{3.5}$$

Each f_K is closed,

$$df_K = 0, \tag{3.5a}$$

due to the Bianchi identity

$$DF = dF + [F, A] = 0. \tag{3.5b}$$

Eq. (3.5b) follows from $d^2 = 0$ and because F is given by

$$F = dA - A^2. \tag{3.5c}$$

Eq. (3.5a) holds in arbitrary dimension because it is an algebraic identity in F and A due to Eq. (3.5b), and therefore the Algebraic Poincaré Lemma (Eq. (4.13) below) implies the existence of the Chern-Simons form q_K^0 whose exterior derivative gives f_K,

$$f_K = dq_K^0(A, F). \tag{3.5d}$$

Chiral solutions follow from a generalization of Eqs. (3.5a) to (3.5d). Each f_K is not only closed, but is also gauge invariant, $sf_K = 0$. Therefore Eq. (3.5a) is generalized to

$$(d + s)f_K = 0. \tag{3.5e}$$

The generalization of Eqs. (3.5b), (3.5c) reads

$$F = (d + s)(A + C) - (A + C)^2, \quad (d + s)F + [F, A + C] = 0. \tag{3.6}$$

Eq. (3.6) follows from

$$sC = C^2, \quad sA = -dC + \{C, A\} \tag{3.6a}$$

which is a consequence of Eqs. (2.3) to (2.7) and (3.4a) (note the minus sign in front of dC which appears in sA and follows from the grading of dx^μ).

Eqs. (3.6) are the same identities for the variables F and $A + C$ and the operator $(d + s)$ as Eqs. (3.5b), (3.5c) for the variables F and A and the operator s. Therefore, if one replaces d in Eq. (3.5d) by $(d + s)$, and A by $A + C$, one gets the algebraic identity

$$f_K = (d + s)\tilde{q}_K(A, C, F) \tag{3.7}$$

where $\tilde{q}_K(A, C, F) = q_K^0(A + C, F)$.

Explicitly, $\tilde{q}_K(A, C, F)$ is given by

$$\begin{aligned} \tilde{q}_K = \textstyle\sum_{l=0}^{m-1} \frac{m!(m-1)!}{(m+l)!(m-l-1)!} \operatorname{Str}(\tilde{A}\tilde{B}^l F^{m_K-l-1}), \\ \tilde{A} = A + C, \quad \tilde{B} = \tilde{A}^2, \\ \operatorname{Str}(M_1 \dots M_n) = \tfrac{1}{n!} \textstyle\sum_\pi \operatorname{tr}(M_{\pi(1)} \dots M_{\pi(n)}). \end{aligned} \tag{3.8}$$

(Str denotes symmetrized traces, the sum $\sum_\pi$ runs over all permutations π of $(1, \dots, n)$.)

Eq. (3.7) contains solutions to Eq. (2.2) with ghost numbers g ranging between 0 and $(2m_K - 1)$ and having form degree $(2m_K - 1 - g)$. In other words, it gives solutions in $(2m_K - 1 - g)$ dimensions. To see this one decomposes $\tilde{q}_K$ into its parts q_K^g with definite ghost number g,

$$\tilde{q}_K = \sum_{g=0}^{2m_K-1} q_K^g(C, A, F). \tag{3.9}$$

q_K^g has form degree $(2m_K - 1 - g)$, because the sum of ghost number and form degree ('total degree') of $\tilde{q}_K$ is $(2m_K - 1)$ (this follows because $(d + s)$ increases the total degree by one and f_K has total degree $2m_K$). Note that the part $q_K^{2m_K-1}$ with highest ghost number appearing in $\tilde{q}_K$ is just Θ_K, Eq. (3.3), which shows the correspondence of the f_K and Θ_K.

Expanding Eq. (3.7) into the parts with different ghost number one gets

$$f_K = dq_K^0, \quad 0 = sq_K^g + dq_K^{g+1} \quad 0 \le q < 2m_K - 1, \quad 0 = sq_K^{2m_K-1}. \tag{3.10}$$

This shows that each q_K^g is a solution of Eq. (2.2) in $(2m_K - 1 - g)$ dimensions (where q_K^g is a volume form).

We note that the existence of a $\tilde{q}_K$ satisfying Eq. (3.7) follows from Eq. (3.5e) and the Algebraic Poincaré Lemma in Eq. (4.3), as Eq. (3.5d) follows from Eq. (3.5a), i.e. we do not need Eqs. (3.6) to prove the existence of $\tilde{q}_K$ (but only to evaluate its explicit form in a convenient way). To show this we apply s to Eq. (3.5d). Because of $\{s, d\} = 0$, $sf_K = 0$, this gives $d(sq_K^0) = 0$, which is an algebraic identity in (derivatives of) C^I and A_μ^I, and holds in arbitrary dimensions. The Algebraic Poincaré Lemma therefore implies the existence of a form q_K^1 such that $sq_K^0 + dq_K^1 = 0$. Applying s to this equation and iterating the argument gives Eqs. (3.10) (of course these arguments alone are not sufficient to prove that the set of q_K^g terminates with a zero-form $q_K^{2m_K-1}$).

Example 2: Chiral Solutions Contained in Monomials Bilinear in the $\tilde{q}_K$ and f_K

We assume the labels K of the Casimir operators ordered such that $K < K'$ implies $m_K \le m_{K'}$. We show that $\tilde{q}_K f_{K'}$, $K \le K'$, contain chiral solutions with ghost numbers ranging from 0 to $(2m_K - 1)$, and that $\tilde{q}_K\tilde{q}_{K'}, K < K'$, contain chiral solutions with ghost numbers ranging fom $2m_{K'}$ to $2(m_K + m_{K'} - 1)$.

Consider a product $f_1 f_2$ of two f_K with $m_1 < m_2$. From the preceeding example it follows that

$$f_1 f_2 = (d+s)(\tilde{q}_1 f_2) = (d+s)(f_1\tilde{q}_2). \tag{3.11}$$

Expanding Eq. (3.11) in ghost numbers, one sees that $\tilde{q}_1 f_2$ gives solutions $q_1^g f_2$ to Eq. (2.2) with ghost numbers $0 \le g \le (2m_1 - 1)$, and $\tilde{q}_2 f_1$ gives solutions $q_2^g f_1$ with ghost numbers $0 \le g \le (2m_2 - 1)$ (i.e. $\tilde{q}_2 f_1$ gives more solutions than $\tilde{q}_1 f_2$). However, the two solutions $q_1^g f_2$ and $q_2^g f_1$ are equivalent, i.e. they differ only by a trivial solution. Especially $q_2^g f_1$ itself is a trivial solution if g exceeds $(2m_1 - 1)$, which is the highest ghost number appearing in $\tilde{q}_1 f_2$. This follows by expanding the following equation into its parts with different ghost number:

$$f_1\tilde{q}_2 - \tilde{q}_1 f_2 = (d+s)(\tilde{q}_1\tilde{q}_2) \tag{3.12}$$

Therefore we drop the solutions we get from $\tilde{q}_2 f_1$, resp. the polynomials P appearing in Eq. (3.4b) can be taken to be independent of monomials $\tilde{q}_2 f_1$ with $m_1 < m_2$.

Eq. (3.12) shows that $\tilde{q}_K f_{K'}$ can be dropped also for $m_K = m_{K'}$, $K' < K$. This eqation also shows that the parts of $\tilde{q}_1\tilde{q}_2$ with ghost numbers exceeding $(2m_2 - 1)$ solve Eq. (2.2), because for these ghost numbers the left-hand-side of Eq. (3.12) vanishes.

We conclude this Section by giving the chiral solutions with ghost numbers zero and one. As in Example 2 we assume the labels K ordered such that $K < K'$ implies $m_K \le m_{K'}$.

Chiral Solutions with $G = 0$:

The polynomials P appearing in Eq. (3.4b) which lead to nontrivial and inequivalent solutions of Eq. (2.2) with $G = 0$ are linear combinations of monomials

$$\tilde{q}_K \prod_{K' \ge K} f_{K'}^{n(K')}, \quad n(K') \ge 0,$$

whose parts with $G = 0$ are

$$q_K^0 \prod_{K' \ge K} f_{K'}^{n(K')},$$

which have form degree $2(m_K + \sum_{K' \ge K} n(K') m_{K'}) - 1$. Chiral solutions with $G = 0$ therefore appear in odd dimensions only. They give just the Chern-Simons actions (recall that the q_K^0, Eq. (3.5d), are the Chern-Simons forms). The result for $G = 0$ therefore contains the proof that the Chern-Simons actions are the only local actions which are gauge invariant (up to boundary terms) but whose Lagrangians are not invariant by themselves but transform into a total derivative.

Chiral Solutions with $G = 1$:

The polynomials P, Eq. (3.4b), which give the inequivalent and nontrivial solutions with $G = 1$, are linear combinations of monomials which either are of the form

$$\tilde{q}_K \prod_{K' \geq K} f_{K'}^{n(K')} \quad n(K') \geq 0 \qquad (typeA)$$

(these monomials appeared in the $G = 0$-case already) or of the form

$$\tilde{q}_K \tilde{q}_{K'} \prod_{K'' \geq K'} f_{K''}^{n(K'')} \quad K < K', \quad m_K = m_{K'} = 1, \quad n(K'') \geq 0. \qquad (typeB)$$

The $G = 1$-parts of the monomials of type A are

$$q_K^1 \prod_{K' \geq K} f_{K'}^{n(K')}, \qquad n(K') \geq 0, \quad q_K^1 = \mathrm{tr}(CF^{m_K - 1} + \ldots).$$

They have form degree $2(m_K + \sum_{K' \geq K} n(K')m_{K'} - 1)$ and therefore appear in even dimensions only.

The $\tilde{q}_K$ with $m_K = 1$ are the abelian ones, $\tilde{q}_K = C^K + A^K$, where C^K and A^K are the ghost and gauge field one-form of the Kth abelian factor of the gauge group. The monomials of type B therefore contain abelian $\tilde{q}_K$ only. Their $G = 1$-parts are given by

$$(C^K A^{K'} - C^{K'} A^K) \prod_{K'' \geq K'} f_{K''}^{n(K'')},$$
$$K < K', \quad m_K = m_{K'} = 1, \quad n(K'') \geq 0$$

and have form-degree $2 \sum_{K''} n(K'')m_{K''} + 1$. Therefore they appear in odd dimensions, and only if the gauge group contains at least two abelian factors.

4. Theorems

Before we sketch in Section 5 the proof of the results described in the preceeding Section we list and comment in this Section the most important theorems needed for that proof.

The first theorem is a useful instrument for the investigation of the cohomology of some nilpotent operator s, $s^2 = 0$ (not necessary the BRS-operator), or for the investigation of an equation

$$sf = dg, \quad s^2 = d^2 = \{s, d\} = 0.$$

4.1 Basic Lemma:

If there exists a linear operator $\mathcal{O}$ with the following properties:

a) its eigenfunctions are complete, i.e. each function f can be uniquely decomposed into

$$f = \sum_\lambda f_\lambda\,, \quad \mathcal{O}f_\lambda = \lambda f_\lambda \quad (f = 0 \Leftrightarrow f_\lambda = 0 \quad \forall \lambda)$$

and the spectrum of $\mathcal{O}$ is isolated around $\lambda = 0$ (i.e. 0 is not an accumulation point of eigenvalues of $\mathcal{O}$)

b) $\mathcal{O}$ can be written as $\mathcal{O} = \{s, r\}$ for some suitable operator r (which implies $[\mathcal{O}, s] = 0$ due to $s^2 = 0$),

then each solution f to $sf = 0$ is of the form $f = f_0 + s(r \sum_{\lambda \neq 0} \frac{1}{\lambda} f_\lambda)$, i.e. trivial if f_0 vanishes,

$$sf = 0 \quad \Rightarrow \quad f = f_0 + s\Big(r \sum_{\lambda \neq 0} \frac{1}{\lambda} f_\lambda\Big). \tag{4.1}$$

The lemma can be extended:

c) If in addition $\{r, d\} = 0$ (and thus $[\mathcal{O}, d] = 0$) then

$$sf = dg \;\Rightarrow\; f = f_0 + s\Big(r \sum_{\lambda \neq 0} \frac{1}{\lambda} f_\lambda\Big) - d\Big(r \sum_{\lambda \neq 0} \frac{1}{\lambda} g_\lambda\Big). \tag{4.2}$$

Proof: We prove the extended version ($g = 0$ gives the proof of the simple version).
Due to (a) f and g can be decomposed into eigenfunctions of $\mathcal{O}$,

$$f = \sum_\lambda f_\lambda \quad \text{and} \quad g = \sum_\lambda g_\lambda .$$

From $sf = dg$ it follows that

$$sf_\lambda = dg_\lambda \quad \text{for all } \lambda \quad \text{because of } [\mathcal{O}, s] = [\mathcal{O}, d] = 0,$$

i.e. each eigenspace of $\mathcal{O}$ is invariant under both s and d.
From (b) it follows that

$$f = \sum_\lambda f_\lambda = f_0 + \{s, r\} \sum_{\lambda \neq 0} \frac{1}{\lambda} f_\lambda = f_0 + sr \sum_{\lambda \neq 0} \frac{1}{\lambda} f_\lambda + rd \sum_{\lambda \neq 0} \frac{1}{\lambda} g_\lambda$$

which proves the lemma, because $\{d, r\} = 0$ due to (c). □

Example 1: BRS-invariant functionals depend on the antighosts $[\bar{C}^I]$ and on the auxiliary fields $[B^I]$ only trivially, i.e.

$$\begin{aligned} s\mathcal{A}([\phi]) + d\mathcal{A}'([\phi]) = 0 \quad &\Rightarrow \\ \mathcal{A}([\phi]) = \mathcal{A}_0([C^I, \psi, A^I_\mu]) + s\mathcal{B}([\phi]) &+ d\mathcal{B}'([\phi]). \end{aligned} \tag{4.3}$$

To prove this we define a graded operator r as follows:

$$rB^I = \bar{C}^I, \quad r\bar{C}^I = rC^I = r\psi = rA^I_\mu = 0, \quad |r| = 1, \quad [r, \partial_\mu] = 0. \tag{4.4}$$

One checks that $\mathcal{O} = \{s, r\} = N_{[\bar{C}]} + N_{[B]}$ is just the counting operator for the variables $[\bar{C}^I, B^I]$. On polynomials of $[\phi]$ $\mathcal{O}$ has a discrete spectrum and its eigenfunctions are complete. By its definition r anticommutes with d and therefore the extended version of the Basic Lemma states that a solution $\mathcal{A}$ of $s\mathcal{A} + d\mathcal{A}' = 0$ depends only trivially on $[\bar{C}^I, B^I]$.

For ghost number zero this implies that $\mathcal{A}_0$ does not depend on the ghosts, i.e. the nontrivial part of a BRS-invariant action depends only on $[\psi, A^I_\mu]$, and hence is BRS-invariant if and only if it is gauge invariant.

For negative ghost number $\mathcal{A}_0$ vanishes, i.e. for $G < 0$ there are only trivial solutions to Eq. (2.2).

Example 2: As a second example we analyse the cohomology of the linearized BRS-operator s_0 ($s_0^2 = 0$) for the algebra given in Section 2, Eqs. (2.3) to (2.5). s_0 acts only on the partial derivatives of the gauge fields,

$$\partial_{\mu_1} \dots \partial_{\mu_{k-1}} A^I_{\mu_k}, \tag{4.5}$$

which we decompose into the parts

$$A_{(k)} := \partial_{(\mu_1} \dots \partial_{\mu_{k-1}} A^I_{\mu_k)}, \tag{4.6}$$

which are totally symmetrized in their Lorentz-indices and transform under s_0 into

$$C_{(k)} := \partial_{\mu_1} \dots \partial_{\mu_k} C^I \tag{4.7}$$

and partial derivatives of the linearized field strengths ${F^o}^I_{\mu\nu}$,

$${F^o}^I_{\mu\nu} := \partial_\mu A^I_\nu - \partial_\nu A^I_\mu, \tag{4.8}$$

which are s_0-invariant. With these quantities the action of s_0 is given by

$$s_0 = \sum_{k\geq 1} C_{(k)} \frac{\partial}{\partial A_{(k)}}. \tag{4.9a}$$

We introduce the operator $r = \sum_{k\geq 1} A_{(k)} \partial / \partial C_{(k)}$ (which does not commute with d). The anticommutator

$$\mathcal{O} = \{s_0, r\} = \sum_{k\geq 1} (C_{(k)} \frac{\partial}{\partial C_{(k)}} + A_{(k)} \frac{\partial}{\partial A_{(k)}}) \tag{4.9b}$$

is the counting operator for all derivatives of the ghosts and the totally symmetrized derivatives of the gauge fields. By means of the Basic Lemma we conclude that the nontrivial part of an s_0-invariant function (polynomial in the $[C^I, A^I_\mu, \psi]$) is independent of $C_{(k)}$, $A_{(k)}$, $k \geq 1$, i.e. can depend only on undifferentiated ghosts C^I and on (derivatives of) the matter fields ψ and the linearized field strengths ${F^o}^I_{\mu\nu}$. This proves the implication '$\Rightarrow$' of the following equation, the implication '$\Leftarrow$' is trivial because the C^I and $[\psi, {F^o}^I_{\mu\nu}]$ are s_0-invariant.

4.2 s_0-Cohomology:

$$s_0 f([C^I, A^I_\mu, \psi]) = 0 \quad \Leftrightarrow \quad f = f_0(C^I, [{F^o}^I_{\mu\nu}, \psi]) + s_0 g([C^I, A^I_\mu, \psi]). \quad (4.9)$$

We note that the sum $f_0(C^I, [{F^o}^I_{\mu\nu}, \psi]) + s_0 g$ appearing in Eq. (4.9) is direct because $s_0 g$ is at least linear in the $C_{(k)}$, $k \geq 1$ due to Eq. (4.9a) and therefore contains no piece which depends only on $C^I, [{F^o}^I_{\mu\nu}, \psi]$.

Example 3: As a third application of the Basic Lemma, we prove that solutions $\mathcal{A}$ of $s\mathcal{A} + d\mathcal{A}' = 0$ are (up to trivial terms) invariant under the global action δ_I of the gauge group (the ghosts transform under the adjoint representation, $\delta_I C^J = -f_{IK}{}^J C^K$). Due to Eq. (4.3) we can assume $\mathcal{A}$ to depend only on $[C^I, \psi, A^I_\mu]$. On these variables δ_I can be expressed by the anticommutator $-\{s, \partial/\partial C^I\}$. As a consequence each Casimir operator $\mathcal{O}_K = g^{I_1 \ldots I_m}\delta_{I_1} \ldots \delta_{I_m}$, $m = m_K$ of the gauge group[2], can be expressed as an anticommutator of s with a suitable operator r_K:

$$\begin{aligned} &\delta_I = -\{s, \frac{\partial}{\partial C^I}\}, \quad \mathcal{O}_K = g^{I_1 \ldots I_m}\delta_{I_1} \ldots \delta_{I_m} = \{s, r_K\}, \\ &r_K = (-)^m g^{I_1 \ldots I_m} \frac{\partial}{\partial C^{I_1}} s \frac{\partial}{\partial C^{I_2}} s \ldots s \frac{\partial}{\partial C^{I_m}}, \quad m = m_K. \end{aligned} \quad (4.10)$$

Each r_K commutes with d (Lorentz-indices are not seen by δ_I). The Casimir operators $\mathcal{O}_K$ of the semisimple part of the gauge group allow a unique decomposition of any $\mathcal{A}$ into eigenfunctions $\mathcal{A}_\lambda$ with eigenvalues $c(K, \lambda)$,

$$\mathcal{O}_K \mathcal{A}_\lambda = c(K, \lambda)\mathcal{A}_\lambda$$

(λ labels the representation according to which $\mathcal{A}_\lambda$ transforms). Due to the Basic Lemma only those $\mathcal{A}_\lambda$ with vanishing eigenvalues to each $\mathcal{O}_K$ can be nontrivial. The only representation with this property is the trivial one. This proves

$$s\mathcal{A} + d\mathcal{A}' = 0 \quad \Rightarrow \quad \mathcal{A} = \hat{\mathcal{A}} + s\mathcal{B} + d\mathcal{B}', \quad \delta_I \hat{\mathcal{A}} = 0. \quad (4.11)$$

[2] $g^{I_1 \ldots I_m}$ are completely symmetric coefficients which are obtained from symmetrized traces $\mathrm{Str}(T_{I_1} \ldots T_{I_m})$ taken in an appropriate matrix representation T_I of the δ_I

The next theorem gives the cohomology of the exterior derivative d which acts on the variables $[\phi]$ as described in Section 2. It states that d-closed forms ω are always exact in the space of forms depending on the $[\phi]$ with the exception of constant forms (which do not depend on $[\phi]$ at all) and of forms of maximal degree, i.e. volume forms $d^D x\mathcal{L}([\phi])$ (which are trivially closed), whose Euler derivative $\hat{\partial}/\hat{\partial}\phi$ with respect to at least one of the fields ϕ does not vanish,

$$\frac{\hat{\partial}}{\hat{\partial}\phi} = \frac{\partial}{\partial\phi} - \partial_\mu \frac{\partial}{\partial(\partial_\mu\phi)} + \dots . \tag{4.12}$$

The proof of this theorem can be found in [3].

4.3 Algebraic Poincaré Lemma:

$$\begin{gathered} d\omega_p([\phi]) = 0 \quad \Leftrightarrow \\ \omega_p([\phi]) = d\omega_{p-1}([\phi]) + \delta_{p,D} d^D x\mathcal{L}([\phi]) + dx^{\mu_1} \dots dx^{\mu_p} c_{\mu_1\dots\mu_p} . \end{gathered} \tag{4.13}$$

The constant forms $dx^{\mu_1} \dots dx^{\mu_p} c_{\mu_1\dots\mu_p}$ ($c_{\mu_1\dots\mu_p}$ are constants) are Lorentz-invariant only if $p = 0$ or $p = D$, i.e. a pure constant or a constant times the volume element $d^D x$ are the only field independent forms that contribute to Eq. (4.13) if the forms ω_p are restricted to be Lorentz-invariant.

An important lemma is the following one. It determines the cohomology of d in the *space of gauge invariant forms* depending on $[\psi, A^I_\mu]$. It states that closed and locally exact gauge invariant forms are also 'covariant exact', i.e. locally exact in the space of gauge invariant forms with the exception that they are (up to a covariant exact form) Chern-forms, i.e. functions of the $f_K = \mathrm{tr}(F^{m_K})$, Eq. (3.5). In other words: if a gauge invariant form ω_p is locally the exterior derivative of some form ω_{p-1} then ω_{p-1} *itself* is gauge invariant, except if ω_p is a function of the f_K.

If ω_p is covariantly exact, i.e. if in each coordinate patch there exists an invariant ω_{p-1}, $\omega_p = d\omega_{p-1}$, then ω_p actually is globally exact because no nontrivial gauge transformations occur which connect the ω_{p-1} of two neighbouring coordinate patches for the ω_{p-1} being group scalars. As a consequence the covariant exact forms do not contribute to topological densities. Therefore the Theorem also states that the Chern-forms are the only invariant and locally exact forms of the $[\phi]$ which may be globally not exact and which can contribute to topological densities.

We note that in curved space an analogous result holds [5] (there are additional f_K in curved space, namely $f_K(R)$ depending on the curvature 2-form R).

4.4 Covariant Poincaré Lemma:

$$\begin{aligned}&\omega_p = \omega_p([\psi, A^I_\mu]), \quad s\omega_p = 0, \quad \omega_p = d\eta_p \\ \Leftrightarrow \quad &\omega_p = d\omega_{p-1} + \hat{\omega}_p(f_K), \quad s\omega_{p-1} = 0, \quad f_K = \mathrm{tr}(F^{m_K}).\end{aligned} \tag{4.14}$$

The Covariant Poincaré Lemma is proven by proving its linearized version [3, 5] (one verifies by means of the s_0-cohomology, Eq. (4.9), that the linearized version in the formulation given below is a necessary condition for the complete version; it is also sufficient because each term appearing in the linearized version is easily extended – by replacing partial derivatives of ψ and $F^{oI}{}_{\mu\nu}$ by symmetrized covariant derivatives and linearized field strengths by the complete nonabelian ones – such that the extended terms satisfy the extended lemma).

4.5 Linearized Covariant Poincaré Lemma:

$$\begin{aligned}&\omega_p = \omega_p([\psi, F^{oI}{}_{\mu\nu}]), \quad \delta_I\omega_p = 0, \quad \omega_p = d\eta_p \\ \Leftrightarrow \quad &\omega_p = d\omega_{p-1}([\psi, F^{oI}{}_{\mu\nu}]) + \hat{\omega}_p(f^o{}_K), \quad \delta_I\omega_{p-1} = 0, \\ &f^o{}_K = \mathrm{tr}(F^{om_K}), \quad F^o = \frac{1}{2} dx^\mu dx^\nu F^{oI}{}_{\mu\nu} T_I = dA.\end{aligned} \tag{4.15}$$

We note that the sum $d\omega_{p-1}([\psi, F^{oI}{}_{\mu\nu}]) + \hat{\omega}_p(f^o{}_K)$ is direct, because $f^o{}_K$ contains as many derivatives as gauge fields A^I_μ, and therefore is not of the form $d\omega_{p-1}([F^{oI}{}_{\mu\nu}])$.

The last theorem we list here determines the Lie algebra cohomology, i.e. the cohomology of the BRS-operator s on functions depending only on the undifferentiated ghosts C^I. It states that all s-invariant functions of the C^I depend (up to trivial contributions) on the ghosts only via the Θ_K which are defined by Eq. (3.3). The proof can be found in [4].

4.6 Lie Algebra Cohomology:

$$sf(C^I) = 0 \Leftrightarrow f(C^I) = \hat{f}(\Theta_K(C^I)) + sg(C^I). \tag{4.16}$$

5. Proof of the Results

We now describe the proof of the results stated in Section 3, up to a stage where it is obvious how they emerge from the theorems of the preceding Section.

We first derive the descent equations for $(G+D-g')$-forms $\mathcal{A}^{g'}$ with ghost number g' which terminate to a s-invariant $(G+D-g)$-form $\mathcal{A}^g$ at some ghost number g (if the form degree has dropped to zero at the latest).

$$\begin{aligned} &s\mathcal{A}^{g'} + d\mathcal{A}^{g'+1} = 0 \quad \text{for } G \le g' < g \le G-D, \\ &s\mathcal{A}^{g} = 0, \quad \mathcal{A}^{g'} \ne s\mathcal{B}^{g'-1} + d\mathcal{B}^{g'} \quad \text{for } G \le g' \le g. \end{aligned} \tag{5.1}$$

Note that we claim that each form $\mathcal{A}^{g'}$ appearing in the descent equations can be taken to be nontrivial in the sense of Eq. (2.9).

Proof: We start with Eqs. (2.2) and (2.9). There are two possible cases:

a) If $\mathcal{A}^{G+1}$ appearing in Eq. (2.2) is trivial or vanishing, i.e. for some $\mathcal{B}^{G+1}, \mathcal{B}^{G}$ (which may vanish), we can express it as $\mathcal{A}^{G+1} = d\mathcal{B}^{G+1} + s\mathcal{B}^{G}$, then $\mathcal{A}'^{G} = \mathcal{A}^{G} - d\mathcal{B}^{G}$, which is equivalent to $\mathcal{A}^{G}$, satisfies $s\mathcal{A}'^{G} = 0$, i.e. the descent equations terminate at $g = G$.
b) If $\mathcal{A}^{G+1}$ is nontrivial and nonvanishing, $\mathcal{A}^{G+1} \ne d\mathcal{B}^{G+1} + s\mathcal{B}^{G}$, then we apply s to Eq. (2.2), which gives $d(s\mathcal{A}^{G+1}) = 0$, i.e. $s\mathcal{A}^{G+1}$ is closed. By the Algebraic Poincaré Lemma, Eq. (4.13), we conclude that $s\mathcal{A}^{G+1}$ is exact or vanishes (because $s\mathcal{A}^{G+1}$ depends on $[\phi]$ or vanishes and is not a D-form), i.e. $\mathcal{A}^{G+1}$ satisfies $s\mathcal{A}^{G+1} + d\mathcal{A}^{G+2} = 0$ for some $\mathcal{A}^{G+2}$ which may vanish.

In case (b) $\mathcal{A}^{G+1}$ satisfies Eqs. (2.2) and (2.9) for G replaced by $G+1$ and, as before, there are the two cases (a) and (b) (for $\mathcal{A}^{G+2}$). We repeat the arguments until we end at an equation $s\mathcal{A}^g = 0$ for some g, which proves Eqs. (5.1). □

A first consequence of Eqs. (5.1) is that each $\mathcal{A}^{g'}$ can be taken to be δ_I-invariant and independent of the antighosts $[\bar{C}^I]$ and the auxiliary fields $[B^I]$ due to Eqs. (4.3) and (4.11),

$$\mathcal{A}^{g'} = \mathcal{A}^{g'}([C^I, \psi, A^I_\mu]), \quad \delta_I \mathcal{A}^{g'} = 0 \quad \forall g'. \tag{5.2}$$

We now concentrate on the equations $s\mathcal{A}^g = 0$ and $s\mathcal{A}^{g-1} + d\mathcal{A}^g = 0$ which are contained in Eq. (5.1). We split these equations into their parts with different degree of homogeneity in the variables $[\phi]$. $\mathcal{A}^g$ splits into parts $\mathcal{A}^g_l$ with degree of homogeneity l in the $[\phi]$ ($\mathcal{A}^{g-1}$ splits analogously). The part $\mathcal{A}^g_\ell$ of $\mathcal{A}^g$ with lowest degree $\ell := l_{\min}$ is called the *head of* $\mathcal{A}^g$. The head will turn out to be characteristic for the whole solution

$$\mathcal{A}^g = \sum_{l \ge \ell} \mathcal{A}^g_l. \tag{5.3}$$

The BRS-operator s splits into s_0, which does not increase the degree of homogeneity in $[\phi]$, and s_1, which increases it by one. The linearized BRS-operator s_0 acts only on derivatives of the gauge fields which are completely symmetrized in their Lorentz indices,

$$s_0 \partial_{(\mu_1} \ldots \partial_{\mu_{k-1}} A^I_{\mu_k)} = \partial_{\mu_1} \ldots \partial_{\mu_k} C^I, \tag{5.4}$$

and vanishes otherwise, Eq. (4.9a). For

$$s = s_0 + s_1 \tag{5.5}$$

the relations $s^2 = [s, \partial_\mu] = 0$ split into

$$s_0^2 = \{s_0, s_1\} = s_1^2 = [s_0, \partial_\mu] = [s_1, \partial_\mu] = 0. \tag{5.5a}$$

(∂_μ does not change the degree of homogeneity in the $[\phi]$). The equations $s\mathcal{A}^g = 0$ and $s\mathcal{A}^{g-1} + d\mathcal{A}^g = 0$ split into

$$\begin{aligned} 0 &= s_0 \mathcal{A}^g_\ell, \quad 0 = s_0 \mathcal{A}^g_l + s_1 \mathcal{A}^g_{l-1} \quad \text{for } l > \ell, \\ 0 &= s_0 \mathcal{A}^{g-1}_l + s_1 \mathcal{A}^{g-1}_{l-1} \quad \text{for } l < \ell, \\ 0 &= s_0 \mathcal{A}^{g-1}_l + s_1 \mathcal{A}^{g-1}_{l-1} + d\mathcal{A}^g_l \quad \text{for } l \geq \ell. \end{aligned} \tag{5.6}$$

Before we work out the consequences which follow from Eqs. (5.6) for the head $\mathcal{A}^g_\ell$ we consider some general aspects of the problem.

We have shown that the forms $\mathcal{A}^{g'}$ appearing in the descent equations (5.1) can be taken to be nontrivial. But up to now we did not mention explicitly that we can consider $\hat{\mathcal{A}}^{g'}$ as equivalent to $\mathcal{A}^{g'}$ if they differ only by trivial terms $s\mathcal{B}^{g'-1} + d\mathcal{B}^{g'}$ (if $g = g'$ in addition we demand $sd\mathcal{B}^g = 0$),

$$\begin{aligned} \mathcal{A}^{g'} &\cong \hat{\mathcal{A}}^{g'} = \mathcal{A}^{g'} + s\mathcal{B}^{g'-1} + d\mathcal{B}^{g'} \quad \text{for} \quad g \neq g', \\ \mathcal{A}^g &\cong \hat{\mathcal{A}}^g = \mathcal{A}^g + s\mathcal{B}^{g-1} + d\mathcal{B}^g, \quad sd\mathcal{B}^g = 0. \end{aligned} \tag{5.7}$$

This definition of equivalence is justified because if we simultaneously change $\mathcal{A}^{g'}$ into $\hat{\mathcal{A}}^{g'}$, $\mathcal{A}^{g'-1}$ into $\hat{\mathcal{A}}^{g'-1}$ and $\mathcal{A}^{g'+1}$ into $\hat{\mathcal{A}}^{g'+1}$ then Eqs. (5.1) are satisfied for the set of forms $(\mathcal{A}^G, \mathcal{A}^{G+1}, \ldots, \mathcal{A}^{g'-2}, \hat{\mathcal{A}}^{g'-1}, \hat{\mathcal{A}}^{g'}, \hat{\mathcal{A}}^{g'+1}, \mathcal{A}^{g'+2}, \ldots, \mathcal{A}^g)$, and $\mathcal{A}^G$ is changed only trivially (if it is changed at all). The condition $sd\mathcal{B}^g = 0$ guarantees that $\hat{\mathcal{A}}^g$ solves $s\hat{\mathcal{A}}^g = 0$ and therefore the set of equivalent forms terminates at the same ghost number g with a nontrivial $\hat{\mathcal{A}}^g$. Note that we already made use of this equivalence to derive Eq. (5.2).

We need an appropriate criterion which allows to select one of two equivalent $\mathcal{A}^g$. The decomposition Eq. (5.3) provides in part such a criterion: if $\mathcal{A}^g \cong \hat{\mathcal{A}}^g$, we pick the one whose decomposition Eq. (5.3) starts at higher degree of homogeneity and drop the other (the criterion fails if $\mathcal{A}^g$ and $\hat{\mathcal{A}}^g$ have heads of the same degree). We make use of this criterion in the following analysis.

5.1 Lemma:

$\mathcal{A}^g$ can be dropped if there exist $\mathcal{B}^{g-1} = \sum_l \mathcal{B}_l^{g-1}$ and $\mathcal{B}^g = \sum_l \mathcal{B}_l^g$ such that

$$\mathcal{A}_\ell^g = s\mathcal{B}^{g-1} + d\mathcal{B}^g + O(\ell+1), \quad sd\mathcal{B}^g = 0,$$

where $O(\ell+1)$ denotes terms which have degree of homogeneity not less than $(\ell+1)$ and $\mathcal{A}_\ell^g$ is the head of $\mathcal{A}^g$.

Proof: If $\mathcal{A}_\ell^g$ satisfies

$$\mathcal{A}_\ell^g = s\mathcal{B}^{g-1} + d\mathcal{B}^g + O(\ell+1), \quad sd\mathcal{B}^g = 0$$

then $\mathcal{A}^g$ is equivalent to

$$\hat{\mathcal{A}}^g = \mathcal{A}^g - s\mathcal{B}^{g-1} - d\mathcal{B}^g = O(\ell+1).$$

Since $\hat{\mathcal{A}}^g = O(\ell+1)$, i.e. since $\hat{\mathcal{A}}^g$ starts at higher degree of homogeneity than $\mathcal{A}^g$ we can drop $\mathcal{A}^g$ by the criterion given above. □

5.2 Lemma:

According to the criterion given above we can choose a representative of the equivalence class containing $\mathcal{A}^g$ such that its head $\mathcal{A}_\ell^g$ depends only on the matter fields, linearized field strengths and their derivatives and on undifferentiated ghosts, i.e. on $C^I, [\psi, F^{\circ I}{}_{\mu\nu}]$.

We abbreviate this lengthy formulation by saying that the head can be taken to depend only on $C^I, [\psi, F^{\circ I}{}_{\mu\nu}]$. As a consequence of Eq. (5.2b) the head also can be taken to be δ_I-invariant (δ_I does not change the degree of homogeneity in the $[\phi]$ and therefore each part $\mathcal{A}_l^{g'}$ of $\mathcal{A}^{g'}$ is separately δ_I-invariant),

$$\mathcal{A}_\ell^g = \mathcal{A}_\ell^g(C^I, [\psi, F^{\circ I}{}_{\mu\nu}]), \quad \delta_I \mathcal{A}_\ell^g = 0. \tag{5.8}$$

Proof: The first equation of Eq. (5.6a) states that $\mathcal{A}_\ell^g$ is s_0-invariant. By means of the s_0-cohomology, Eq. (4.9), we conclude

$$\mathcal{A}_\ell^g = \hat{\mathcal{A}}_\ell^g(C^I, [\psi, F^{\circ I}{}_{\mu\nu}]) + s_0\mathcal{B}_\ell^{g-1} \quad \text{respectively}$$
$$\hat{\mathcal{A}}^g := \mathcal{A}^g - s\mathcal{B}_\ell^{g-1} = \hat{\mathcal{A}}_\ell^g + O(\ell+1).$$

In case of $\hat{\mathcal{A}}_\ell^g = 0$ we get

$$\mathcal{A}_\ell^g = s\mathcal{B}_\ell^{g-1} + O(\ell+1),$$

i.e. by Lemma (5.1) $\mathcal{A}_\ell^g$ can be dropped. In case of $\hat{\mathcal{A}}_\ell^g \neq 0$,

$$\hat{\mathcal{A}}^g = \mathcal{A}^g - s\mathcal{B}_\ell^{g-1} = \hat{\mathcal{A}}_\ell^g + O(\ell+1).$$

$\hat{\mathcal{A}}^g$ is equivalent to $\mathcal{A}^g$ and its head $\hat{\mathcal{A}}_\ell^g$ satisfies Eq. (5.8). □

5.3 Lemma:

The following conditions are necessary (but not sufficient) for $\mathcal{A}^g_\ell$ to be head of a solution $\mathcal{A}^g$ of $s\mathcal{A}^g = 0$ (the second condition gives no restrictions on $\mathcal{A}^g_\ell$ if $g = G$ because in this case it is trivially fulfilled):

$$s^c \mathcal{A}^g_\ell = 0, \quad s^c \mathcal{A}'^{g-1}_{\ell-1} - \bar{d}\mathcal{A}^g_\ell = 0. \tag{5.9}$$

Here $\mathcal{A}'^{g-1}_{\ell-1}$ is the part of $\mathcal{A}^{g-1}_{\ell-1}$ which does not contain derivatives of the ghosts. The operator s^c is the part of s which acts only on the ghosts C^I and $\bar{d}$ differentiates only the ghost number zero variables,

$$s^c = \frac{1}{2} f_{JK}{}^I C^J C^K \frac{\partial}{\partial C^I}, \quad \bar{d} = dx^\mu \bar{\partial}_\mu, \quad \bar{\partial}_\mu C^I = 0, \quad \bar{\partial}_\mu[\psi, A^I_\nu] = \partial_\mu[\psi, A^I_\nu]. \tag{5.10}$$

Proof: We look at those equations (5.6) which contain the head $\mathcal{A}^g_\ell$ and are not already satisfied by Eq. (5.8), namely

$$s_0 \mathcal{A}^g_{\ell+1} + s_1 \mathcal{A}^g_\ell = 0 \quad \text{and} \quad s_0 \mathcal{A}^{g-1}_\ell + s_1 \mathcal{A}^{g-1}_{\ell-1} + d\mathcal{A}^g_\ell = 0.$$

Eqs. (5.9) are those parts of these equations which are independent of differentiated ghosts $[\partial C]$. To see this we look for the parts of s_1 and d which do not contain $[\partial C]$. The relevant part of d of course is just $\bar{d}$. On functions which are independent of $[\partial C]$ the relevant part of s_1 is $s_{1,0} = -s^c - C^I \delta_I$ (this follows because $sC^I = -\frac{1}{2} C^J \delta_J C^I$). On $\mathcal{A}^g_\ell$ and the part of $\mathcal{A}^{g-1}_{\ell-1}$ which is independent of $[\partial C]$ $s_{1,0}$ therefore is given by $-s^c$ due to Eq. (5.2) and this proves Eq. (5.9) (recall that s_0 has no part independent of $[\partial C]$). □

5.4 Lemma:

The head can be taken to depend on the ghosts only via the Θ_K,

$$\mathcal{A}^g_\ell = \mathcal{A}^g_\ell(\Theta_K, [\psi, F^{o}{}^I_{\mu\nu}]). \tag{5.11}$$

Proof: By the first equation (5.9) and the Lie algebra cohomology, Eq. (4.16) (which can be used here because s^c treats all variables as constants apart from the C^I), we conclude

$$\mathcal{A}^g_\ell = \hat{\mathcal{A}}^g_\ell(\Theta_K, [\psi, F^{o}{}^I_{\mu\nu}]) + s^c \mathcal{B}^{g-1}_{\ell-1}.$$

$\mathcal{B}^{g-1}_{\ell-1}$ is s_0-invariant due to Eq. (5.8) (it depends only on $C^I, [\psi, F^{o}{}^I_{\mu\nu}]$). Therefore

$$s\mathcal{B}^{g-1}_{\ell-1} = (s_0 + s_{1,0})\mathcal{B}^{g-1}_{\ell-1} + X = s^c \mathcal{B}^{g-1}_{\ell-1} + X.$$

X is a term at least linear in derivatives $[\partial C]$ of C^I which has degree of homogeneity ℓ. It follows that X is s_0-invariant because $s\mathcal{B}^{g-1}_{\ell-1}$ and $s^c \mathcal{B}^{g-1}_{\ell-1}$ are s_0-invariant due to $\{s, s_0\} = \{s^c, s_0\} = 0$ and $s_0 \mathcal{B}^{g-1}_{\ell-1} = 0$. By the s_0-cohomology,

Eq. (4.9), we conclude $X = s_0 Y$ because X is at least linear in $[\partial C]$. Y has the same degree of homogeneity as X, namely ℓ. Inserting $s_0 Y$ into the equation for $s\mathcal{B}^{g-1}_{\ell-1}$ we get

$$s^c \mathcal{B}^{g-1}_{\ell-1} = s\mathcal{B}^{g-1}_{\ell-1} - s_0 Y = s\mathcal{B}^{g-1}_{\ell-1} - sY + O(\ell+1).$$

Therefore we get $\mathcal{A}^g_\ell = \hat{\mathcal{A}}^g_\ell + s(\mathcal{B}^{g-1}_{\ell-1} - Y) + O(\ell+1)$. Eq. (5.11) now follows by the arguments used in the proof for Eq. (5.8). □

We now investigate separately the cases $g = G$ and $g > G$. The difference of the two cases arises from the special role played by the volume forms in the Algebraic Poincaré Lemma, Eq. (4.13), which ultimately leads to the distinction between chiral and Lagrangian solutions.

5.5 The Case $g > G$:

We first treat the case $g > G$ and show that in this case $\mathcal{A}^g_\ell$ can be taken to be independent of the matter fields ψ and to depend on $[F^{o}{}^I_{\mu\nu}]$ only via the linearized Chern-forms $f^o{}_K$:

$$\begin{gathered}\mathcal{A}^g_\ell = \mathcal{A}^g_\ell(\Theta_K, f^o{}_K), \quad g \neq G, \quad f^o{}_K = \mathrm{tr}(F^{o\, m_K}), \\ F^o = \tfrac{1}{2} dx^\mu dx^\nu F^{o}{}^I_{\mu\nu} T_I = dA.\end{gathered} \tag{5.12a}$$

Proof: Inserting Eq. (5.11) into the second equation (5.9) we get

$$\bar{d}\mathcal{A}^g_\ell(\Theta_K, [\psi, F^{o}{}^I_{\mu\nu}]) = s^c \mathcal{A}'^{g-1}_{\ell-1}.$$

$\bar{d}\mathcal{A}^g_\ell$ depends on the C^I only via the Θ_K (recall that $\bar{d}$ treats the ghosts as constants) and therefore is not s^c of any $\mathcal{A}'^{g-1}_{\ell-1}$ due to Eq. (4.16). We conclude that $\bar{d}\mathcal{A}^g_\ell$ and $s^c\mathcal{A}'^{g-1}_{\ell-1}$ have to vanish separately, so $\bar{d}\mathcal{A}^g_\ell = 0$. Because $\mathcal{A}^g_\ell$ is not a volume form the Algebraic Poincaré Lemma Eq. (4.13) states that $\mathcal{A}^g_\ell$ is $\bar{d}$-exact or $\bar{d}$-constant. The constants with respect to $\bar{d}$ are the Θ_K in this case and Lorentz-invariance and $g \neq G$ therefore restrict the $\bar{d}$-constant part to be a zero-form depending only on the Θ_K. This proves Eq. (5.12a) if $\mathcal{A}^g_\ell$ is a zero-form.

If $\mathcal{A}^g_\ell$ is not a zero-form then it is $\bar{d}$-exact, $\mathcal{A}^g_\ell = \bar{d}\eta$. Because $\mathcal{A}^g_\ell$ also is δ_I-invariant we apply the linearized Covariant Poincaré Lemma, Eq. (4.15), and conclude

$$\mathcal{A}^g_\ell = \bar{d}\omega^g_\ell(\Theta_K, [\psi, F^{o}{}^I_{\mu\nu}]) + \hat{\omega}^g_\ell(\Theta_K, f^o{}_K), \quad \delta_I \omega^g_\ell = 0.$$

Using $\bar{d} = d - d^c$, where d^c differentiates only the ghosts, we get

$$\bar{d}\omega^g_\ell = d\omega^g_\ell - d^c\omega^g_\ell.$$

Since $s_0 d^c \omega^g_\ell = 0$ (because $d^c\omega^g_\ell$ depends not on $A_{(k)}$, Eq. (4.9a)) and $d^c\omega^g_\ell$ depends linearly on derivatives of the ghosts (due to the action of d^c) it is s_0-trivial due to Eq. (4.9),

$$d^c\omega^g_\ell = s_0\eta^{g-1}_\ell .$$

Therefore we get

$$\mathcal{A}^g_\ell = d\omega^g_\ell(\Theta_K, [\psi, F^{oI}{}_{\mu\nu}]) - s_0\eta^{g-1}_\ell + \hat{\omega}^g_\ell(\Theta_K, f^o{}_K),\ \delta_I\omega^g_\ell = 0.$$

ω^g_ℓ is easily extended to an s-invariant $\omega^g_{\rm inv}$ following the prescription we gave in the text above Eq. (4.15), $\omega^g_{\rm inv} = \omega^g_\ell + O(\ell+1)$. We conclude that

$$\hat{\mathcal{A}}^g := \mathcal{A}^g - d\omega^g_{\rm inv} + s\eta^{g-1}_\ell = \hat{\omega}^g_\ell(\Theta_K, f^o{}_K) + O(\ell+1).$$

Eq. (5.12a) now follows by the arguments used in the proof for Eq. (5.8). □

5.6 The Case $g = G$:

In the case $g = G$ the result is that $\mathcal{A}^G_\ell$ is either of the form $\mathcal{A}^G_\ell(\Theta_K, f^o{}_K)$, or it is of the form $d^Dx\mathcal{L}$ where $\mathcal{L}$ is a superfield in the Θ_K with component fields which are either δ_I-invariant functions of $[\psi, F^{oI}{}_{\mu\nu}]$ with nonvanishing Euler derivative with respect to ψ or A^I_μ or constant:

$$\mathcal{A}^G_\ell = d^Dx\mathcal{L}^G(\Theta_K, [\psi, F^{oI}{}_{\mu\nu}]) + \hat{\mathcal{A}}^G_\ell(\Theta_K, f^o{}_K) + d^Dx\omega^G(\Theta_K). \tag{5.12b}$$

Proof: $\mathcal{A}^G_\ell$ is a volume form, i.e. $\mathcal{A}^G_\ell = d^Dx\hat{\mathcal{L}}^G$. $\hat{\mathcal{L}}^G$ is a superfield in the Θ_K. Those component fields of $\hat{\mathcal{L}}^G$ which have nonvanishing Euler derivatives with respect to ψ or A^I_μ contribute to $\mathcal{L}^G$ in Eq. (5.12b). $\tilde{\mathcal{L}}^G := \mathcal{L}^G - \hat{\mathcal{L}}^G$ has vanishing Euler derivatives with respect to ψ or A^I_μ. $d^Dx\tilde{\mathcal{L}}^G$ is trivially $\bar{d}$-closed because it is a volume form. By means of the Algebraic Poincaré Lemma we conclude

$$d^Dx\tilde{\mathcal{L}}^G(\Theta_K, [\psi, F^{oI}{}_{\mu\nu}]) = \bar{d}\eta(\Theta_K, [\psi, A^I_\mu]) + d^Dx\tilde{\eta}(\Theta_K).$$

By the Covariant Poincaré Lemma it follows

$$\bar{d}\eta(\Theta_K, [\psi, A^I_\mu]) = \bar{d}\omega(\Theta_K, [\psi, F^{oI}{}_{\mu\nu}]) + \hat{\omega}(\Theta_K, f^o{}_K).$$

As in the proof of Eq. (5.12a) it follows from $\bar{d} = d - d^c$ and

$$\omega(\Theta_K, [\psi, F^{oI}{}_{\mu\nu}]) = \omega_{\rm inv}(\Theta_K, \{\psi, F^I_{\mu\nu}\}) + O(\ell+1)$$

that

$$\mathcal{A}^G_\ell = d\omega_{\rm inv}(\Theta_K, \{\psi, F^I_{\mu\nu}\}) + sY + \hat{\omega}(\Theta_K, f^o{}_K) + \\ + \tilde{\eta}(\Theta_K) + \mathcal{L}^G(\Theta_K, [\psi, F^{oI}{}_{\mu\nu}]) + O(\ell+1).$$

By the arguments used in the proof for Eq. (5.8) this leads to Eq. (5.12b). □

5.7 Final Results Emerging from Equations (5.12):

$d^Dx\mathcal{L}^G(\Theta_K, [\psi, F^{o}{}^I_{\mu\nu}])$ is easily completed to a solution $\mathcal{A}^G$ of $s\mathcal{A}^G = 0$: One simply has to replace $F^{o}{}^I_{\mu\nu}$ by the complete nonabelian field strength $F^I_{\mu\nu}$ and the partial derivatives of ψ and $F^I_{\mu\nu}$ by symmetrized covariant ones.

$$d^Dx\mathcal{L}^G(\Theta_K, [\psi, F^{o}{}^I_{\mu\nu}]) + d^Dx\omega^G(\Theta_K)$$

are just the heads of the Lagrangian solutions (see Section 3).

The heads for $g > G$ given by Eq. (5.12a) and $\hat{\mathcal{A}}^G_\ell(\Theta_K, f^o{}_K)$ lead to the chiral solutions. They are also easily completed to solutions $\mathcal{A}^g$ of $s\mathcal{A}^g = 0$ by replacing the linearized fieldstrengths-2-forms by the corresponding completed nonabelian forms. The completed $\mathcal{A}^g$ are then functions of the f_K and Θ_K which are s-invariant by themselves.

However, among these solutions $\mathcal{A}^g(\Theta_K, f_K)$ to $s\mathcal{A}^g = 0$, there are some which do not correspond to a solution of Eqs. (5.1), either because they are trivial, or because some of the $\mathcal{A}^{g'}$ needed to satisfy Eqs. (5.1) do not exist. The determination of those $\mathcal{A}^g(\Theta_K, f_K)$ which correspond to (inequivalent) solutions of Eqs. (5.1) follows from an analysis of the action of $(d + s)$ on functions of the f_K and $\tilde{q}_K$ which are given in Eq. (3.8). This analysis is a generalization of Example 2 in Section 3 and can be found in [4]. We note that if there is a solution to Eqs. (5.1) corresponding to $\mathcal{A}^g(\Theta_K, f_K)$ then one obtains it from $\mathcal{A}^g(\Theta_K, f_K)$ by replacing all Θ_K appearing in $\mathcal{A}^g$ by the corresponding $\tilde{q}_K$. This shows how the chiral solutions emerge from Eqs. (5.12a), (5.12b).

6. Conclusion

We end here the presentation of our solution of the consistency equation. The result is that there are no anomalies besides the well-known ones. For ghost number zero our solution proves that gauge invariant local actions stem from gauge invariant Lagrangians with precisely one exception: In odd dimensions there exist the Chern-Simons forms which are not gauge invariant though their integrals are. In even dimensions the building blocks of chiral anomalies are the Chern-forms $f_K = trF^{m(K)}$ and the piece q^1_K (with ghost number one) of the generalized Chern-Simons form $\tilde{q}_K$. In $D = 2n$ dimensions all chiral anomalies are products

$$\mathcal{A}^1_{D=2n} = q^1_{K_1} f_{K_2} \cdots f_{K_r}$$

which are restricted by the requirement that $\mathcal{A}^1$ is a volume form (D-form).

q_K^1 is a $(2m(K)-2)$-form, so

$$\sum_{s=1}^{r} m(K_s) = n+1.$$

In $D = 4$ dimensions this leaves the following possibilities for nonvanishing m_{K_s}:

a) $(m(K_1), m(K_2), m(K_3)) = (1,1,1)$,
b) $(m(K_1), m(K_2)) = (1,2)$,
c) $m(K_1) = 3$.

The solutions corresponding to (a) or (b) occur only if the gauge group contains a $U(1)$ (simple groups have $m(K) \geq 2$) and are called *abelian anomalies*. They are slightly special because they are BRS-invariant, i.e. the descent equations terminate at ghost number 1. The solution corresponding to (c) is the nonabelian anomaly. It contains the connection form explicitely (as the evaluation of $\tilde{q}_K$ in Eq. (3.8) shows) and the descent equations terminate only at ghost number $2m(K_1) - 1 = 5$. In higher dimensional models many combinatorical solutions of $\sum_s m(K_s) = \frac{D+2}{2}$ exist. If, e.g., $D = 10$ and the gauge group is $SO(1,9) \times E_8$ (a popular supergravity example, $SO(1,9)$ is the Lorentz group), then the degrees $m(K)$ can be taken from

$$m(K)_{SO(1,9)} \in \{2,4,6,8,5\} \quad \text{and}$$
$$m(K')_{E_8} \in \{2,8,12,14,18,20,24,30\}, \quad \text{see [1].}$$

All possible chiral anomalies correspond to the following seven decompositions of

$$\frac{10+2}{2}: \quad 2+2+2 = 2+2+2' = 2+2'+2' = 2'+2'+2' = 2+4 = 2'+4 = 6.$$

The descent equations terminate at ghost number 3 (in case of the first six decompositions) or 11 (in the last case).

The list would increase considerably if E_8 was replaced by $SU(8)$, which has Casimir invariants of degree $m(K)_{SU(8)} \in \{2,3,4,5,6,7,8\}$.

In odd dimensions chiral anomalies can occur only if the gauge group contains two abelian factors at least. In $D = 2n+1$ dimensions all chiral anomalies are products

$$\mathcal{A}^1_{D=2n+1} = (C^K A^{K'} - C^{K'} A^K) f_{K_1} \cdots f_{K_r},$$
$$m(K) = m(K') = 1, \quad \textstyle\sum_{s=1}^{r} m(K_s) = n.$$

$C^K, C^{K'}$ and $A^K, A^{K'}$ are the ghosts and connection forms belonging to the Kth and K'th abelian factor of the gauge group.

We note that there is an apparent irregularity in the degrees $m(K)$ of $SO(D)$ if D is even: The last $m(K)$ is not D but $D/2$. Explicitly, this corresponds to the topological densities

$$\begin{aligned} f_{K=D/2} &= F^{[a_1 b_1]} \dots F^{[a_K b_K]} \varepsilon_{a_1 b_1 \dots a_K b_K} \\ &= \frac{1}{2^{D/2}} d^D x \varepsilon^{\mu_1 \nu_1 \dots \mu_K \nu_K} F^{[a_1 b_1]}_{\mu_1 \nu_1} \dots F^{[a_K b_K]}_{\mu_K \nu_K} \varepsilon_{a_1 b_1 \dots a_K b_K} . \end{aligned}$$

(Recall that the Lie algebra of $SO(D)$ can be labeled by an antisymmetric pair of vector indices). The curvature tensor $R_{\mu\nu}{}^{\rho\sigma}$ can be considered as the field strength of $SO(D)$. For $D = 2$ f_1 is nothing but $\sqrt{g}R$, so $\sqrt{g}R$ is a topological density in two dimensions and Einstein gravity in $D = 2$ is trivial.

For $D = 4$, f_2 coincides with $\sqrt{g}(R^{\mu\nu\rho\sigma} R_{\mu\nu\rho\sigma} - 4R^{\mu\nu} R_{\mu\nu} + R^2)$, which again is a topological density (in 4 dimensions).

Acknowledgements

This project was supported in part by the Deutsche Forschungsgemeinschaft, by the Fonds zur Förderung der wissenschaftlichen Forschung under Grant No. J0398-PHY and by the National Science Foundation under Grant No. PHY82-17853, supplemented by funds from the National Aeronautic and Space Administration. F.B. was supported by the Deutsche Forschungsgemeinschaft.

References

1. A. Borel and C. Chevalley, Mem. Amer. Math. Soc. **14**, 1 (1955)
2. F. Brandt, N. Dragon and M. Kreuzer, Phys. Lett. **231B**, 263 (1989)
3. F. Brandt, N. Dragon and M. Kreuzer, Nucl. Phys. **332B**, 224 (1990)
4. F. Brandt, N. Dragon and M. Kreuzer, Nucl. Phys. **332B**, 250 (1990)
5. F. Brandt, N. Dragon and M. Kreuzer, Nucl. Phys. **340B**, 187 (1990)
6. L.D. Faddeev, Phys. Lett. **145B**, 81 (1984)
7. L. O'Raifertaigh, "Group Structure of Gauge Theories" (Cambridge U.P., Cambridge 1986)
8. B. Zumino, in ***Relativity, Groups and Topology II***, eds. B.S. DeWitt and R. Stora (North-Holland, Amsterdam 1984)
9. J. Wess and B. Zumino, Phys. Lett. **37B**, 95 (1971)

Modular Invariance, Causality and the PCT-Theorem

M. Reuter

Institut für Theoretische Physik der Universität,
Appelstraße 2, D(W)-3000 Hannover 1,
Federal Republic of Germany

Abstract

We describe the $IOSp(D,2|2)$-extension of the Poincaré group in the BRST-quantization of the (spinning) relativistic point particle. The Batalin-Fradkin-Vilkovisky method is used to construct the corresponding field theory, and its dimensional reduction by the Parisi-Sourlas mechanism is proven. We show that a certain element in the identity component of the $SO(D,2)$ subgroup of $IOSp(D,2|2)$ induces the PCT-transformation in the physical subspace. We clarify the role of modular transformations (i.e., of world-line orientation-reversing diffeomorphisms) and argue that the PCT-transformation is the same as a modular transformation seen in an $SO(D,2)$-rotated frame.

In theories with chiral fermions, $OSp(D,2|2)$ is typically broken down to $O(D-1,1) \otimes OSp(1,1|2)$, but modular invariance still seems to be at the heart of causality and PCT-invariance.

1. Introduction

During the past few years the covariant quantization of reparametrization-invariant systems such as strings, membranes or Einstein-gravity has received much attention. The simplest system of this kind is the relativistic point particle, which has been studied in a variety of ways [1 – 14]. Among them, BRST-methods, in particular the path-integral formulation of Batalin, Fradkin and Vilkovisky [15, 16], proved to be very powerful and convenient. One of the most intriguing results obtained in this context was the discovery of Neveu and West [6] that the BRST- and the anti-BRST-operator of the relativistic point particle in D-dimensional Minkowski space can be unified with the Poincaré group $IO(D-1,1)$ in an inhomogeneous orthosymplectic supergroup $IOSp(D,2|2)$. This group acts linearly on a $(D+4)$-dimensional space consisting of the D-dimensional Minkowski space together with an "internal space" coordinatized by two additional bosonic coordinates (the evolution parameter and the Lagrange multiplier of the constraint) and two fermionic coordinates (the Batalin-Fradkin-Vilkovisky [BFV] ghosts). However, Parisi-Sourlas dimensional reduction guarantees that the physical states of the theory "live" in D dimensions only.

In the following we give a brief discussion of the origin of the $IOSp(D,2|2)$-symmetry and of its physical interpretation. In particular we shall discuss its relation to the PCT-theorem of relativistic quantum field theory. An important role in this discussion will be played by the "modular transformations", i.e., by the orientation-reversing diffeomorphisms of the world line. We shall see that modular invariance not only implies causal boundary conditions for the propagator in the physical sector of the theory, but is also responsible for its PCT-invariance or, equivalently, for the fulfillment of the microcausality condition. The latter is usually considered to be an axiom of relativistic quantum field theory, but here, starting from first quantization, it can be derived in a certain sense.

2. BRST-Quantization

In this section we briefly review the BRST-quantization of a spinless point particle in the framework of the BFV-approach [18, 19]. More details can be found in [20] and [21].

We consider world-lines $X^\mu(\tau)$, $\tau \in [\tau_1, \tau_2]$, $\mu = 0,\ldots,D-1$, of particles propagating in a D-dimensional Minkowski space. The classical action is given by

$$S_{cl} = -m \int_{\tau_1}^{\tau_2} d\tau \sqrt{\dot{X}_\mu^2(\tau)}, \tag{2.1}$$

and the canonical momenta are $P^\mu = m\dot{X}^\mu / \sqrt{\dot{X}_\mu^2}$. They give rise to the constraint

$$\phi(\tau) \equiv P_\mu^2(\tau) - m^2 \approx 0. \tag{2.2}$$

The classical Hamiltonian vanishes identically, so that the first order action is given by

$$S_1 = \int_{\tau_1}^{\tau_2} d\tau [\dot{X}^\mu P_\mu - \lambda(\tau)(P_\mu^2 - m^2)], \tag{2.3}$$

where $\lambda(\tau)$ is a Lagrange multiplier enforcing the constraint $\phi \approx 0$. The action in Eq. (2.3) is invariant under the infinitesimal world line diffeomorphisms [22] generated by ϕ,

$$\begin{aligned} \delta X^\mu(\tau) &= 2\varepsilon(\tau) P^\mu(\tau) \\ \delta P_\mu(\tau) &= 0 \\ \delta\lambda(\tau) &= \dot{\varepsilon}(\tau), \end{aligned} \tag{2.4}$$

provided the parameter $\varepsilon(\tau)$ vanishes at the end points: $\varepsilon(\tau_1) = \varepsilon(\tau_2) = 0$. The BFV-quantization [18, 19] of (2.3) consists of extending the phase space by promoting $\lambda(\tau)$ and its conjugate momentum $\pi(\tau)$ to dynamical degrees of freedom. Furthermore, one associates two ghosts $\eta^i(\tau)$, $(i = 1, 2)$ and two antighosts $Q_i(\tau)$ with the constraints $\phi \approx 0$ and $\pi \approx 0$. The additional constraint $\pi \approx 0$ has to be imposed in order to preserve the dynamics of the theory in the presence of $\lambda(\tau)$ as a new dynamical variable. The canonical (anti-)commutation relations are

$$\begin{aligned} [X^\mu, P_\nu] &= -i\delta^\mu_\nu \\ [\lambda, \pi] &= i \\ [Q_i, \eta^j] &= -i\delta^j_i . \end{aligned} \tag{2.5}$$

Here [,] denotes the $\mathbb{Z}_2$-graded commutator. Introducing the standard notation

$$\begin{aligned} \eta^i &= (c, -i\mathcal{P}) \\ Q_i &= (\bar{\mathcal{P}}, i\bar{c}), \end{aligned} \tag{2.6}$$

the BRST-operator [19] reads

$$\Omega = c(P_\mu^2 - m^2) - i\mathcal{P}\pi. \tag{2.7}$$

In order to arrive at an $IOSp(D, 2|2)$-invariant formulation we now perform two canonical transformations, i.e., transformations preserving the commutator relations (2.5). First we make the replacements

$$\begin{aligned} c \to \lambda c, \quad \mathcal{P} \to \lambda\mathcal{P}, \quad \bar{c} \to \lambda^{-1}\bar{c}, \\ \bar{\mathcal{P}} \to \lambda^{-1}\bar{\mathcal{P}}, \quad \pi \to \pi + (\bar{c}P - c\bar{\mathcal{P}}), \end{aligned} \tag{2.8}$$

and then we define the following combinations of the ghosts:

$$\begin{aligned} \tilde{x}_1 &= -i\sqrt{2}(c + \tfrac{1}{2}\bar{c}) \\ \tilde{p}^1 &= \frac{i}{\sqrt{2}}(\tfrac{1}{2}\bar{\mathcal{P}} + \mathcal{P}) \\ \tilde{x}_2 &= -i\sqrt{2}(c - \tfrac{1}{2}\bar{c}) \\ \tilde{p}^2 &= \frac{i}{\sqrt{2}}(\tfrac{1}{2}\bar{\mathcal{P}} - \mathcal{P}). \end{aligned} \tag{2.9}$$

The variables $\tilde{x}_a$ and $\tilde{p}^a$ $(a = 1, 2)$ have been introduced, since, as suggested by their names, they obey a canonical anticommutation algebra. In terms of the new variables the BRST-charge reads

$$\Omega = \Omega_1 + \Omega_2, \tag{2.10}$$

where

$$\Omega_a = \frac{\lambda}{\sqrt{2}}[\tfrac{i}{2}\tilde{x}_a(P_\mu^2 - m^2 + i\varepsilon_{ab}\tilde{p}^a\tilde{p}^b) - i\tilde{p}^a\pi] \tag{2.11}$$

with $[\Omega_a, \Omega_b] = 0$. Similarly, the anti-BRST-operator [19] can be written as $\bar{\Omega} = \Omega_1 - \Omega_2$. It can be shown [8] that the pair (Ω_1, Ω_2) transforms according to the spinor representation of an $Sp(2)$-group generated by certain bilinears in $\tilde{x}_a$ and $\tilde{p}^a$.

Since the classical Hamiltonian vanishes identically in our case, the BRST-invariant Hamiltonian [19] is completely determined by the gauge fixing function χ:

$$H = [\chi, \Omega]. \tag{2.12}$$

We shall make the choice

$$\chi = \frac{i}{2}\mathcal{P} = \frac{1}{\sqrt{2}}(\tilde{p}^1 + \tilde{p}^2), \tag{2.13}$$

which implements the proper-time gauge condition $\dot{\lambda} = 0$ in the BFV-path integral [22]. The resulting Hamiltonian reads

$$H = \tfrac{1}{2}\lambda[P_\mu^2 - m^2 + i\varepsilon_{ab}\tilde{p}^a\tilde{p}^b]. \tag{2.14}$$

Now we introduce a Schrödinger picture representation of the operators X^μ, P_μ, λ, π, $\tilde{x}_a$ and $\tilde{p}^a$. The position variables are realized as multiplicative operators, whereas the conjugate momenta are represented by the derivative operators

$$P_\mu = i\partial_\mu, \quad \pi = -i\frac{\partial}{\partial\lambda}, \quad \tilde{p}^a = -i\frac{\partial}{\partial\tilde{x}_a}. \tag{2.15}$$

They act on wave functions $\psi = \psi(x^\mu, \lambda, \tilde{x}^a, \tau)$, where τ is the evolution parameter (proper-time). The τ-evolution of ψ is governed by the Schrödinger equation

$$i\frac{\partial\psi}{\partial\tau} = H\psi = \tfrac{1}{2}\lambda[P_\mu^2 - m^2 + i\varepsilon_{ab}\tilde{p}^a\tilde{p}^b]\psi. \tag{2.16}$$

It can be derived from the action

$$S = i\int d^D X \int_{-\infty}^{\infty} dp_+ \int_{-\infty}^{\infty} dx_- \int d\tilde{x}_1\, d\tilde{x}_2 \psi^*[2ip_+\partial_\tau - \Box + 2i\partial_{\tilde{x}_1}\partial_{\tilde{x}_2}]\psi, \tag{2.17}$$

where $\Box \equiv -\partial_\mu\partial^\mu - m^2$. Here we have separated off a factor $\lambda^{\frac{1}{2}}$ from ψ and replaced λ by $p_+ \equiv \lambda^{-1}$ as the argument of the wave function. Eq. (2.17) is the action of the BRST-invariant field theory studied by Neveu and West [6]. To make its symmetries more explicit, we introduce the Fourier transform of ψ with respect to p_+. The conjugate variable will be called x_-. It will also prove convenient to set $x_+ \equiv \tau$:

$$\Phi(X^\mu, x_+, x_-, \tilde{x}_a) = \int_{-\infty}^{\infty} \frac{dp_+}{2\pi} e^{ip_+x_-}\psi(X^\mu, x_+, p_+, \tilde{x}_a). \tag{2.18}$$

Using the "Cartesian" linear combinations

$$X_D \equiv \frac{1}{\sqrt{2}}\,(x_+ - x_-) \quad \text{and} \quad X_{D+1} \equiv \frac{1}{\sqrt{2}}\,(x_+ + x_-), \tag{2.19}$$

the action can be written as

$$\mathcal{S} = -2\pi \int d^{D+4}X\; \Phi^*(X^A)\,\Box_{OSp}\,\Phi(X^A)), \tag{2.20}$$

where $d^{D+4}X \equiv d^DX\,dX^D\,dX^{D+1}i d\tilde{x}_1 d\tilde{x}_2$ and

$$\begin{aligned}\Box_{OSp} &= -\Box + 2\partial_{x_+}\partial_{x_-} + 2i\partial_{\tilde{x}_1}\partial_{\tilde{x}_2}\\ &= \eta^{AB}\partial_A\partial_B + m^2.\end{aligned} \tag{2.21}$$

The indices $A, B, \ldots$ refer to coordinates $X^A = (X^\mu, X^D, X^{D+1}, \tilde{x}^a)$ of a $(D+4)$-dimensional "super Minkowski space", obtained from the Minkowski space by adding two bosonic coordinates X^D and X^{D+1} and two fermionic coordinates $\tilde{x}^a$. We shall also use the notation $X^\alpha = (X^\mu, X^D, X^{D+1})$ for the bosonic coordinates alone. The metric tensor η_{AB} (or its inverse η^{AB}, respectively) appearing in Eq. (2.21) is defined as

$$\begin{aligned}\eta_{\alpha\beta} &= \text{diag}[+1, -1, -1, \ldots, -1, +1]\\ \eta_{ab} &= i\varepsilon_{ab}\\ \eta_{a\alpha} &= 0.\end{aligned} \tag{2.22}$$

The tensor $\eta_{\alpha\beta}$ is the metric of a $(D+2)$-dimensional Minkowski space with two time-like coordinates, X^0 and X^{D+1}. The metric η_{AB} is left invariant by the transformations of the supergroup $OSp(D,2|2)$ or, including translations, by $IOSp(D,2|2)$ [23]. The operator $\Box_{OS\tilde{p}}$ is the corresponding $IOSp(D,2|2)$-invariant Laplacian; hence the $IOSp(D,2|2)$-invariance of the action displayed in Eq. (2.20) is manifest. The superfield Φ transforms according to the scalar representation of $OSp(D,2|2)$,

$$\Phi^\Lambda(X^A) = \Phi(\Lambda^A{}_B X^B), \tag{2.23}$$

where $\Lambda^A{}_B$ are the representation matrices in the vector-representation. The generators in the scalar representation are

$$J_{AB} = X_A P_B - (-)^{[A][B]} X_B P_A, \tag{2.24}$$

where $P_A = (P_\mu, P_D, P_{D+1}, \tilde{p}_a)$ with $[X_A, P_B] = -i\eta_{BA}$ ($[A]$ equals zero (one) if A is bosonic (fermionic)). Using the Schrödinger equation, the BRST-operator $\Omega = \Omega_1 + \Omega_2$ can be identified with elements of the super Lie algebra of $OSp(D,2|2)$. One finds that

$$\Omega_a = \frac{-i}{\sqrt{2}}\,J_{a-}. \tag{2.25}$$

The nilpotency of Ω is a consequence of the $OSp(D,2|2)$-Lie algebra structure.

At this point a remark about the gauge fixing function χ might be in order. Applied to our model, the BFV-theorem states that, for suitable boundary conditions, the path integral

$$Z_\chi = \int [dX^\mu dP_\mu d\lambda d\pi d\eta^i dQ_i] \exp\left\{iS_1 + \int_{\tau_1}^{\tau_2} d\tau(\dot{\lambda}\pi + \dot{\eta}^i Q_i - \{\chi, \Omega\}\right\} \tag{2.26}$$

is independent of the gauge fixing functional χ and is identical to the standard path-integral for constrained systems. This requires, however, that χ is admissable in the following sense [22]: Let us consider the action of the gauge group defined in Eq. (2.4) in the space of functions $\lambda(\tau)$ and let us define (in analogy with Riemann surface theory) the *Teichmüller space* to be the space of orbits of the connected part of the gauge group. The modular group is defined as the quotient of the complete gauge group (including "large" transformations) by its connected component. Finally, the quotient of the Teichmüller space by the modular group is called *modular space*: the gauge inequivalent configurations of the system are characterized by the points in the modular space. For each point in the modular space, an admissable gauge fixing selects one and only one representative from the orbits in the equivalence class of orbits belonging to the same point in the modular space (and which therefore are related by "large" gauge transformations). Since the BFV-method takes care only of the "small" gauge transformations continuously connected to the identity, there will be no such gauge fixing χ in general. The best one can achieve is to find a "good" [22] gauge fixing such that the path-integral reduces to an integral over the Teichmüller space (rather than moduli space). This integral then has to be restricted "by hand" to a fundamental domain of the modular group. These remarks apply to any reparametrization invariant system. For the point particle they have the following concrete meaning: the orbits of the gauge group are classified by the *Teichmüller parameter*

$$C = \int_{\tau_1}^{\tau_2} d\tau\, \lambda(\tau). \tag{2.27}$$

Because of the condition $\varepsilon(\tau_1) = \varepsilon(\tau_2) = 0$, C cannot be changed by the transformation shown in Eq. (2.4). Hence the Teichmüller space consists of the entire real line $\mathbb{R}$. However, the crucial point to note is that the full symmetry group of the action in Eq. (2.1) is doubly connected: besides being the transformations in Eq. (2.4) which are connected to the identity, it contains orientation-reversing diffeomorphisms which exchange the end-points $X^\mu(\tau_1)$ and $X^\mu(\tau_2)$, i.e., the modular group is $\mathbb{Z}_2$. It is generated by the reflection $\tau \to -\tau$. Invariance of the equations of motion resulting from Eq. (2.3) requires that $\lambda(\tau)$ transforms as

$$\lambda(\tau) \to -\lambda(-\tau). \tag{2.28}$$

Thus the Teichmüller parameter changes its sign under $\mathbb{Z}_2 : C \to -C$. Consequently, the modular space is the real half-line $\mathbb{R}/\mathbb{Z}_2 = \mathbb{R}^+$.

It has been shown [5, 22] that the gauge fixing used above,

$$\dot{\lambda}(\tau) = 0, \quad \lambda(\tau) = \lambda_0 = const., \tag{2.29}$$

is indeed a "good" one. The corresponding Teichmüller parameter is simply

$$C = \lambda_0 \, (\tau_2 - \tau_1). \tag{2.30}$$

The transition amplitude from an initial point X_1^μ to a final point X_2^μ, obtained by evaluating the integral in Eq. (2.26), reads

$$G(X_2^\mu, X_1^\mu) = \int_{-\infty}^{+\infty} dC \, C^{-D/2} \exp\left\{ -\frac{i}{2} \left[\frac{(x_2 - x_1)^2}{C} + m^2 C \right] \right\}. \tag{2.31}$$

It is expressed in terms of an integral over the Teichmüller space, i.e., over the whole real C-axis. The correct final answer is obtained by restricting the C-integration to a fundamental domain of $\mathbb{Z}_2$, e.g. to $0 \le C < \infty$. Then the right-hand-side of Eq. (2.31) has a unique interpretation as a distribution and coincides with the standard Feynman (or causal) propagator. This is a first indication of the very close relation between modular invariance and causality. We shall make this connection more precise later on.

3. Second Quantization and Dimensional Reduction

As usual, the physical states of our model could be determined by computing the cohomology of the BRST-operator Ω. Using Eq. (2.7), it is an easy exercise [20] to prove that the cohomology classes of Ω (at ghost number zero) are in a one-to-one correspondence with the solutions of the Klein-Gordon equation. Here we shall adopt another strategy. We shall consider the term in Eq. (2.20) as the free part of an $OSp(D, 2|2)$-invariant classical field theory, which is now quantized (or, more appropriately, "second quantized") by a path-integral over the superfield Φ. In this language the projection onto the physical subspace amounts to invoking the celebrated Parisi-Sourlas dimensional reduction [24, 25]. The field theory we are going to discuss is defined by the generating functional

$$Z[J] = \int [d\Phi] \exp\left(\frac{i}{2\pi} \int d^{D+4}X \{ \tfrac{1}{2} \Phi \Box_{OSp} \Phi - V(\Phi) + J\Phi \} \right). \tag{3.1}$$

Besides the free action in Eq. (2.20) it contains an $OSp(D, 2|2)$-invariant interaction term $V(\Phi)$ (the normalization of the action is chosen for later convenience). Let us now describe the Parisi-Sourlas reduction of the functional $Z[J]$. We shall see that, by suitably restricting the source functions J, $Z[J]$ will describe a purely D-dimensional theory of a real scalar boson. The usual Parisi-Sourlas dimensional reduction is based upon the "magic formula"

$$\int dx_1 dx_2 d\tilde{x}_1 d\tilde{x}_2 g(x_1^2 + x_2^2 + 2i\tilde{x}_1\tilde{x}_2) = (2\pi i) g(0), \tag{3.2}$$

which is valid for all functions g vanishing at infinity and depending on the bosonic variables (x_1, x_2) and the fermionic variables $(\tilde{x}_1, \tilde{x}_2)$ only through the $OSp(D, 2|2)$-invariant combination $x_1^2 + x_2^2 + 2i\tilde{x}_1\tilde{x}_2$. Heuristically we can interpret Eq. (3.2) by saying that "negative-dimensional" anticommuting coordinates compensate the commuting ones and thus effectively reduce the dimensionality of the space. In our case Eq. (3.2) is not applicable, since we are dealing with the pseudo-orthosymplectic group $OSp(1,1|2) \subset OSp(D,2|2)$, whose invariant length-square is

$$-X_D^2 + X_{D+1}^2 + 2i\tilde{x}_1\tilde{x}_2 = 2(x_+x_- + i\tilde{x}_1\tilde{x}_2) \equiv 2y. \tag{3.3}$$

It can be shown that Eq. (3.2) does not hold for a Minkowskian signature ("Euclidean" and "Minkowskian" always refer to (X_D, X_{D+1})-space). However, provided the function $g(y)$ fulfills certain analyticity requirements, one can prove [20] the following Minkowski space analogue of Eq. (3.2):

$$\int dx_+ dx_- d\tilde{x}_1 d\tilde{x}_2 g(x_+x_- + i\tilde{x}_1\tilde{x}_2) = (\pm 2\pi) g(0) \tag{3.4}$$

This relation is valid if $g(y)$ is holomorphic in the upper (resp. lower) complex y-plane.

Equipped with the Minkowski space reduction formula in Eq. (3.4) we can prove that for sources restricted to the physical subspace according to

$$J(X^\mu, X^D, X^{D+1}, \tilde{x}^a) = j(X^\mu)\delta(X^D)\delta(X^{D+1})\delta(\tilde{x}^a), \tag{3.5}$$

the functional $Z[J]$ of Eq. (3.1) becomes identical to that of a purely bosonic D-dimensional theory:

$$Z[J] = Z_D[j], \tag{3.6}$$

where

$$Z_D[j] = \int [d\phi] \exp\left(i \int d^D X [\tfrac{1}{2}\phi(-\partial_\mu^2 - m^2)\phi - V(\phi) + j\phi] \right). \tag{3.7}$$

In Eq. (3.7), the integration is only over $\phi(X^\mu) \equiv \Phi(X^\mu, 0, 0, \tilde{0})$, i.e., over the lowest component of the superfield Φ, with X^D and X^{D+1} put equal to zero. The proof of Eq. (3.7) consists of integrating out all field variables which do not couple to $j(X^\mu)$ [20]; it proceeds in analogy with either its perturbative [24] or its non-perturbative [25] analogue in Euclidean Parisi-Sourlas field theory. In either case one has to make sure that Eq. (3.4) is indeed applicable. As we showed in Ref. [20], the prerequisite analyticity conditions are fulfilled only if one uses the Feynman propagator as the Green's function of $\Box_{OSp}$ or of $(-\partial_\mu^2 - m^2)$, respectively. This is another indication that the BRST-method, when applied carefully, not only gives rise to the correct *formal* path-integral in Eq. (3.7) for the dynamics in the physical sector of the theory, but also "knows" about causality: only the (D-dimensional) theory with *causal* boundary conditions can be obtained as the dimensional reduction of a BRST-field theory.

Next we present another argument in favour of the Feynman boundary conditions, which illustrates more clearly the connection between modular invariance and causality. We recall that a modular transformation is given by $\tau \to -\tau, \lambda \to -\lambda$, or, using $p_+ \equiv \lambda^{-1}$, by $p_+ \to -p_+$. On the wave functions it acts as

$$\Psi_{\mathrm{mod}}(X^\mu, p_+, \tau, \tilde{x}_a) = \Psi(X^\mu, -p_+, -\tau, \tilde{x}_a), \tag{3.8}$$

or, in terms of the new variables, as

$$\Phi_{\mathrm{mod}}(X^\mu, X^D, X^{D+1}, \tilde{x}_a) = \Phi(X^\mu, -X^D, -X^{D+1}, \tilde{x}_a). \tag{3.9}$$

Let us now try to construct a Green's function for the Schrödinger equation Eq. (2.16). In ordinary non-relativistic Schrödinger theory it would be chosen as the position-space representation of

$$(i\partial_\tau - H - i\varepsilon)^{-1}. \tag{3.10}$$

The $i\varepsilon$-term introduced in this way has the effect of propagating the wave function Ψ forward in τ, wheras its complex conjugate Ψ^* is propagated backward. Inserting our Hamiltonian in Eq. (2.14) and rescaling the propagator by $2p_+$, which is the same redefinition leading to Eq. (2.17), we obtain from Eq. (3.10):

$$(P_\mu^2 - m^2 + i\varepsilon_{ab}\tilde{p}^a\tilde{p}^b - 2ip_+\partial_\tau + 2i\varepsilon p_+)^{-1}. \tag{3.11}$$

As it stands, this propagator is incompatible with modular invariance, because it propagates $\Psi(\Psi^*)$ forward (backward) in τ irrespective of the sign of p_+. Modular invariance requires that $\Psi(p_+ > 0)$ and $\Psi^*(p_+ < 0)$ are propagated forward and that $\Psi(p_+ < 0)$ and $\Psi^*(p_+ > 0)$ are propagated backward in τ. This means that for negative values of p_+ the direction of the propagation has to be changed. This is easily implemented by the replacement $2i\varepsilon p_+ \to i\varepsilon$. The new propagator, seen from a D-dimensional point of view (the ghost term and the $p_+\partial_\tau$-piece disappear upon dimensional reduction), is precisely the Feynman propagator! The reversed propagation for $p_+ < 0$ or $\lambda < 0$ was already suggested by one of the equations of motion following from the action in Eq. (2.3):

$$\frac{dX^0(\tau)}{d\tau} = 2\lambda(\tau)P^0(\tau). \tag{3.12}$$

A sign-change of λ without a concomitant sign-change of $d/d\tau$ would destroy the causal propagation pattern of positive (negative) P^0 propagating forward (backward) in the physical time X^0. Thus we find that causal propagation in the physical D-dimensional world can be considered as a consequence of modular invariance.

4. Ward Identities of $OSp(4,2|2)$

Next we derive the Ward identities resulting from the $OSp(4,2|2)$-invariance of the action and show that one of them can be identified with the PCT-theorem [26] (in this section we put $D = 4$ for simplicity). The only slightly unusual point in obtaining these Ward identities is that the sources are restricted to the special form in Eq. (3.5). This is a reflection of the fact that, despite the $OSp(4,2|2)$-invariance of the action, the physical state condition $\Omega|\text{phys}>= 0$ is *not* $OSp(4,2|2)$-invariant. In the path-integral formalism this manifests itself in the δ-functions appearing in Eq. (3.5), which select certain directions (in the $(D+4)$-dimensional super-Minkowski space) along which the Parisi-Sourlas reduction acts. Thus our starting point is

$$Z[j] = \int [d\Phi^{\Lambda}] \exp\left(iS_{OSp}[\Phi^{\Lambda}] + i\int d^4X\, j(X^\mu)\Phi^{\Lambda}(X^\mu,0,0,\tilde{0})\right), \qquad (4.1)$$

where "$\tilde{0}$" indicates that both ghost variables have been set equal to zero. In Eq. (4.1) we wrote the integration variable as Φ^{Λ}, which is the field obtained from Φ by acting on it according to Eq. (2.23) with some $\Lambda \in OSp(4,2|2)$. Furthermore, S_{OSp} denotes the $OSp(4,2|2)$-invariant action entering the path-integral in Eq. (3.1). Hence we have $S_{OSp}[\Phi^{\Lambda}] = S_{OSp}[\Phi]$. Excluding the possibility of anomalous $OSp(4,2|2)$-breaking, we assume that in Eq. (4.1) also the measure is invariant: $[d\Phi^{\Lambda}] = [d\Phi]$. Consequently, it follows that for all $\Lambda \in OSp(4,2|2)$:

$$\begin{aligned} Z[j] &= \int [d\Phi] \left(iS_{OSp}[\Phi] + i\int d^4X\, j(X^\mu)\Phi(X^\mu,0,0,\tilde{0})\right) \\ &= \int [d\Phi] \left(iS_{OSp}[\Phi] + i\int d^4X\, j(X^\mu)\Phi^{\Lambda}(X^\mu,0,0,\tilde{0})\right). \end{aligned} \qquad (4.2)$$

The Ward identities are now easily derived by taking functional derivatives with respect to $j(X^\mu)$. Here we are only interested in the special case $\Lambda = \Lambda_{PCT}$, where Λ_{PCT} is defined to act on the bosonic variables as

$$(\Lambda_{PCT})^{\alpha}{}_{\beta} X^{\beta} = -X^{\alpha} \qquad (4.3)$$

and to leave the ghosts unchanged. Obviously, the transformation Λ_{PCT} changes the sign of all six bosonic coordinates; it can be visualized as a sequence of three rotations of angle π in the (X^0, X^{D+1})-, (X^1, X^2)- and (X^3, X^D)-planes, respectively. The (X^0, X^{D+1})-plane consists of two time-like directions. Due to the presence of the new X^{D+1}-coordinate the time-reversal transformation $X^0 \to -X^0$ can be obtained by a continuous rotation, rather than by a discrete reflection, as is the case in $(3+1)$-dimensional Minkowski space. Similarly, because of the extra space-like X^D-direction, the reflection $X^3 \to -X^3$ is a simple rotation in the extended bosonic space. Performed together with a trivial π-rotation in the (X^1, X^2)-plane it has the effect of inducing a parity transformation in the physical X^μ-space. We stress that in the sense of

$OSp(4,2|2)$ the group element Λ_{PCT} is continuously connected to the identity: it consists of three special $SO(2)$-rotations. The additional time-like (space-like) coordinates $X^{D+1}(X^D)$ makes it possible to unify time-reflection T (parity P) with the connected component of the Lorentz group. Differentiation of Eq. (4.2) with respect to $j(X^\mu)$ yields for $\Lambda = \Lambda_{PCT}$:

$$\langle 0|T\phi(X_1^\mu)\ldots\phi(X_n^\mu)|0\rangle = \langle 0|T\phi(-X_1^\mu)\ldots\phi(-X_n^\mu)|0\rangle. \tag{4.4}$$

Here $\phi(X^\mu) \equiv \Phi(X^\mu, 0, 0, \tilde{0})$ is the physical lowest component of Φ. By integrating out the other fields, which are decoupled from $j(X^\mu)$, one ends up with the conventional T-ordered (i.e. X^0-ordered) n-point functions. From Eq. (4.4) it follows that

$$\langle 0|\phi(X_1^\mu)\ldots\phi(X_n^\mu)|0\rangle = \langle 0|\phi(-X_1^\mu)\ldots\phi(-X_n^\mu)|0\rangle. \tag{4.5}$$

This equation is correct for any temporal ordering of the arguments, since all commutator terms (resulting from commuting the arguments to the T-ordered form) mutually cancel. Eq. (4.5) is precisely the PCT-theorem [26] for an interacting scalar theory, expressed in terms of Wightman functions. It states that in any relativistic quantum field theory (defined by a normal-ordered Hermitean Lagrangian invariant under the connected part of the Lorentz group and quantized with the correct connection between spin and statistics), the Hamiltonian commutes with the PCT-operator. In the present case of a Hermitean scalar field we find from Eq. (4.5) that the operator $\Theta \equiv PCT$ acts on the field operators as

$$\Theta\phi(X^\mu)\Theta^{-1} = \phi(-X^\mu). \tag{4.6}$$

The phases entering the definitions of the separate P-, C- and T-transformations are chosen here in such a way that there appears no phase or sign factor on the right-hand-side of Eq. (4.6). We shall come back to this point later on when we discuss fields of higher spin.

Before closing this section we describe an intriguing relation between the PCT-transformation and the modular transformation. In fact, we shall see that in an $OSp(4,2|2)$-invariant theory the PCT-transformation is the same as a modular transformation, but "seen" in an $SO(4,2)$-rotated frame. Here $SO(4,2)$ is the subgroup of $OSp(4,2|2)$ acting on the commuting coordinates $X^\alpha = (X^\mu, X^D, X^{D+1})$. We know from Eq. (3.9) that a modular transformation acts on the superfield Φ according to

$$\begin{aligned}\Phi_{\text{mod}}(X^\mu, X^D, X^{D+1}, \tilde{x}^a) &= \Phi(X^\mu, -X^D, -X^{D+1}, \tilde{x}^a)\\ &= \Phi(\Lambda^\alpha_{\text{mod}\ \beta}X^\beta, \tilde{x}^a),\end{aligned} \tag{4.7}$$

where $\Lambda_{\text{mod}} \equiv \text{diag}[1,1,1,1,-1,-1]$. Note that $\Lambda_{\text{mod}} \in SO(4,2)$, but that Λ_{mod} is not continuously connected to the identity. Next observe that, by a special $SO(4,2)$-rotation, we can interchange the two time-variables X^0 and X^{D+1}, namely by a $\pi/2$-rotation in the (X^0, X^{D+1})-plane. Similarly we can exchange the new spatial coordinate X^D with one of the X^i, say X^3. Denoting these rotations by Λ_{ex}, we have

$$\Lambda_{\text{ex}}(X^0, X^1, X^2, X^3, X^D, X^{D+1}) = (X^{D+1}, X^1, X^2, X^D, -X^3, -X^0). \quad (4.8)$$

If we now apply a modular transformation in Eq. (4.7) on $\Lambda^{\alpha}_{\text{ex}\,\beta}X^{\beta}$, it changes $(-X^3, -X^0)$ to $(+X^3, +X^0)$. Finally we apply the inverse exchange transformation Λ_{ex}^{-1} to bring X^3 and X^0 back to their original positions:

$$\Lambda_{\text{ex}}^{-1}\Lambda_{\text{mod}}\Lambda_{\text{ex}}(X^0, X^1, X^2, X^3, X^D, X^{D+1}) = (-X^0, X^1, X^2, -X^3, X^D, X^{D+1}) \quad (4.9)$$

Obviously, this is a PCT-transformation combined with a trivial π-rotation $\mathcal{R}_{1,2}(\pi)$ in the $1-2$-plane:

$$\Lambda_{\text{ex}}^{-1}\Lambda_{\text{mod}}\Lambda_{\text{ex}} = \tilde{\Lambda}_{PCT}\mathcal{R}_{1,2}(\pi). \quad (4.10)$$

Here $\tilde{\Lambda}_{PCT} \in SO(4,2)$ is a "genuine" PCT-transformation, changing only the signs of X^{μ} but leaving X^D and X^{D+1} unaltered. Like Λ_{mod} (but unlike Λ_{PCT}) it is *not* in the identity component of $SO(4,2)$.

This simple argument shows that in an $OSp(4,2|2)$-invariant theory the PCT-transformation and the modular transformation are basically the same thing; they just differ by a conjugation with the rotation matrix Λ_{ex}. We also recall that according to the analysis of Jost [26], the PCT-theorem is essentially equivalent to microcausality, or, more precisely, to "weak local commutativity". Taking both statements together, we again find that modular invariance is related to causality, this time, however, in a slightly different form than in connection with the Feynman propagator.

5. The Spinning Particle

We now generalize the preceeding discussion by including particles with spin. We shall not present here all the technical details, but only the most important results. A more complete account may be found in [21].

The massive spinning particle in D-dimensional Minkowski space is described by commuting position variables $X^{\mu}(\tau)$ and $(D+1)$ anticommuting variables $\xi^{\mu}(\tau)$ and $\xi_5(\tau)$ [27]. The theory is defined by the two first class constraints

$$\phi_1 \equiv P_{\mu}^2 - m^2, \quad \phi_2 \equiv P_{\mu}\xi^{\mu} - m\xi_5. \quad (5.1)$$

Hence the action, in Hamiltonian form, reads

$$S_1 = \int_{\tau_1}^{\tau_2} d\tau[-P_{\mu}\dot{X}^{\mu} - \tfrac{i}{2}\{\xi_{\mu}\dot{\xi}^{\mu} - \xi_5\dot{\xi}_5\} - \lambda_1\phi_1 - \lambda_2\phi_2]. \quad (5.2)$$

Since the constraint ϕ_1 (ϕ_2) is commuting (anticommuting), also the Lagrange multiplier λ_1 (λ_2) has to be chosen commuting (anticommuting). The action S_1 is invariant under the gauge transformations

$$\delta(\cdot) = [(\cdot), \varepsilon_1\phi_2 + \varepsilon_2\phi_1], \quad \delta\lambda_1 = \dot{\varepsilon}_1 + i\varepsilon_2\lambda_2, \quad \delta\lambda_2 = \dot{\varepsilon}_2, \tag{5.3}$$

provided $\varepsilon_i(\tau_1) = \varepsilon_i(\tau_2) = 0$, $i = 1, 2$. The constraint ϕ_1 generates world-line reparametrizations, wheras ϕ_2 yields local supersymmetry transformations [1, 2]. The system of Eq. (5.2) can be quantized along the lines of the BFV-method, and one again finds that the BRST- and anti-BRST-operators can be unified with the Lorentz group in the $OSp(D,2|2)$-supergroup [10, 21]. The $OSp(D,2|2)$-invariant action replacing the expression in Eq. (2.20) reads

$$\mathcal{S} = \int d^{D+4}X\, \bar{\Psi}[i\Gamma^A\partial_A - m]\Psi. \tag{5.4}$$

The matrices $\Gamma^A \equiv (\Gamma^\mu, \Gamma^D, \Gamma^{D+1}, \Gamma^a)$ belong to an orthosymplectic Clifford algebra

$$\Gamma_A\Gamma_B + (-1)^{[A][B]}\Gamma_B\Gamma_A = 2\eta_{BA}. \tag{5.5}$$

Its representations are necessarily infinite-dimensional. A convenient representation is given by

$$\begin{aligned} \Gamma_\mu &= \gamma_\mu \otimes I \otimes I, \quad \Gamma_D = \Gamma_5 \otimes i\sigma_2 \otimes I \\ \Gamma_{D+1} &= \gamma_5 \otimes \sigma_2 \otimes I, \quad \Gamma_a = -\gamma_5 \otimes \sigma_3 \otimes A_a \cdot K, \end{aligned} \tag{5.6}$$

where the operators A_a $(a = 1, 2)$ can be expressed in terms of the creation and annihilation operators $a^\dagger$ and a of the harmonic oscillator:

$$A_1 = a + a^\dagger, \quad A_2 = i(a - a^\dagger), \quad [a, a^\dagger] = 1. \tag{5.7}$$

Thus the Γ_A's are tensor products of the matrices γ_μ forming a Clifford algebra in D dimensions, the 2-dimensional matrices (I, σ_i) and operators (I, A_a) acting on the infinite dimensional Fock space of the harmonic oscillator $(a, a^\dagger)$ (K denotes a Klein factor whose form is irrelevant here). Because the matrix multiplication involves infinite sums, special techniques are necessary to prove the Parisi-Sourlas reduction of the action in Eq. (5.4), see [21]. As expected, the action in Eq. (5.4) is invariant if the wave function Ψ transforms as a spinor under $OSp(D,2|2)$:

$$\Psi'(X^A) = S(\Lambda)\Psi(\Lambda^A{}_B X^B) \quad \text{with} \quad S(\Lambda) = \exp[-\tfrac{i}{4}\omega^{AB}\Sigma_{AB}]. \tag{5.8}$$

Here ω^{AB} are the parameters of the transformation Λ and $\Sigma_{AB} \equiv \frac{i}{2}[\Gamma_A, \Gamma_B]$, where (as always) [,] denotes the graded commutator. It can be shown that the modular transformation is given by

$$\begin{aligned} \Psi_{\text{mod}}(X^A) &= S_{\text{mod}}\,\Psi(\Lambda_{\text{mod}}{}^A{}_B X^B) \\ &\equiv [I \otimes \sigma_3 \otimes I]\,\Psi(\Lambda_{\text{mod}}{}^A{}_B X^B). \end{aligned} \tag{5.9}$$

The action in Eq. (5.4) is invariant under this transformation. If we also require that the associated Green's function is modular invariant, we again find that only Feynman boundary conditions are admissable.

We can easily calculate the spinor matrix $S_{PCT} \equiv S(\Lambda_{PCT})$. In the representation given by Eq. (5.9) we find (again putting $D = 4$):

$$S_{PCT} \equiv \exp[\tfrac{i\pi}{2}\{\Sigma^{0,D+1} + \Sigma^{1,2} + \Sigma^{3,D}\}] = i\gamma_5 \otimes \sigma_3 \otimes I. \tag{5.10}$$

Now we would like to know which transformation the matrix in Eq. (5.10) induces on the physical subspace. Quite generally, the dimensional reduction can be described as follows: For higher spin fields, the dimensional reduction of some tensor or spinor field consists not only of the reduction $X^A \to X^\mu$ of the argument of the field, it also entails a reduction in the number of field components. A field $A^B(X^C)$, say, which transforms as a vector under $OSp(4,2|2)$, is reduced to a field $A^\mu(X^\nu)$, i.e. to a vector under $O(3,1)$. Due to a kind of Parisi-Sourlas mechanism *in field space* the extra fermionic components A^a compensate for the extra bosonic fields A^D and A^{D+1}. In the same way the spinor under $OSp(4,2|2)$, $\Psi_{\delta_1\delta_2 n}(X^A)$ with $\delta_1 = 1,\ldots,4$, $\delta_2 = 1,2$ and $n = 0,1,2,3,\ldots$, is reduced to a spinor $\psi_{\delta_1}(X^\mu)$ of $O(3,1)$ (or its covering group). For this to be true one has to use the representation of the Γ-matrices in Eq. (5.6); the indices δ_1, δ_2 and n then refer to the first, second and third factor of the tensor product, respectively. Consequently, in the physical subspace of the $\psi_{\delta_1}(X^\mu)$'s, only the first factor of the Γ- or Σ-matrices has a non-trivial action. This implies that Eq. (5.10) gives rise to the transformation

$$\psi(X^\mu) \to -i\gamma_5\psi(-X^\mu), \tag{5.11}$$

which is exactly the standard PCT-transformation for a Dirac fermion! In fact, we can derive the associated Ward identities as in Sect. 4.

Let us consider a rather general $OSp(4,2|2)$-invariant theory containing a complex scalar Φ, a charged fermion Ψ and a vector field A^C. Then the action is $S_{OSp} = S_{OSp}[\Phi,\Phi^*,\Psi,\bar{\Psi},A^C]$. The rotation Λ_{PCT} gives rise to the transformations

$$\begin{aligned}
\Phi^{PCT}(X^\alpha,\tilde{x}^a) &= \Phi(-X^\alpha,\tilde{x}^a)\\
\Phi^{*PCT}(X^\alpha,\tilde{x}^a) &= \Phi^*(-X^\alpha,\tilde{x}^a)\\
\Psi^{PCT}(X^\alpha,\tilde{x}^a) &= S_{PCT}\,\Phi(-X^\alpha,\tilde{x}^a)\\
\bar{\Psi}^{PCT}(X^\alpha,\tilde{x}^a) &= \bar{\Psi}(-X^\alpha,\tilde{x}^a)\,S_{PCT}^{-1}\\
A_C^{PCT}(X^\alpha,\tilde{x}^a) &= -(-)^{[C]}A_C(-X^\alpha,\tilde{x}^a).
\end{aligned} \tag{5.12}$$

In the same way as in Sect. 4 one can derive the corresponding Ward identity:

$$\begin{aligned}&\langle 0|\hat{\phi}(v^\mu)\dots\hat{\phi}^\dagger(w^\mu)\dots\hat{\psi}_\alpha(x^\mu)\dots\hat{\bar{\psi}}_\beta(y^\mu)\dots\hat{A}_\nu(z^\mu)\dots|0\rangle\\ &=\langle 0|\dots(-\hat{A}_\nu)(-z^\mu)\dots[-i\hat{\bar{\psi}}(-y^\mu)\gamma_5]_\beta\dots[-i\gamma_5\hat{\psi}(-x^\mu)]_\alpha\dots\\ &\qquad\dots\hat{\phi}^\dagger(-w^\mu)\dots\hat{\phi}(-v^\mu)|0\rangle\end{aligned}\tag{5.13}$$

(The caret is to indicate that we are dealing with field operators rather than integration variables).

The relations in Eq. (5.13) coincide with the well-known form of the PCT-theorem [26]. Assuming PCT-invariance of the vacuum, Eq. (5.13) corresponds to the following operatorial transformation laws under $\Theta \equiv PCT$:

$$\begin{aligned}\Theta\hat{\phi}(x)\Theta^\dagger &= \hat{\phi}^\dagger(-x),\quad \Theta\hat{\phi}^\dagger(x)\Theta^\dagger = \hat{\phi}(-x)\\ \Theta\hat{\psi}_\alpha(x)\Theta^\dagger &= i(\gamma_5)_{\alpha\beta}\hat{\psi}^\dagger_\beta(-x)\\ \Theta\hat{\bar{\psi}}_\alpha(x)\Theta^\dagger &= -i\hat{\psi}_\beta(-x)(\gamma_5\gamma_0)_{\beta\alpha}\\ \Theta\hat{A}_\mu(x)\Theta^\dagger &= -\hat{A}_\mu(-x)\end{aligned}\tag{5.14}$$

It is important to realize the difference between these equations and the $SO(4,2)$-transformation laws in Eq. (5.12).

The rotation caused by Λ_{PCT} transforms $\Phi(X^\mu,0,0,\tilde{0})$ into $\Phi(-X^\mu,0,0,\tilde{0})$ and $\Phi^*(X^\mu,0,0,\tilde{0})$ into $\Phi^*(-X^\mu,0,0,\tilde{0})$, wheras Θ, acting on operators, maps $\hat{\phi}(x)$ onto $\hat{\phi}^\dagger(-x)$ and $\hat{\phi}^\dagger(x)$ onto $\hat{\phi}(-x)$. This difference is due to the antilinear nature of Θ: when it acts on a field operator it gives the Hermitean adjoint, while when it acts on a state (a one-particle wave function for instance) it does not involve a complex conjugation. The same remark applies to ψ and $\bar{\psi}$. The path integral integration variables in Eq. (5.12) behave like one-particle wave functions. It is only because these are not complex conjugated by Θ that the PCT-transformation can be represented by an $SO(4,2)$-rotation. On the other hand, both the separate C- and the separate T-transformations involve a complex conjugation of the one-particle wave functions, and they therefore cannot be represented by an $SO(4,2)$-rotation. The situation is different for P and the product CT, which involve no complex conjugation of ϕ and ψ. Parity is given simply by two π-rotations in the (X^3,X^D)- and (X^1,X^2)-planes, respectively, and CT by a single π-rotation in the time-plane (X^0,X^{D+1}). By dimensional reduction we obtain the standard transformation laws in the physical subspace. Therefore we arrive at the important conclusion that not only PCT, but also P and CT separately are symmetries in the physical sector of every $OSp(4,2|2)$-invariant model. This means that in any theory in which P and CT are broken, at the higher dimensional level also $OSp(4,2|2)$ must be broken. This can be the case in theories of 4-dimensional Weyl fermions or of Dirac fermions with γ_5 couplings, for instance. A natural question is whether there exists a higher-dimensional generalization of the four-dimensional γ_5-matrix. Is it possible to find a matrix $\hat{\Gamma}$ such that a Yukawa term $\bar{\Psi}\hat{\Gamma}\Psi\Phi$, say, is $OSp(4,2|2)$ invariant and dimensionally reduces to $\bar{\psi}\gamma_5\psi\phi$? In [21] we have shown that the answer is negative. The reason is not difficult to understand: we argued that a low-dimensional parity transformation can be obtained as the physical projection of some $SO(4,2)$-rotation. Hence, from a low-dimensional point of view, the Σ_{AB}-spinor representation contains both left-handed and right-handed particles. Even in the massless case, no $OSp(4,2|2)$-covariant separation of a Dirac spinor into two Weyl spinors is possible. However, this does not imply that it is impossible to obtain a parity-violating theory from BRST-quantization. The full $OSp(4,2|2)$-symmetry is not really necessary for a consistent BRST-field theory, i.e., we could allow $\hat{\Gamma}$ to break $OSp(4,2|2)$ down to some subgroup of transformations commuting with $\hat{\Gamma}$. The only indispensible requirements are that this subgroup contains the Lorentz group $O(3,1)$ and the BRST-transformation. However, breaking down $OSp(4,2|2)$ to a subgroup we have a priori no guarantee that the PCT-rotation S_{PCT} is contained in this subgroup. Here again modular invariance plays a crucial rôle: It turns out [21] that modular invariance together with Lorentz invariance is sufficient to guarantee PCT-invariance! In this way we are again led to the by now familiar connection between modular invariance on the one hand and (micro-)causality or PCT-invariance on the other. Moreover, in this situation where $OSp(4,2|2)$ is partly broken, we also realize that modular invariance is a more fundamental requirement than $OSp(4,2|2)$-invariance. To close with, we give an explicit example of a matrix $\hat{\Gamma}$ which is Lorentz-, BRST- and modular invariant:

$$\hat{\Gamma} = \gamma_5 \otimes I \otimes I. \tag{5.15}$$

It leaves $O(3,1) \otimes OSp(1,1|2)$ unbroken. Generically, the transformations of

$$\frac{OSp(4,2|2)}{O(3,1) \otimes OSp(1,1|2)} \tag{5.16}$$

are no longer symmetries of the higher dimensional theory, except for the isolated element Λ_{PCT}, which is guaranteed to be a symmetry by modular invariance.

6. Conclusion

We have shown that for both scalar and spinor particles the BRST-field theory can be made invariant under the supergroup $IOSp(4,2|2)$, which unifies the Poincaré transformations with the BRST-transformation. In this framework the PCT-transformation can be considered either as a continuous rotation in an extended bosonic space or as a modular transformation seen in a rotated frame. We repeatedly stressed the rôle played by modular invariance in obtaining a causal reduced theory, where "causal" refers both to causal (or Feynman) boundary conditions for the propagator and to microcausality, which is related to PCT-invariance.

Acknowledgement

It is a pleasure to thank the organizers of the workshop for their kind hospitality at Bad Honnef and for the excellent atmosphere there. I would also like to thank Ennio Gozzi for an enjoyable collaboration on the subject presented here.

References

1. C.A.P. Galvao, C. Teitelboim: J. Math. Phys. **21**, 1863 (1980)
2. M. Henneaux, C. Teitelboim: Ann. Phys. **143**, 127 (1982)
3. C. Teitelboim: Phys. Rev. **D 25**, 3159 (1982)
4. P.D. Mannheim: Phys. Lett. **B 137**, 385 (1984); **B 166**, 191 (1986)
5. S. Monaghan: Phys. Lett. **B 178**, 231 (1986); **B 181**, 101 (1986)
6. A. Neveu, P. West: Phys. Lett. **B 182**, 343 (1986)
7. A. Barducci, R. Casalbuoni, D. Dominici, R. Gatto: Phys. Lett. **B 187**, 135 (1987)
8. R. Casalbuoni: Talk given at the *12th Edicion De Los Encuentros relativistas* (La Laguna, 1987)
9. T. Hori, C.B. Kim: Phys. Lett. **B 207**, 44 (1988)
10. A. Barducci, R. Casalbuoni, D. Dominici, R. Gatto: Phys. Lett. **B 194**, 257 (1987)
11. R. Casalbuoni, D. Dominici, R. Gatto, J. Gomis: Phys. Lett. **B 198**, 177 (1987); J. Gomis, J. Roca: Phys. Lett. **B 207**, 309 (1988)
12. A. Aratyn, R. Ingermanson, A.J. Niemi: Phys. Rev. Lett. **58**, 965 (1987)
13. E.S. Egarian: Phys. Lett. **B 202**, 535 (1988); R.P. Manvelyan: Phys. Lett. **B 205**, 504 (1988)
14. T. Filk, H. Römer: Z. Phys. **C 39**, 203 (1988)
15. P. Thomi: J. Math. Phys. **30**, 470 (1989)
16. A. Neveu, P. West: Nucl. Phys. **B 293**, 266 (1987)
17. W. Siegel, *Introduction to String Field Theory* (World Scientific, Singapore 1988) and references therein
18. E.S. Fradkin, G.A. Vilkovisky: Phys. Lett. **B 55**, 244 (1975); I.A. Batalin, G.A. Vilkovisky: Phys. Lett. **B 69**, 309 (1977); E.S. Fradkin, T.E. Fradkin: Phys. Lett. **B 72**, 343 (1978)
19. M. Henneaux: Phys. Rep. **126**, 1 (1985)
20. E. Gozzi, M. Reuter: Nucl. Phys. **B 320**, 160 (1989)
21. E. Gozzi, M. Reuter: Nucl. Phys. **B 325**, 356 (1989)
22. J. Govaerts: preprints CERN-TH. 4950/88 and 5010/88
23. A.C. Hirshfeld: In *The Fundamental Interaction – Geometrical Trends*, Proceedings of the 1987 Bad Honnef meeting, ed. by J. Debrus and A.C. Hirshfeld (Plenum Press, New York 1988)
24. G. Parisi, N. Sourlas: Phys. Rev. Lett. **43**, 744 (1979); Nucl. Phys. **B 206**, 321 (1982)
25. J.L. Cardy: Phys. Lett. **B 125**, 470 (1983)
26. B. Zumino (unpublished); G. Lüders: Ann. Phys. **2**, 1 (1957); W. Pauli: in *Niels Bohr and the Development of Physics*, ed. by W. Pauli (McGraw-Hill, New York 1955); G. Lüders, B. Zumino: Phys. Rev. **106**, 385 (1957); R. Jost: Helv. Phys. Acta **30**, 409 (1957); R.F. Streater, A.S. Wightman: *PCT, Spin and Statistics, and all that*, (Benjamin, New York, 1964)
27. A. Barducci, R. Casalbuoni, L. Lusanna: Nuovo Cimento **35A**, 377 (1976); F.A. Berezin, M.S. Marinov: Ann. Phys. **104**, 336 (1977); L. Brink, S. Deser, B. Zumino, P. diVecchia, P. Howe: Phys. Lett. **B 64**, 435 (1976)

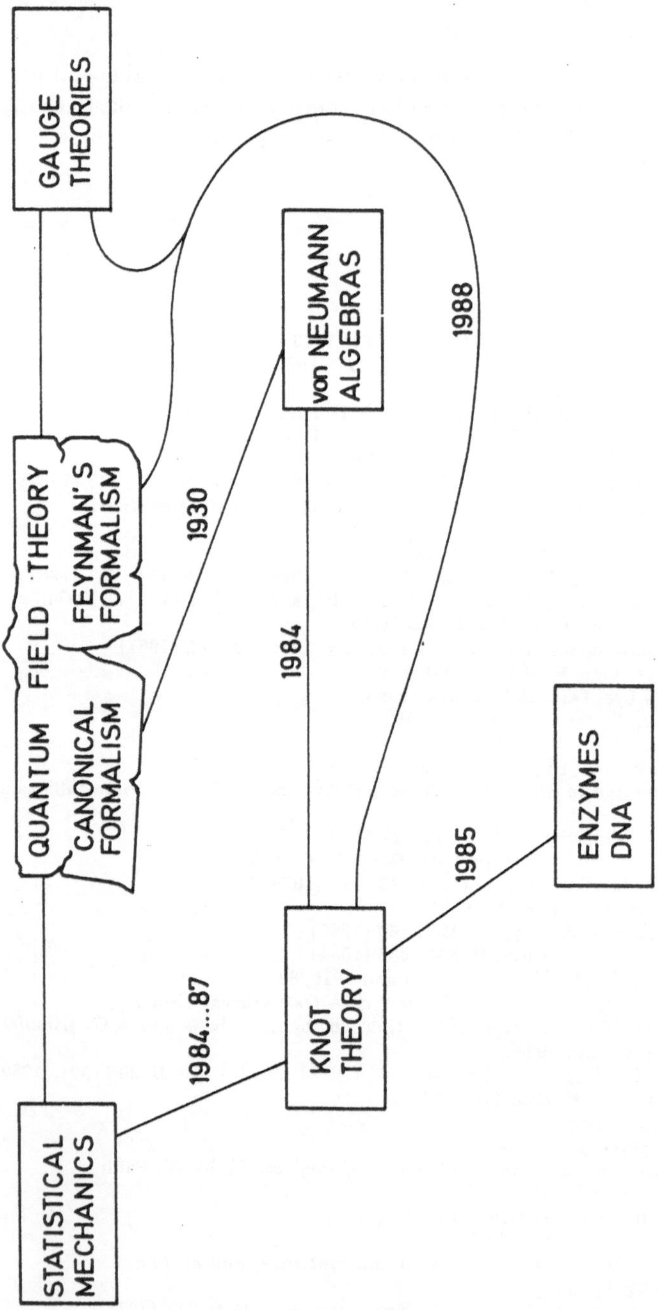

Fig. 1. Ramifications

Knots and Their Links with Biology and Physics

Thomas Schücker

Institut für Theoretische Physik der Universität
Philosophenweg 16, D(W)-6900 Heidelberg,
Federal Republic of Germany

Abstract

The discovery of a link between apparently unrelated fields is always a particular highlight in the development of natural sciences. An entire wave of such discoveries was triggered in 1984 and is still rolling. It all started with a bridge between knot theory and the theory of von Neumann algebras: the Jones polynomials. Within one year biologists recognized the usefulness of these polynomials for the classification of the enzymes transforming our DNA. At about the same time, mathematicians discovered a most remarkable connection between statistical mechanics and knot theory. Again, the mathematician and physicist Vaughan Jones played a central role. Finally, in 1988 Ed Witten successfully used techniques of quantum field theory and gauge theory to arrive at the Jones polynomial.

1. Knot Theory

About a hundred years old, knot theory is a quite recent field of mathematics. Its main goal, a complete classification of all knots, is still unachieved today. Fig. 11 shows such a classification going back to the year 1932, which includes all knots with nine or less crossings [1]. Today with the help of computers this table has been pushed to thirteen crossings.

To be sure not to lose the knot while manipulating it, mathematicians always "seal" it: after the knot is tied, its loose ends are glued together. The fundamental problem in classifying knots is that in general it is difficult to tell whether two knots are equivalent, i.e. continuous deformations of each other. In more technical terms, a knot is a differentiable map $K(s)$ from the circle into $\mathbb{R}^3$, which is self-avoiding (injective).

Fig. 2. The unknot in different diguises

Two knots $K(s)$ and $K'(s)$ are called *equivalent*, if there is a differentiable isotopy

$$f : [0,1] \to \mathrm{Diff}(\mathbb{R}^3), \qquad t \mapsto f_t, \tag{1.1}$$

with f_0 the identity map of $\mathbb{R}^3$ and such that $K'(s) = f_1(K(s))$.

Fig. 2 shows three equivalent knots, the so-called *unknots*. Fig. 3 shows two unequivalent knots, the positive and negative *trefoil knot*. The positive one consists of three positive crossings. To determine the sign of a crossing, we orient the knot. If the vector product of the upper with the lower arrow of a crossing points out of the page towards us the crossing is positive. If we reverse the orientation of the knot, both arrows change sign, but their vector product does not.

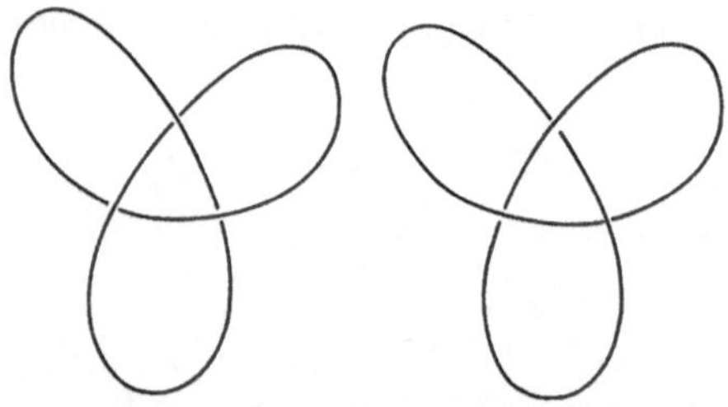

Fig. 3. Non-equivalent knots: The positive and negative trefoil knot

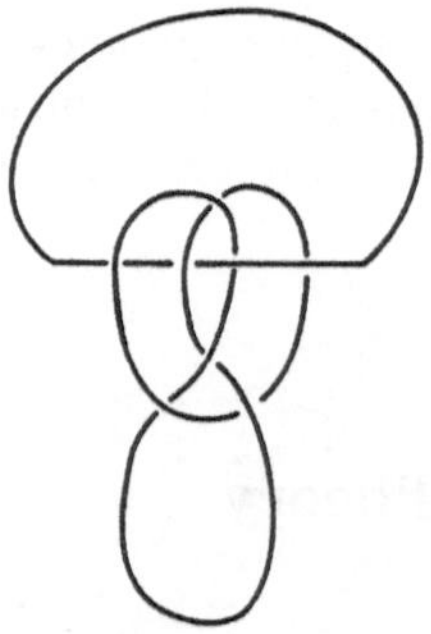

Fig. 4. A link: The clove hitch.

An isolated non-trivial result about knots was already found in 1833 by Carl Friedrich Gauß [2]. He asked himself what magnetic field a knotted wire would produce if it carried an electric current. This consideration led him to an analytic formula for the linking number. A link consists of a finite number of knots, called *components* of the link. These components can be interlaced. Fig. 4 shows a link of two unknots, the well-known *clove hitch*, whose linking number is two.

A particular interest in the classification of knots came from chemistry following the discovery of the periodic table of the elements at the end of the last century. Lord Kelvin and other scientists tried to understand atoms as knots, hoping that the table of knots would reproduce the regularities of the table of elements. In this picture the different masses of the atoms would be explained by the resistance felt by the knots when moving through a viscous ether. Molecules would be represented by links. Although the classification of knots is still uncomplete, this attractive picture was refuted rather quickly. However, it reappeared frequently in nuclear and particle physics. Its most recent variant, string theory, has just passed its remarkable climax. There have also been attempts to apply knot theory to the chemistry of polymers [3]. In 1988 Christine Dietrich-Buchecker and Jean Pierre Sauvage in Strassbourg succeeded to synthesize the trefoil knot with molecular threads [4]. They now have the world's smallest knot.

Independently of possible applications, mathematicians continued to work on the classification of knots. The introduction of abstract topological and group-theoretical methods in particular by Alexander 1928 [5], Reidemeister 1932 [1] and Seifert 1934 [6] brought substantial progress.

2. The Jones Polynomials

The theory of von Neumann algebras is an even younger branch of mathematics. In quantum mechanics observables are described by operators acting on a Hilbert space. These operators form algebras, whose systematic analysis was started in the thirties by John von Neumann [7]. The classification of these algebras still fascinates both mathematicians and physicists today [8], among them Vaughan Jones. In 1974 he came from his native New Zealand to Geneva to begin a thesis on axiomatic quantum mechanics. Due to unfortunate personnel politics at the physics institute he switched to mathematics, where he began his thesis on subfactors of von Neumann algebras with André Haefliger in 1975. Subfactors describe subsystems of quantum mechanical systems. What particularly intrigued Jones was that subfactors can have non-integer dimensions. In 1980 he went to Philadelphia, where he was appointed professor. Despite depreciating remarks of some colleagues he continued to work on the classification of subfactors. In a rather technical proof he found the relations:

$$\sigma_i \sigma_j = \sigma_j \sigma_i, \quad |i-j| \geq 2 \tag{2.1}$$

$$\sigma_i \sigma_{i+1} \sigma_i = \sigma_{i+1} \sigma_i \sigma_{i+1}, \tag{2.2}$$

which are Artin's presentation of the braid group. In Rehren's lectures at this School you find details about a similar connection between operator algebras and the braid group. These relations were the key to Jones' definition in 1984 of a new class of polynomials [9] which allow a giant step towards the classification of knots and links.

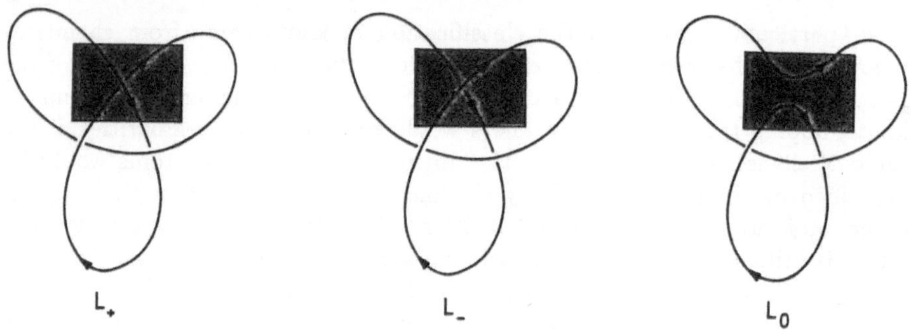

Fig. 5. The reduction procedure to calculate the Jones polynomial of the positive trefoil knot, first step.

To every link L Jones associates a Laurent polynomial $V_L(t)$ in the variable $\sqrt{t}$, which is calculated inductively in the number of crossings according to the following scheme: First choose an orientation of the link and project it into the plane. Any link can be transformed into a simpler link (with less crossings) by replacing a suitably chosen overcrossing by an undercrossing. This procedure is illustrated in Fig. 5 for the positive trefoil $L =: L_+$, where the L_- produced in this way is already equivalent to the unknot. We must also consider a third link L_0, obtained from L_+ by replacing the chosen overcrossing by the uncrossing as in Fig. 5. Obviously, also L_0 has less crossings than L_+. The Jones polynomial of L_+ is obtained from the polynomials of the simpler links L_- and L_0 by the so-called *skein relation:*

$$V_{L_+}(t) = t^2 V_{L_-}(t) + t(\sqrt{t} - \frac{1}{\sqrt{t}})V_{L_0}(t). \tag{2.3}$$

Iterating this procedure we finally arrive at a link with no crossings at all, i.e., an non-interlaced collection of n unknots, whose polynomial is defined to be

$$V_n(t) = [-(\sqrt{t} + \frac{1}{\sqrt{t}})]^{n-1}. \tag{2.4}$$

In particular, the Jones polynomial of the unknot is equal to one. Therefore the Jones polynomial of the positive trefoil knot is

$$V_{L_+}(t) = t^2 \cdot 1 + t(\sqrt{t} - \frac{1}{\sqrt{t}})V_{L_0}. \tag{2.5}$$

Applying the skein relation once more to $L_0 =: L_{0+}$ as in Fig. 6, we obtain

$$V_{L_0}(t) = V_{L_{0+}} = t^2[-(\sqrt{t} + \frac{1}{\sqrt{t}})] + t(\sqrt{t} - \frac{1}{\sqrt{t}}) \cdot 1 \tag{2.6}$$

and finally, for the positive trefoil knot:

$$V_{L_+}(t) = -t^4 + t^3 + t. \tag{2.7}$$

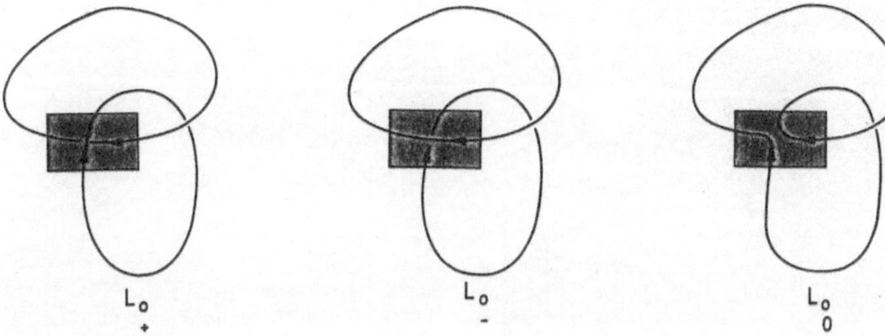

Fig. 6. Second step

For this algorithm to make sense, it must still be shown that different choices of orientation and projection of the link lead to the same polynomial. Even more: equivalent links have identical Jones polynomials. The proofs are complicated, but can be simplified using statistical mechanics as sketched below. To summarize: The Jones polynomials help us to classify links. If two links have different polynomials, as for instance the positive and negative trefoil knot, they cannot be equivalent. However, this classification is not complete; there are non-equivalent knots with identical polynomials, as the two shown in Fig. 7.

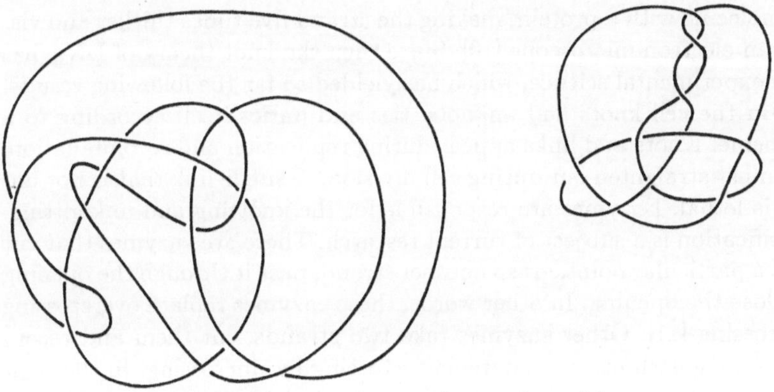

Fig. 7. Non-equivalent knots with the same Jones polynomial

Let us mention three more properties of the Jones polynomial.

1. The Jones polynomial of a link with an odd number of components is a Laurent polynomial in t.
2. The Jones polynomial of the mirror image $-L$ of a link L is given by

$$V_{-L}(t) = V_L(t^{-1}). \tag{2.8}$$

3. The number of components c of a link L is obtained from the value of its Jones polynomial at $t = 1$, $V_L(1) = (-2)^{c-1}$. (2.9)

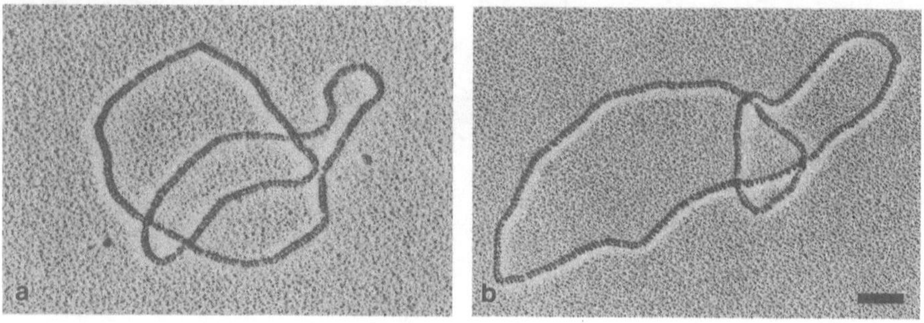

Fig. 8. Knots tied by Nature: The positive and negative trefoil of DNA under the electron microscope [10]. The scale bar measures $10^{-7}m$.

3. Enzymes and DNA

It is a long standing problem in biology to understand the mechanisms responsible for the knotting and unknotting of DNA molecules. DNA is normally packed very tightly into cells. The total length of all a person's DNA molecules is about 100 times the distance between the earth and the sun. Until recently biologists could only speculate about the geometry of DNA. This changed in 1983, when Cozzarelli of Berkeley and Stasiak of Zürich found a way to coat the DNA molecule with a protein, making the strand five times thicker and visible under an electron microscope [10]. Since then the knot theory of DNA has become an experimental science, which has yielded so far the following results. The DNA in the cell knots and unknots, ties and unties itself according to a definite scheme. Knots and links appear during replication and recombination. The DNA must straighten out during cell division. A single link that is not undone then is lethal. Enzymes are responsible for the knotting and unknotting. Their classification is a subject of current research. There are enzymes that cut a strand at a particular point, grasp another strand, pass it through the opening and then close the opening. In other words, these enzymes replace overcrossing by undercrossing [11]. Other enzymes take two strands, cut them and reconnect them among each other, they replace crossing by uncrossing. In the light of these results, it is no surprise that Jones polynomials offer a practical algorithm for classifying these enzymes. Experimentally, the inductive procedure described above is reversed. First unknots are formed artificially from DNA strands. Then one particular enzyme is allowed to operate on these unknots. The links thereby produced are observed under the electron microscope, Fig. 8. Their Jones polynomials help to characterize the enzyme.

4. Statistical Mechanics and Knot Theory

Remarkably enough, a proof of invariance of the Jones polynomial under continuous deformations can be given using methods from statistical mechanics [12, 13]. Furthermore, these methods lead to new link invariants.

To start, we checker-board shade the graph of a projected link L as in Fig. 9, and construct its *dual lattice* by assigning a vertex to each shaded region and an edge with sign to each crossing. This sign has nothing to do with the sign of an oriented crossing introduced before, and is defined as follows. Lay a pencil over the upper line of the crossing and rotate the pencil clockwise until it covers the lower line. If the pencil has swept over a shaded region, the sign of the crossing is negative. Now we put a spin $\frac{1}{2}$ on each vertex, taking values $\sigma = 1, -1$, and construct an Ising model over the dual lattice with *Boltzmann weights* $w_+(\sigma, \sigma')$ and $w_-(\sigma, \sigma')$ according to the sign of the edge.

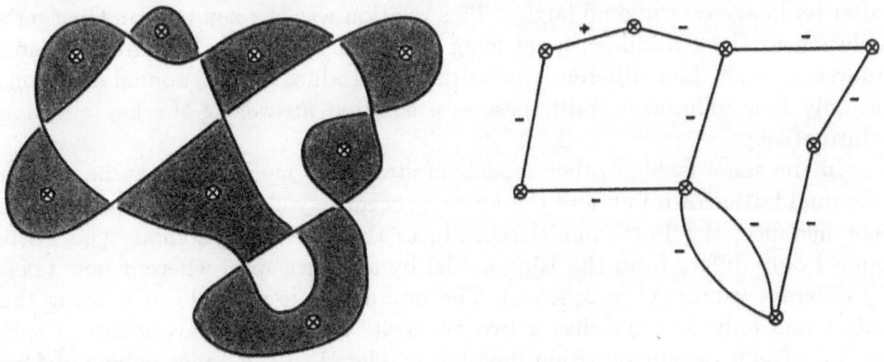

Fig. 9. The dual lattice of a knot

Let us recall that the Ising model was introduced to describe phase transitions in ferromagnets where σ represents the spin of an electron in an iron atom. In this case, the Boltzmann weights are related to the interaction energy of two neighbouring electron spins $E_{\sigma\sigma'} = J_{\sigma\sigma'}$ by

$$w(\sigma, \sigma') = \exp\left(\frac{-E_{\sigma\sigma'}}{kT}\right) \tag{4.1}$$

Returning to our more general Ising model over the dual lattice of the link L, we consider its partition function

$$Z_L = \sum_{\text{configurations}} \prod_{\text{edges}} w_\pm(\sigma, \sigma'). \tag{4.2}$$

The key question is: Can we choose the four Boltzmann weights

$$w_+(1,1) = w_+(-1,-1) \qquad w_+(1,-1) = w_+(-1,1) \tag{4.3}$$
$$w_-(1,1) = w_-(-1,-1) \qquad w_-(1,-1) = w_-(-1,1) \tag{4.4}$$

such that the partition function does not change when we deform the link L continuously? There are infinitely many continuous deformations, but the following circumstance simplifies our task: Every continuous deformation of L can be decomposed into a sequence of elementary deformations on the projection of L, the so-called *Reidemeister moves.* A Reidemeister move deforms the link only at one place and in a simple fashion. For instance, we can simplify the knot of Fig. 9 in the upper left-hand corner using the first Reidemeister move of Fig. 10. Under a Reidemeister move the dual lattice changes only locally, as shown in Fig. 10 for two typical moves. Accordingly, the partition function changes under a Reidemeister move in a transparent fashion, and its invariance yields polynomial equations for the Boltzmann weights. The equation belonging to the second Reidemeister move of Fig. 10 is called "star triangle relation" after its image on the dual lattice. This relation was the key to Lars Onsager's solution of the two-dimensional Ising model in 1944 [14]. The Reidemeister moves with all their different configurations produce 32 polynomial equations in only four unknowns. Still there is a solution answering the key question affirmatively.

In the same fashion other models of statistical mechanics may be put on the dual lattice of a link and the above procedure yields other link invariants. For instance, the Potts model reproduces the Jones polynomial. The Potts model only differs from the Ising model by a higher spin, where σ now takes Q different values ($Q = 3,4,5...$). The interaction is again located along the edges and only distinguishes if two "nearest" neighbours have identical spin or not. The invariant partition function is related to the Jones polynomial by $Q = 2 + t + \frac{1}{t}$.

To summarize, we can say that knot theory has successfully generalized techniques which were developed in statistical mechanics for regular lattices, e.g. of crystals to lattices stemming from links. This situation belongs to the rather unfrequent examples in which the flow of information is from physics to mathematics. More frequently, mathematical techniques are utilized to solve physical problems. Another example of this rather unfrequent category is the subject to the next section.

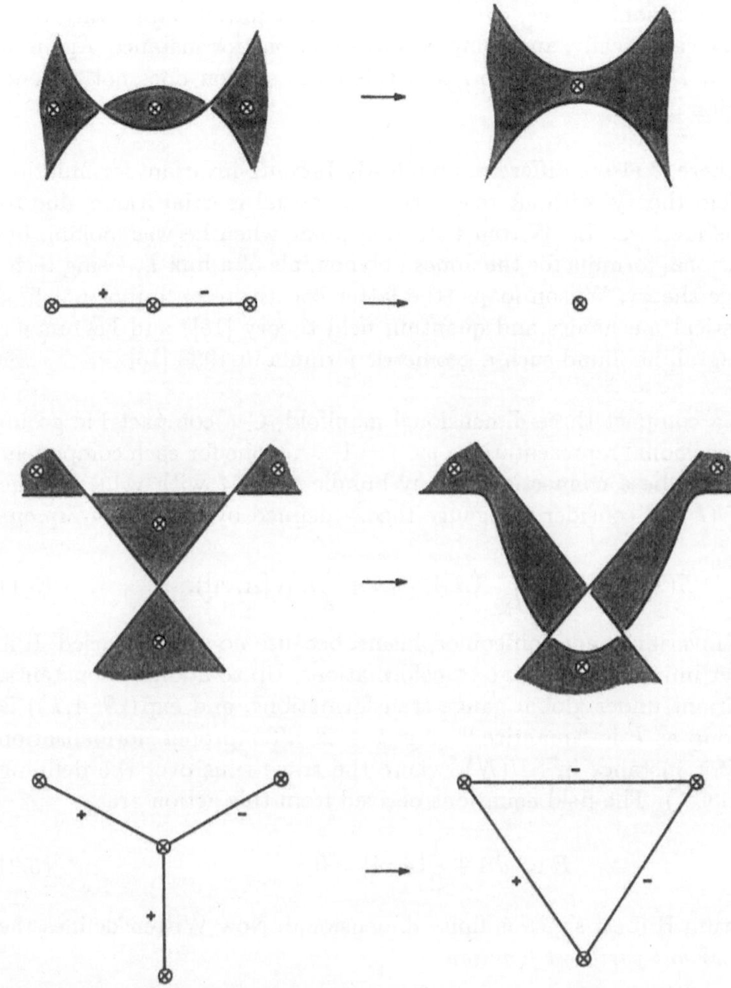

Fig. 10. Two Reidemeister moves and their representation on the dual lattice

5. Quantum Field Theory and the Jones Polynomials

Since their discovery in 1984, mathematicians have tried to find a three-dimensional geometric interpretation of the Jones polynomials. Indeed, the calculation of the polynomial takes an uncomfortable detour via a two-dimensional projection, necessitating a posteriori the complicated proof that the polynomial does not depend on the projection chosen. We encounter the same problem in relativistic quantum field theory. Let us recall that in quantum mechanics space and time play fundamentally different roles: time is not subject to an uncertainty relation, only space is quantized, in sharp contradistinction with Lorentz invariance. We are therefore forced to project the four-dimensional spacetime onto a three-dimensional space, a Cauchy surface of initial data. Only then can we quantize canonically, and compute cross sections for instance. Again, a complicated proof shows a posteriori that this cross section does not depend on the projection chosen.

However, there is also a different, manifestly Lorentz-invariant formulation of quantum field theory, without reference to a special inertial frame, due to Feynman. This idea was Ed Witten's starting point when he was looking for a three-dimensional formula for the Jones polynomials of a link L. Using techniques of gauge theory, Wilson loops (the latter constitute an important link between statistical mechanics and quantum field theory [15]) and Feynman's functional integral, he found such a geometric formula in 1988 [16].

Let M be a compact three-dimensional manifold, G a compact Lie group with finite dimensional representations ρ_j, $j = 1,\ldots,c$, one for each component of the link. Let A be a connection on any bundle over M with values in the Lie algebra of G and consider the gauge theory defined by the *Chern-Simons action:*

$$S(A,k) := \frac{k}{4\pi}\int_M \mathrm{Tr}(A \wedge dA + \frac{1}{3}A \wedge [A,A]). \tag{5.1}$$

This action is invariant under diffeomorphisms, because no metric is used. It is invariant under infinitesimal gauge transformations. Up to additive constants, it is also invariant under global gauge transformations, and $\exp(iS(A,k))$ is invariant as long as k is "quantized", e.g. $k \in \mathbb{Z}$ after proper normalisation of the trace (for instance in $SU(N)$, where the trace runs over the defining representation $\mathbb{C}^N$). The field equations derived from this action are

$$F := dA + \frac{1}{2}[A,A] = 0, \tag{5.2}$$

and the quantum Hilbert space is finite dimensional. Now Witten defines the *manifestly invariant partition function*

$$Z_L(k) := \int \mathcal{D}A \, \exp(iS(A,k)) \prod_{K_j} \mathrm{Tr}_{\rho_j} P\exp(\textstyle\int_{K_j} A), \tag{5.3}$$

where $\mathcal{D}A$ denotes the functional integral over all connections on the fixed bundle, K_j are the components of the link and P is the path-ordering along the knots K_j. Witten's partition function reproduces the Jones polynomial

$$Z_L(k) = V_L\left(\exp\left(\frac{2\pi i}{k+2}\right)\right), \tag{5.4}$$

if we choose $M = S^3$, $G = SU(2)$, the trivial bundle and $\rho_j = \mathbb{C}^2$ for all link components. A further attractive feature of Witten's formula is that it remains valid in arbitrary compact three manifolds M, and for $L = \emptyset$, the empty knot, it yields an invariant of three manifolds themselves. The main problem is that the functional integral is still not well defined, a situation comparable to the status of the delta function before Laurent Schwartz invented distributions. Quantum field theory is not a consistent theory, it is plagued by divergences and ambiguities [17]. On the other hand, we must admit that quantum field theory has had impressive phenomenological success in high energy physics, where in the context of Feynman diagrams reduction procedures similar to the one described for the Jones polynomial play an important role.

Now that mathematicians use tricks of quantum field theory, is there better hope for a consistent quantum field theory at last, or even for a quantum theory of general relativity, which in its geometric aspects somehow resembles knot theory [18]? Maybe we can at least hope for an improvement of the dialogue between mathematicians and physicists, that has been so badly neglected in the last decades.

References

1. K. Reidemeister: ***Knotentheorie*** (Springer, Berlin 1932)
2. C. F. Gauß: Königliche Gesellschaft der Wissenschaften zu Göttingen **5**, 602 (1877)
3. H. L. Frisch, E. Wasserman: J. Am. Chem. Soc. **83**, 3789 (1961)
4. Ch. O. Dietrich-Buchecker, J P. Sauvage: Angew. Chem., Int. Ed. **28**, 189 (1989)
5. J. W. Alexander: Trans. AMS **30**, 275 (1928)
6. H. Seifert: Math. Ann. **110**, 571 (1934)
7. F. J. Murray, J. von Neumann: Ann. Math. **37**, 116 (1936)
8. A. Connes: Proc. Symp. Pure Math. **38**, 43 (1982)
9. V. Jones: Bull. AMS **12**, 103 (1985); Ann. Math. **126**, 335 (1987)
10. M. A. Krasnow, A. Stasiak, S. J. Spengler, F. Dean, T. Koller, N. R. Cozzarelli: Nature **304**, 559 (1983)
11. J. D. Griffith, H. A. Nash: Proc. Natl. Acad. Sci. USA **82**, 3124 (1985)
12. L. Kauffman: ***Topology*** **26**, 395 (1987)
13. V. Jones: Pac. J. Math. **137**, 311 (1989)
14. L. Onsager: Phys. Rev. **65**, 117 (1944)
15. K. Symanzik: J. Math. Phys. **7**, 510 (1966);
    F. Wegner: J. Math. Phys. **12**, 2259 (1971);
    K. Wilson: Phys. Rev. **D10**, 2445 (1974)
16. E. Witten: Comm. Math. Phys. **121**, 351 (1989)
17. P. A. M. Dirac: Scien. Am. **208**, 45 (1963)
18. C. Rovelli, L. Smolin: Phys. Rev. Lett. **61**, 1155 (1988)

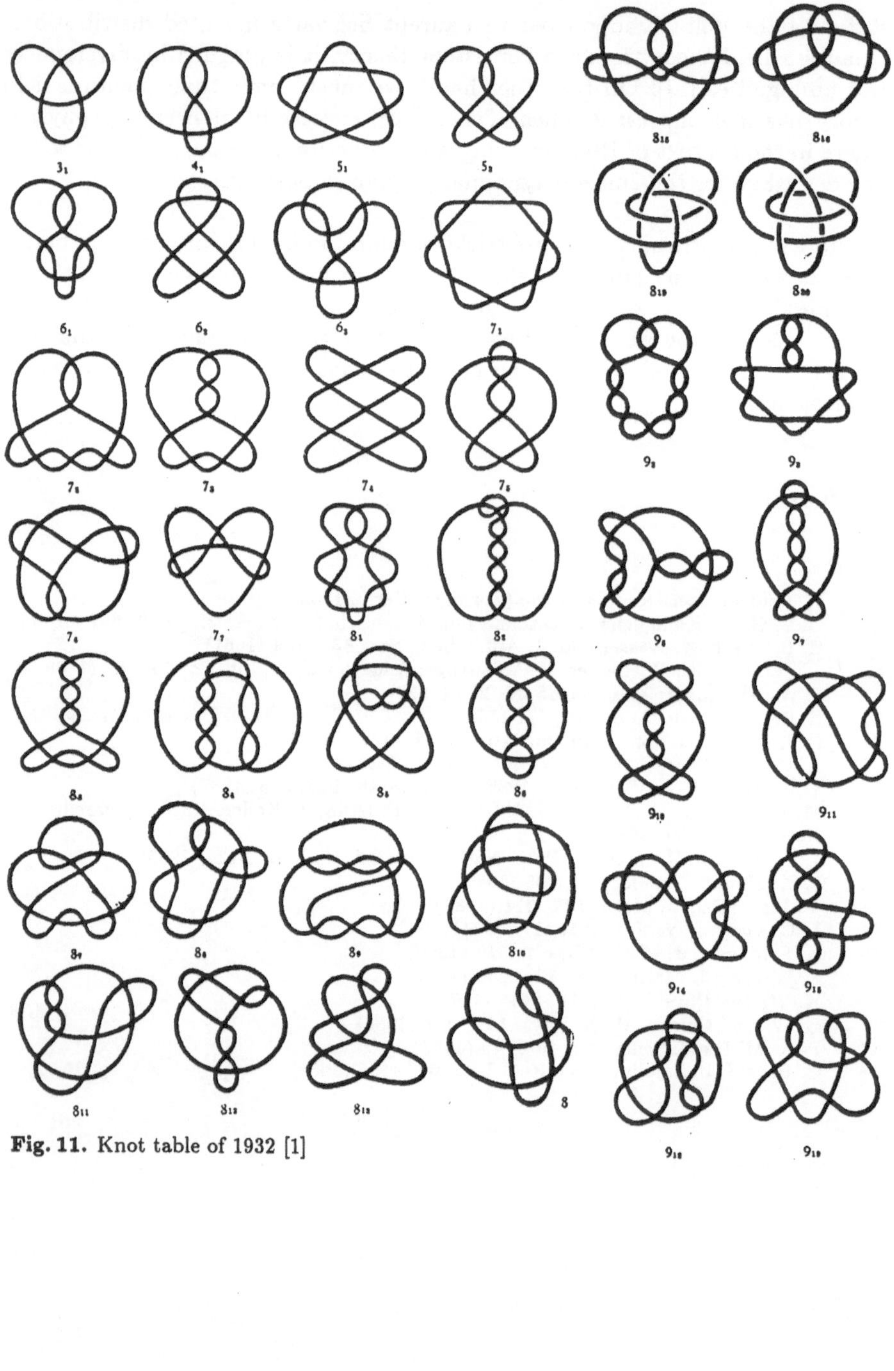

Fig. 11. Knot table of 1932 [1]

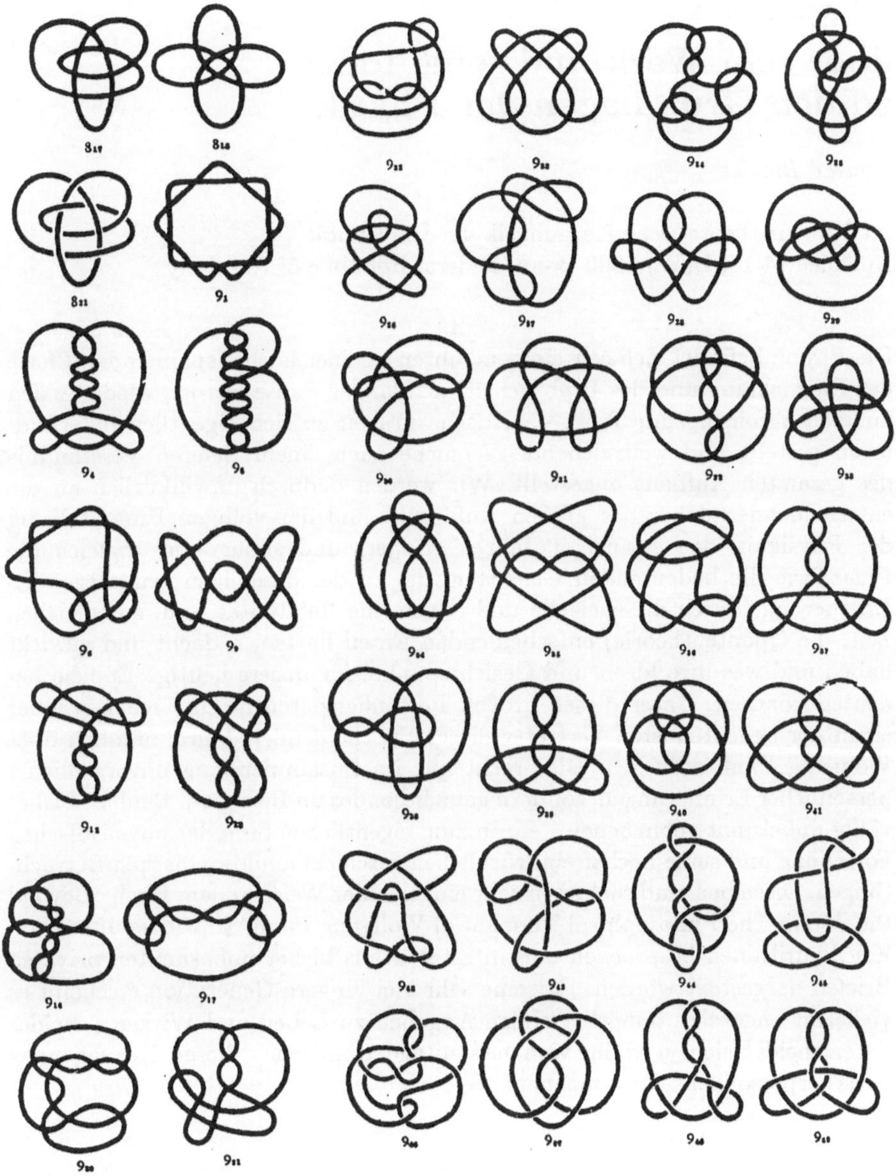

Wolfgang Pauli – über sein Werk und seine Ideen zu den Grundlagen der Physik

Konrad Bleuler

Institut für theoretische Kernphysik der Universität
Nußallee 14-16, D(W)-5300 Bonn, Federal Republic of Germany

I.

Die Physik befindet sich seit einigen Jahren in einer höchst spannenden Phase der Entwicklung und des Umbruchs: Nach langer Pause werden wieder – den äußeren Erfolgen zum Trotz – kritische Fragen zu den eigentlichen Grundlagen gestellt und weitreichende Versuche nach einem tieferen Verständnis des gesamten Aufbaus angestellt. Wir werden dadurch unwillkürlich an die entscheidenden Jahre des großen Aufbruchs und der völligen Umgestaltung der Physik in den zwanziger Jahren erinnert und stellen uns sogleich die Frage, wie die bedeutenden Gelehrten, die an der damaligen grundlegenden Erneuerung (durch die spezielle und allgemeine Relativität und, zur gleichen Zeit, die Quantentheorie) entscheidenden Anteil hatten, gedacht und gewirkt haben und was ihre Ideen und Gesichtspunkte für unsere heutige Epoche bedeuten könnten. Unter diesen großen Persönlichkeiten möchte ich hier über einen der bedeutendsten Vertreter dieser entscheidenden Jahre, nämlich über Wolfgang Pauli sprechen. Dies nicht nur im Zusammenhang unvergeßlicher persönlicher Erinnerungen, sondern ganz besondes im Bezug auf Paulis – bisher völlig unbekannt gebliebenen – Ideen zum eigentlichen Sinn der physikalischen Forschung und seine höchst eindrücklichen geschichtsphilosophischen Betrachtungen. Dieselben sind soeben in sehr eingehender Weise in dem Buch „Beyond the Atom: The Philosophical Thought of Wolfgang Pauli“ (Springer 1988) von K.V.Laurikainen besprochen und in Zitaten aus bisher unbekannten privaten Briefen dargestellt worden. Für eine sehr viel jüngere Generation erscheint es vielleicht angezeigt, zunächst einige Angaben zu Leben und Wirken – beides unterscheidet sich ja völlig vom heutzutage Üblichen – dieser überragenden Forscherpersönlichkeit zu machen.

II.

Wolfgang Pauli wurde also im Jahre 1900 – er bemerkte gelegentlich, daß dieses runde Datum sehr praktisch sei – in Wien geboren. Der ursprüngliche, für seine tiefe Begabung bezeichnende Familienname „Pacheles“ wurde von seinem Vater (Professor in Wien und „nobelpreisverdächtig“ bezgl. seiner chemisch-pharmakologischen Forschung) in „Pauli“ abgeändert, wie dies wohl für viele aus dem Osten ins K. und K. Kaiserreich Zugewanderte üblich gewesen war. Man sollte in diesem Zusammenhang vielleicht bemerken, daß das „geistige *und* künstlerische Wien“ ohne die vielen großen Namen wie Kafka, Freud, Mahler u.a.m. wohl kaum denkbar wäre. Aus Paulis Schulzeit in Wien wäre unter anderem zu berichten, daß er sehr viel später, aber eher mit einem bitteren Unterton, beiläufig erwähnte: „Ich war halt eben ein (mathematisches) Wunderkind“ und daß er auf den Rand eines Buches, sehr im Gegensatz zu seinem Patenonkel Mach, schrieb: „Es ist wichtig, neben der Physik *auch* Sprachen und Geschichte zu betreiben“, eine Bemerkung, die Späteres, wie wir sehen werden, vorweg nimmt.

Die Studienzeit wurde durch einen unglaublichen Glücksfall, der wohl für die Entwicklung der ganzen modernen Physik von entscheidender Bedeutung ist, gezeichnet: Zusammen mit Werner Heisenberg und Gregor Wentzel – es bildete sich daraus eine freundschaftliche und geistige Verbindung für alle späteren Jahre – studierte er bei Arnold Sommerfeld in München: Man kann in diesem Zusammenhang die wissenschaftsgeschichtliche Bedeutung eines großen Lehrers und Förderers der Forschung kaum überschätzen. Dasselbe gilt in gleicher Weise für die spätere und wieder gemeinsame Arbeit unter der geistigen, aber auch menschlich verständigen Leitung von Nils Bohr in Kopenhagen. Über Paulis späteren, in jeder Beziehung höchst originellen Chef, Prof. Lenz in Hamburg, erzählte Pauli auch noch in Zürich (er kam 1928 dahin und blieb – mit einer Unterbrechung wärend des Krieges – dort bis zu seinem Tode 1958) die ergötzlichsten Anekdoten.

III.

Paulis wissenschaftliche Laufbahn begann 1919 „kometenhaft“ mit dem bis heute grundlegenden „Handbuchartikel“ über spezielle und allgemeine Relativitätstheorie, zu welchem er von Sommerfeld, in gewissem Sinne als „Lückenbüßer“ für einen anderen Autor, angeregt worden war. Albert Einstein war über dieses Werk so tief beeindruckt, daß er in der Folge Pauli zu seinem geistigen (aber auch beruflichen) Nachfolger ernannte. Dies bedeutete für Pauli, wie mir scheint, eine große Verpflichtung, in gewissem Sinne sogar eine Bürde: In plastischer Erinnerung ist mir in diesem Zusammenhang ein sehr viel späterer Vortrag (aus Anlaß der feierlichen Einweihung einer Einstein-Büste), in welchem Pauli Einsteins völlige Ablehnung der Quantentheorie mit den Worten „es wäre doch alles in Einsteins Sinn gewesen“ zu begründen oder irgendwie zu entschuldigen suchte: Bei der Verlesung seines – in unüblicher Weise – wohlvorbereiteten Manuskripts vor „auserkorenem“ Publikum holte Pauli – in Vertiefung über seinem schwierigen Text – aus seiner Westentasche einen kleinen Stift und setzte in aller Ruhe die notwendigen – beim Atemholen natürlich auffallenden – Kommas in sein Papier ein. Der junge Pauli, der die Einstein'sche Theorie aufgrund seiner mathematischen Bildung, aber auch aufgrund deren großartiger logischer Geschlossenheit wohl mit einem Schlage überblickte, stürzte sich aber sogleich auf die damals in logischer Hinsicht eben noch sehr unvollständige ältere Bohr'sche Quantenmechanik und schrieb auch darüber einen ersten Handbuchartikel. Es erscheint deshalb umso verständlicher, daß Pauli von Anfang an in ganz entscheidender Weise in Briefen und persönlichen Gesprächen zuerst mit Heisenberg und später mit Bohr an der Entstehung und, hauptsächlich, an der endgültigen Formulierung der heutigen Quantentheorie mitgewirkt hat. Die absolut grundlegende Bedeutung wie auch die Endgültigkeit dieses vielleicht größten Umbruchs in der Geschichte der Physik wurde von Pauli auch in späteren Jahren (z.T. in Verbindung mit anderen „Interpretationsversuchen") oft betont. Der Eintritt in diese völlig neue und wirklich großartige Welt der Quantentheorie muß für einen maßgeblich beteiligten und noch sehr jungen Forscher wie Wolfgang Pauli ein tiefes und entscheidendes Erlebnis gewesen sein: Es erscheint mir als einer der Beweggründe für Paulis spätere tiefschürfende Betrachtungen über Sinn und Zweck der physikalischen Forschung wie auch für seine Betrachtungen zur Ideengeschichte der physikalischen Grundbegriffe: Paulis entscheidendste Beiträge während dieser „Gründerjahre" sind wohlbekannt: Sein für den Gesamtaufbau der Materie, d.h. für Atome, Kerne und (neuerdings!) Hadronen grundlegenden Ausschließungsprinzip, das zusammen mit der Pauli'schen unrelativistischen Spintheorie die Grundlage für deren mathematische Behandlung bildet; seine – von heute aus gesehen – visionäre Neutrinohypothese und, als krönenden Abschluß (zusammen mit Heisenberg), die allgemeine Quantenfeldtheorie (d.h. *die* relativistische Partikeltheorie), aus der er in der Folge die – in neuester Zeit erweiterte – „Spin-Statistik-Relation" als, wie er sich ausdrückte, wichtigste Folge der speziellen Relativität ableitete.

IV.

Nachdem seit der Formulierung dieser drei allgemeingültigen Grundstrukturen der Quantentheorie mehr als ein halbes Jahrhundert vergangen ist, können wir heutzutage deren Bedeutung (als *die* formale Grundlage aller seither durchgeführten quantentheoretischen Rechnungen) in gebührender Weise einschätzen. Zugleich – so möchte ich vermuten – hatte aber Pauli selbst im Falle der Quantenfeldtheorie schon früh seine tiefen Zweifel in Bezug auf deren endgültige und auch mathematisch befriedigende Grundlage für eine allgemeine Theorie der Elementarpartikel. Er wollte sich, wie er sich später (d.h. nach '46) ausdrückte, *nicht* „renormieren" lassen und stellte sich damit (allerdings zusammen mit Dirac, Bohr und anderen) zeitweise in einen schroffen Gegensatz zu einer „begeisterten" und zunächst höchst erfolgreichen jüngeren Generation. Ein (teilweises?) Eingeständnis von dieser Seite erfolgte wohl erst lange nach Paulis Tode, z.T. im Zusammenhang mit dem (erfolglosen?) „Rettungsversuch" im Rahmen der „Strings" und des neuesten, sehr tief- und weitreichenden Versuches einer „Nicht-kommutativen Geometrie" nach A.Connes, D.Kastler u. a.. Mir scheint, daß mit dieser jüngsten, noch kaum allgemein bekannten Entwicklung vielleicht eine „Vision" in Erfüllung geht, die Pauli in einem langen und unvergessenen Gespräch wenige Wochen vor seinem Tode äußerte: „Zur wirklichen Lösung des Singularitätenproblems (d.h. der Renormierungsfragen) ist nochmals ein grundlegender Schritt von derselben Bedeutung und Größe wie der vorangegangene (soeben erwähnte) in den zwanziger Jahren notwendig."

V.

Pauli schloß diese große Periode der zwanziger Jahre, in der die Physik in relativ kurzer Zeit von Grund auf erschüttert und zugleich völlig erneuert wurde, mit seinem berühmten Handbuchartikel „Die Prinzipien der Quantentheorie", der bis heute seine Gültigkeit bewahrt hat, ab. Fortan galt er weitherum als oberste Instanz (oder höchster Richter) in der Beurteilung neuer Beitrage oder der vielen, unvermeidlichen „Verbesserungsversuche". In steter Erinnerung bleibt sein allgemeines „Verdikt" in solchen Fällen: Aus den vier Möglichkeiten sah er immer nur zwei, nämlich: „Neu und falsch", oder „Alt und richtig". Neben dem absolut vernichtenden „Alt *und* falsch" sah er aber die Notwendigkeit einer noch tieferen Klassifikation, nämlich „Nicht einmal falsch". Neben vielen solchen, z.T. gefürchteten „globalen Verurteilungen", die gelegentlich, insbesondere gegenüber Vortragenden, allzu hart ausfielen, gibt es eine, wie mir scheint, unbegrenzte Zahl sog. „Pauli-Anekdoten", die „mehr oder weniger wahr", „gut erfunden" oder geradezu „visionär" waren. Pauli, der wie Einstein oft mit der entsprechenden Geste von „Ihm" sprach, prophezeite z.B.: *Er* hat von dieser (mathematischen) Möglichkeit (es handelte sich um die von Pauli bekanntlich abgelehnte – weil paritätsverletzende – Weyl'sche Neutrinogleichung) ***keinen*** Gebrauch gemacht. Im Gegensatz zu dieser nichterfüllten ‚Vision' (Pauli macht sich später wegen dieses Fehlurteils über sich selbst lustig: „Gerne gebe ich jetzt einen Teil meines ‚Renommés' ab, denn ich besitze ja genug davon") ging eine andere „gut erfundene", nach der Pauli schließlich auch mit „Ihm" selbst nicht einverstanden ist, später in ganz anderer, tragischer Weise, wie wir sehen werden, in Erfüllung. Zunächst möchte ich aber ganz kurz an seine nun schon weit zurückliegenden Vorlesungen erinnern: Es gab oft sehr lange Momente der Stille und der völligen Abwendung von den Zuhörern, die ich heute gerne als ein „Zwiegespräch" oder eher als einen „Zwist" mit „Ihm" interpretieren würde.

VI.

Trotz der vielen fröhlichen und ungezwungenenen Stunden des Zusammenseins in all den späteren Jahren in Zürich, auf kleineren munteren Ausflügen, gemeinsamen Konzertbesuchen und heiteren Gesprächen beim gemeinsamen Lunch, fühlte man doch, daß Pauli damit zugleich ein Gefühl der Einsamkeit, der Zweifel und der geistigen Spannung zu überbrücken versuchte. Die waren z.T. gewiß die „Nachwirkungen" einer schweren inneren Krise (nach diesen besprochenen ersten 10 Jahren seiner wirklich großartigen Leistungen), die ihn in entscheidender Weise auch mit Fragen der Psychoanalyse und damit auch mit C.G.Jung in Verbindung brachte. Dies war für Pauli wohl ein weiterer äußerer Anlaß, sich mit tiefgreifenden Fragestellungen über die geistigen, psychologischen und auch geschichtlichen Hintergründe der, von außen gesehen, so erfolgreichen und so zielbewußten Forschung zu beschäftigen. Ein zentraler (heute vielfach übersehener oder besser, verdrängter) Gesichtspunkt Paulis lautet: Zum wirklichen Erfassen der Erscheinungen und zur Aufstellung, d.h. besser gesagt, zur Entdeckung grundlegender Gesetze braucht es (gleichberechtigt) zwei verschiedene Voraussetzungen: 1. Eine ausreichende Gesamtheit empirischer Daten *und* 2. vielleicht noch wichtiger, d.h. unumgänglich, eine Intuition oder einen besonderen Gedanken, z.B. einen Hinweis aus einem anscheinend völlig verschiedenen Forschungsgebiet. Ein modernes und sehr eindrückliches Beispiel wäre – neben dem viel zu oft erwähnten fallenden Apfel Newtons – Hermann Weyls Idee des geometrisch ungemein natürlichen (lokalen) Eichprinzips aus dem Jahre '29, das nach der (sozusagen zwangsweisen) Erweiterung nach Yang und Mills fast ein halbes Jahrhundert später *die* Grundstruktur für alle drei die ganze Erscheinungswelt der Physik überdeckenden Grundgesetze (d.h. G.R., GWS, String) wurde.

Um der Frage der Herkunft oder des Entstehens der entscheidenden Intuitionen (oder der inneren Sicht einiger Forscher) näher zu kommen, vertiefte sich Pauli in (Jung'sche) Fragen zur menschlichen Seele und deren Erscheinungsformen in der Wissenschaftsgeschichte: Er fand einen erlösenden Hinweis in dem philosophischen, oder besser mystischen Hauptwerk Keplers: Es ist der für ihn entscheidende Begriff des „Archetypus", eines Ausdrucks, der ihn zugleich dem Jung'schen Kreis näher brachte. Er bedeutet in gewissem Sinn tiefe und unbewußte, uns allen eingeprägte abstrakte Vorstellungen oder, wenn ich mich so ausdrücken darf, bereits vorhandene „Ur-Intuitionen". In demselben Werk Keplers fand Pauli aber auch die Vorstellungen über (Sphären-) Harmonien, die uns heute – nach dieser großartigen „Geometrisierung" der physikalischen Grundgesetze – eigentlich nicht mehr allzu ferne stehen. Diese Tatsache entspricht dem von Pauli aus dem Dänischen (wohl auf Grund der vielen und tiefgreifenden Gespräche mit Nils Bohr) übernommenen Begriff des „Gangere", der so viel wie geistige Wiederkehr in neuer, sehr verschiedener Gestalt bedeutet.

Diese, wie wir heute abwertend zu sagen pflegen, „mittelalterlichen" Vorstellungen, nach denen der Raum belebt war und die Partikel an die vorgegebenen Stellen leitete, wurden von Pauli auch in ihrer Nachwirkung bis in unsere Zeit analysiert und damit in völlig neuer Gestalt wiederentdeckt: Mit großem Verständnis spricht er (sehr im Gegensatz zu üblicher Schulweisheit!) über Newtons tiefe Sorge bezüglich seines Fernwirkungsgesetz für die Gravitation, das den Raum in gewissem Sinne „entseelt" und damit gegen religiöse Vorstellung zu verstoßen scheint. Pauli spricht humorvoll und geradezu poetisch von dessen Erhebung auf den „Olymp" abstrakter und absoluter Wahrheiten, der überdies von Kant mit (Pauli nahestehenden!) Verbotstafeln umstellt worden wäre: „Man hätte dann", so fährt Pauli fort, wobei er Einstein meint, „alle Mühe gehabt, den Raum von da wieder herunter zu kriegen". In der Tat kann man Pauli verstehen, wenn er die Riemann'sche Geometrie als eine „geistige" Wiederkehr älterer Vorstellungen empfunden hat.

Zugleich geht Pauli aber in der Geschichte weit zurück, und zwar bis zu der vergeistigten Welt Platos, auf die er seinen Wahlspruch „Das Alte ist auch das Neue" mit Überzeugung anwendet. Entscheidend beeindruckt ist er aber von einem der Hauptbegriffe der Neuplatoniker, der „Anima Mundi" oder „Weltseele", deren Wirkung er, wie oben angedeutet, über viele Jahrhunderte hinweg bis zur Zeit Newtons nachweist und verfolgt.

VII.

Ohne hier auf Paulis tiefgründigen und weiterführenden Überlegungen eingehen zu können, möchte ich – wenn ich mir einen solchen Hinweis erlauben darf – auf einen ganz anderen Zugang eingehen, den Pauli zu diesem Grundbegriff der Neuplatoniker gewiß auch hatte: Dies war sein tiefes (aber keineswegs technisches) Erlebnis und Verständnis der ‚großen' Musik (von Bach bis Schubert). Ich erinnere mich noch gut, wie er (aus Anlaß der von Paulis Assistent, Charles Ens, zu organisierenden, gemeinsamen Konzertbesuche) tief versunken einer Aufführung des späten Schubert'schen C-dur Quintetts lauschte: Er sah gerade in dieser Musik – bestätigt durch eine bezeichnende Bemerkung, die er in Paumgartners Schubertbiographie (die er eingehend studierte) fand – die Rolle der Intuition und der Eingebung, die Pauli, vielleicht unbewußt, mit der „Anima Mundi" in Verbindung brachte. Auch diese berührt ja die Seelen der Menschen über alle Zeiten und Erdteile hinweg. Unvergeßlich bleibt besonders das gemeinsame Anhören des späten Beethoven-Quartetts – in aller Stille – mit dem „Dankgebet eines Genesenden an die Gottheit" in Paulis Heim.

Damit kommen wir auch zu Paulis letztem Lebensjahr: Er machte – schon von einem unabänderlichen Krebsleiden gezeichnet – einen späten, doch mit allergrößter Anstrengung verfolgten Versuch, in erneuter (aber von einem Mißverständnis über das gesteckte Ziel verdunkelten) Zusammenarbeit mit Werner Heisenberg die mathematischen Schranken in der Quantenfeldtheorie zu durchbrechen. Ohne hier auf das ungemein tragische und in seinen Gründen weitgehend mißverstandene Ende dieser Zusammenarbeit einzugehen, komme ich gleich zu einem letzten, ausgedehnten und ungemein eindrücklichen Gespräch im Rahmen eines mehrstündigen gemeinsamen Spazierganges durch die tiefen Wälder des Jura (Chaumont) oberhalb von Neuchâtel: Es war dies eher Paulis wissenschaftliche Selbstbiographie mit der harten Selbstkritik, sein eigentliches und letztes Ziel (der mathematischen Konsistenz einer geometrisch zu erweiternden Quantenfeldtheorie) *nicht* erreicht zu haben. In einer zufällig am Wege stehenden, verlassenen Kapelle trat er ein und hielt eine kurze Weile der stillen Einkehr.

Wenige Wochen darauf wurde Pauli – nach viel zu später Entdeckung seines Leidens – in Eile zu einer (nicht mehr durchführbaren) Operation in das dem Institut nächstgelegen Spital gebracht. Als ihn Charles Enz auf die Zimmernummer, sie hieß 137, hinwies, verstand Pauli sogleich, daß er jetzt vor „Ihm" stehen würde, um auch das große Rätsel dieser Zahl zu erfahren. Ohnehin bestand ja für Pauli (nach einer etwas erweiterten Deutung einer Stelle in einem späteren persönlichen Brief) der eigentliche Sinn und das wirkliche Streben der Forschung darin, den weiten Weg zu „Ihm" zu finden.

Wolfgang Pauli: His Scientific Work and His Ideas on the Foundations of Physics

Konrad Bleuler

Institut für Theoretische Kernphysik der Universität
Nußallee 14-16, D(W)-5300 Bonn, Federal Republic of Germany

1. Introduction

Physics finds itself in recent years in an exciting and revolutionary phase of development: after a long intermission – and despite practical successes – critical questions about the proper foundations are again being asked, and far-reaching attempts are being made to gain a deeper understanding of the whole structure of the theory of our time. Like it or not, we are reminded of the decisive years of the revolution in physics in the 1920's, and we ask ourselves what was in the minds of the great scholars of that time, who participated in the re-structuring of physics made necessary by the advent of the special and the general theory of relativity and – simultanously – of quantum theory, and what relevance their thoughts and ideas could have for our time. From amongst these extraordinary personalities I wish to speak here about one of the foremost: Wolfgang Pauli. I would like to do so not just in view of my own unforgettable personal recollections, but especially in view of Pauli's – to this day largely unknown – ideas on the deeper meaning of physical research and his far-reaching philosophical and historical considerations. Just recently these ideas have been extensively discussed by K.V. Laurikainen in his book "Beyond the Atom: The Philosophical Thought of Wolfgang Pauli" (Springer Press 1988), based on quotations from previously inaccessable private correspondence. For the benefit of a new generation it may, however, be appropriate to first discuss some details of the life and work of this deepthinking scientific personality – both of which were quite unusual by present-day standards.

2. Youth and University Years

Wolfgang Pauli was born in Vienna in the year 1900 – he sometimes remarked on the practicality of this round number. His father (a professor in Vienna and, some thought, deserving of a Nobel prize for his work in chemical-pharmaceutical research) changed the family name from the original "Pacheles" to "Pauli", a step which was taken by many of the Eastern immigrants to the Austrian Empire. This origin may shed some light on his talent; one should perhaps recall in this connection that the "spiritual and artistic" Vienna of that period is hardly imaginable without the names of Kafka, Freud, Mahler and many others. From Pauli's school years in Vienna it should be noted that he causally remarked in later years, with an undertone of irony, "I was just a (mathematical) prodigy", and that he wrote in the margin of a book the following opinion, which is in marked contrast to the view of his godfather Mach: "It is important, besides physics, *also* to study history and languages." This remark presages, as we shall see, later far-reaching developments in his understanding of science.

Pauli's years at university were marked by a signal coincidence, which was probably of central importance for the entire subsequent development of modern physics: he studied, together with Werner Heisenberg and Gregor Wentzel (the three remained life-long friends) under Arnold Sommerfeld in Munich. In this context one can hardly overestimate the scientific importance of a great teacher and patron of research. The same may be said of his later collaboration in Copenhagen with Niels Bohr, who was his spiritual and human mentor. As regards his subsequent stay in Hamburg, Pauli told delightful anecdotes of his superior, the singular Professor Lenz, immortalized by the famous "Pauli-Lenz vector". As early as 1928 Pauli was – rather young but well-known – nominated a full professor at the ETH in Zurich, where he remained, except for an interval during the war, until his death in 1958.

3. Scientific Career

Pauli's scientific career began with a meteoric rise to fame in 1919, when he wrote the article on special and general relativity for the "Handbuch der Physik". This article is regarded even today as the authoritative review of the subject. Sommerfeld had encouraged him to write the article, after the first author he asked had declined. Albert Einstein was so deeply impressed by this work that he proclaimed Pauli as his spiritual (and professional) heir. Pauli regarded this, I believe, as a great duty, in a sense even a burden: in this connection I have a lively memory of a lecture held much later (on the occasion of the unveiling of a bust of Einstein), in which Pauli attempted to explain, or perhaps to somehow excuse, Einstein's rejection of the quantum theory: "Quantum mechanics was completely in the spirit of Einstein," he said. While reading his (for once) well-prepared manuscript in front of the "illustrious" audience, Pauli, deeply absorbed in his difficult text, calmly drew a small pencil from his vest-pocket and proceeded to insert the commas necessary to the natural breath spots for reading aloud. The young Pauli, who was able to immediately grasp Einstein's theory not only because of his own mathematical sophistication, but also because of the theory's great logical consistency, next plunged himself into the study of Bohr's "old quantum theory", at that time perfectly new but logically incomplete, and also wrote the first "Handbuch" article on this subject. It therefore seems only natural that Pauli contributed in an essential way, through correspondence and personal conversation first with Bohr and then with Heisenberg, to the emergence of the final formulation of modern quantum mechanics. The profound meaning, as well as the finality of this perhaps greatest revolution in the history of physics, was often emphasized by Pauli in later years in connection with "alternative interpretations". Entering into this fully new and indeed breathtaking abstract world of modern quantum theory must have made a deep and indelible impression on the principal architects of the theory, especially Pauli, who was the first (even before Schrödinger) to solve the hydrogen-atom-problem, using 'modern' algebraic techniques. It seems to me that this may well have been one of the motivations for Pauli's later profound reflections on the meaning and aim of physical research as such, as well as his deliberations on the intellectual history of physics' fundamental concepts. Pauli's outstanding contributions of those pioneering days are well known: his exclusion principle, which is fundamental for the structure of atoms, nuclei and (nowadays!) hadrons, and together with Pauli's non-relativistic spin theory forms the very basis for their mathematical treatment; his – from a modern vantage point – visionary neutrino hypothesis; and, as crowning achievement (together with Heisenberg), the general theory of quantum fields (i.e. *the* relativistic theory of elementary particles), from which he later deduced the – recently generalized – "spin-statistics theorem", a principle which he characterized as the most important consequence of the theory of special relativity. Very much the same holds for his famous CPT-theorem.

4. Doubts and Open Questions

Now that more than half a century has passed since the formulation of Pauli's decisive contributions to quantum theory, we can correctly estimate their importance as *the* formal foundation for all following discussions and calculations. At the same time I would like to conjecture that Pauli himself already had, in an early stage of the development of quantum field theory, profound doubts as to its adequacy and mathematical consistency as a general theory of elementary particles. He refused to be "renormalized", as he expressed himself in relation to that famous principle of renormalisation, thus setting himself (together, however, with Dirac, Bohr and others) for a time into strict opposition with an enthusiastic, and at first very successful, younger generation. A (partial?) concession in this respect came only long after Pauli's death, firstly in connection with the (unsuccessful?) attempts to "save" local quantum field theory by seeking recourse to "strings", and secondly in the recent profound and far-reaching attempts at a "non-commutative geometry" of A. Connes, D. Kastler and others. It seems to me that this most recent (as yet not generally recognized) development is perhaps the fulfilment of a "vision", which Pauli expressed in a long and unforgettable discussion a few weeks before his death: "For a real solution of the problem of singularities (i.e. the question of renormalization) a step of the same size and significance as that which was taken once before in the twenties might be necessary."

5. The Final Authority

Pauli closed this great period of the twenties, during which physics was in short order completely destroyed and then fully rebuilt, with his famous "Handbuch" article "The Principles of the Quantum Theory", which retains its validity to this day. Subsequently he was widely considered to be the final authority (or highest arbiter) for evaluating new contributions, or the many inavoidable "improvements". His general "verdict" in such cases was notorious: out of four possible outcomes he always found himself confronted with two: "new and wrong", or "old and right". Besides the absolutely annihilating "old *and* wrong", he invented an even lower assessment, namely "not even wrong". Besides many true instances of such "global judgements", which were occassionally all too severe, especially for younger lecturers, it seems to me that there exists an inexhaustible store of Pauli anecdotes, which are "more or less true", "easily imaginable", or simply "visionary". Like Einstein, Pauli spoke, often with accompanying gesticulation, of "Him", and prophesied, for example: "*He* has made no use of this (mathematical) possibility" (this concerned Weyl's neutrino equation, which Pauli, as is well known, disfavored because it violates parity). In contrast to this "vision", which was not fulfilled (Pauli later indulged, in connection with this wrong judgement, in a bit of self-satire, saying "I would gladly relinquish some of my reputation, I have enough of it"), another "easily imaginable" anecdote, according to which Pauli disagreed even with "Him", was, as we shall later see, fulfilled in a tragic way. Before that, however, I would like to remind you briefly of his, by now long past, lectures. There were often long moments of silence, during which he seemed to be completely unaware of his audience, which I would today interpret as *dialogues*, or more likely *arguments*, with "Him".

6. Pauli and Jung

Despite the many happy and seemingly carefree hours spent together in the later years in Zurich, during spontaneous outings, concerts or animated discussions over a shared lunch, one felt that Pauli was struggling to overcome a feeling of loneliness, of doubt or of spiritual tension. This was undoubtably in part due to the after-effects of a difficult inner crisis (after the first ten years of his really great achievements), which brought him into intimate contact with questions of psychoanalysis, and thereby with C.G. Jung. This was certainly an additional external motivation for Pauli to concern himself with profound questions regarding the spiritual, psychological and historical roots of scientific research, which, seen from the outside, seems so consistent and successful. A central point of Pauli's, today often neglected or, one might say, repressed, was the idea that (for a real grasp of the phenomena, i.e. for the establishment, or better, discovery of fundamental laws) two equally important prerequisites are necessary: (1) An adequate fund of empirical data, *and* (2), perhaps even more important, i.e. indispensable, an intuition, or a particular insight, as, for example, the hint of a connection to an apparently completely disconnected area of research. A modern, and very impressive example – besides the overworked case of Newton's falling apple – is Hermann Weyl's idea of the geometrically natural (local) gauge principle in the year 1929, which later became – after the extension by Yang and Mills nearly a half-century later – *the* underlying structure for all three fundamental physical laws, covering the complete spectrum of physical phenomena (i.e. general relativity, the Glashow-Weinberg-Salam model of electroweak interactions and Quantum Chromodynamics in strong, i.e. Hadronic, interactions).

In order to approach the question of the origin or genesis of this decisive intuition (or the researcher's inner sight), Pauli immersed himself in Jung's questions concerning the human soul and its appearances in the history of science. He found an enlightening hint in Kepler's philosophical works: Kepler's decisive concept is that of the "archetype"; this concept brought Pauli closer to Jung's circle. It meant, in a certain sense, a deep and unconscious abstract idea embedded in our common consciousness, or, if I may so say, a "primeval intuition". In the same work of Kepler Pauli found the concept of the harmonies (of the spheres), which today, after the great "geometrization" of the fundamental physical laws, no longer appears to us so far-fetched: It corresponds, in a way, to the concept of "Gangere", which Pauli had adopted from the Danish (certainly as a result of his many deep discussions with Niels Bohr), representing a kind of spiritual reincarnation, in a new and very different form.

Such ideas, which we today often characterize deprecatingly as "mediaeval", according to which space is animate and particles are guided to their ordained positions, were analyzed by Pauli with respect to their after-effects on our time, and thereby rediscovered in a new guise: he speaks with great sympathy (in contrast to the conventional wisdom) of Newton's deep concern over gravity's action-at-a-distance, which in a sense despiritualizes space, and thereby seemed at odds with religious precepts. Pauli speaks humorously, and even poetically, of its elevation to the "Olympus" of abstract and absolute truths, and even surrounded, according to Kant, by "No tresspassing" signs (This especially pleased Pauli). "One had", Pauli continues (referring to Einstein), "great trouble to get space back down from this pinnacle". We can thus understand how Pauli felt Riemannian geometry to be a "spiritual" reincarnation of older concepts.

At the same time Pauli goes even further back in history, to the spiritualized world of Plato, for whom his motto that "the old is also the new" is singularly approprate. However, he was most deeply impressed by one of the concepts of the neo-Platonists, the "Anima Mundi", or the "world spirit", whose influence he traced through the centuries, as indicated above, up to the time of Newton.

7. Conclusion

Without being able here to go into Pauli's profound further considerations, I would like – if I can allow myself such a hint – to indicate a completely different approach, which Pauli also had to this fundamental concept of the neo-Platonists: this was his deep (but not "technical") experience and appreciation of "great" music (from Bach to Schubert). I recall very distinctly, how (on the occasion of a concert which Pauli's assistant Charles Enz had arranged for us to attend) Pauli sat deeply absorbed in a performance of Schubert's late C-major quintet: He saw revealed in this music (confirmed by a revealing remark in Paumgarten's biography of Schubert which he had carefully studied) the rôle of intuition and the inspiration, which he, perhaps subconsciously, associated with the "Anima Mundi": Music also touches the souls of men at all times and in all places. Another unforgettable experience was listening together – in absolute silence – to the late Beethoven-quartet "Thanksgiving Prayer of a Reconvalescent to God" in Pauli's home.

With this we arrive at Pauli's last year: In close collaboration with Werner Heisenberg he made a late attempt, with all the power left at his command (he was already marked by the signs of an incurable cancer), to penetrate the mathematical boundaries of quantum field theory. This brave effort was plagued from its conception by a misunderstanding concerning its envisaged aim, i.e., a theory of particles on the one hand, a mathematically consistent field theory on the other! I cannot go here into the largely misunderstood reasons for the tragic end of this collaboration. I come instead to speak of one of my last extended and profoundly touching discussions with Pauli in the course of a long walk through the deep woods of the Jura (Chaumont) above Neuchâtel. Pauli expounded his scientific autobiography, accompanied by harsh self-criticism for not having achieved his own final aim: the mathematical consistency of a geometrically extended quantum field theory. In an abandoned chapel we happened to pass he entered and kept for a short time a silent revery.

A few weeks later Pauli was rushed – after a far too late diagnosis of his disease – to the hospital nearest to the Institute for a proposed operation which turned out to be no longer possible. When Charles Enz pointed out to him the number of the room – it was 137 – Pauli understood at once that he would soon stand before "Him", in order to finally learn the great secret of this number. In any case Pauli's own aim (according to a somewhat generalized interpretation of a passage in one of his late personal letters), *and* the true aim of real research, was to find the way to "Him".

Editor's Note: In order to appreciate these last remarks the reader has to know the famous Pauli anecdote, to which the author alludes at several points but never quite reveals: Upon arriving in Heaven Pauli finally gets the chance to find out the mystery of the fine-structure constant, which has eluded him all his life. He asks God for the explanation. God goes to the blackboard, the chalk clicks and clicks ... God stands back from the blackboard. Pauli's response: "I'm sorry, I'm still not convinced."

Participants of the Meeting

Ackermann, Thomas	Abteilung Mathematik der Universität, Naturwissenschaftliche Fakultät, Universitätsstraße 31, W-8400 Regensburg
Behrndt, Klaus	Sektion Physik 01 der Humboldt-Universität, Postfach 1297, O-1086 Berlin
Bettge, Lutz	Theoretische Physik III der Universität, Postfach 50 05 00, W-4600 Dortmund 50
Bieber, Robert	Abteilung Mathematik der Universität, Naturwissenschaftliche Fakultät, Universitätsstraße 31, W-8400 Regensburg
Binz, Ernst	Lehrstuhl für Mathematik I der Universität, Seminargeb.A5, Schloß, W-6800 Mannheim
Bischoff, Wolfgang	Goebenstraße 25, W-2350 Neumünster
Bleyer, Ulrich	Einstein-Laboratorium der ehem. Akademie der Wissenschaften der DDR, Rosa-Luxemburg-Straße 17a, O-1590 Potsdam
Blumenhagen, Ralph	Endenicher Straße 331, W-5300 Bonn 1
Brandt, Friedemann	Institut für Theoretische Physik der Universität, Appelstraße 2, W-3000 Hannover 1
Cell, B.	Sektion Physik der Universität, Karl-Marx-Straße 10/11, O-7010 Leipzig
Damm, Anke	Sundgauallee 39, W-7800 Freiburg
Debrus, Joachim	Physikzentrum Bad Honnef Hauptstraße 5, W-5340 Bad Honnef
Diekmann, Burkhard	Händelstraße 15, W-5300 Bonn
Eberlein, Claudia	Sektion Physik der Universität, WB Quantenfeldtheorie, Karl-Marx-Straße 10/11, O-7010 Leipzig
Flohr, Michael	Physikalisches Institut der Universität, Nußallee 12, W-5300 Bonn
Föll, Marc	Dreikreuzenstraße 12, W-3000 Hannover 91
Förste, Stefan	Sektion Physik 01 der Humboldt-Universität, Postfach 1297, O-1086 Berlin
Forger, Michael	Physikalisches Institut der Universität, Hermann-Herder-Straße 3, W-7800 Freiburg
Fröbe, Holger	Sektion Physik der Universität, WB Quantentheorie, Max-Wien-Platz 1, O-6900 Jena
Gebert, Reinhold	Neureuther Straße 39, W-8000 München 40

Gotzes, Siegfried	Theoretische Physik III der Universität, Postfach 50 05 00, W-4600 Dortmund 50
Groote, Stefan	Theoretische Physik III der Universität, Postfach 50 05 00, W-4600 Dortmund 50
Hammerschmitt, Achim	Institut für Theoretische Physik der Universität, Philosophenweg 16, W-6900 Heidelberg
Hehl, Friedrich W.	Institut für Theoretische Physik der Universität, Zülpicher Straße 77, W-5000 Köln 41
Heinrich, Olaf	Zentralinstitut für Astrophysik der ehem. Akademie der Wissenschaften der DDR, Rosa-Luxemburg-Straße 17a, O-1561 Potsdam
Hellmund, Meik	Sektion Physik der Universität, WB Hochenergiephysik, Karl-Marx-Platz 10/11, O-7010 Leipzig
Herrmann, Stefan	Weberstraße 99, W-5300 Bonn
Hinrichsen, Haye	Physikalisches Institut der Universität, Nußallee 12, W-5300 Bonn
Hirshfeld, Allen C.	Theoretische Physik III der Universität, Postfach 50 05 00, W-4600 Dortmund 50
Hoppe, Jens	Institut für Theoretische Physik der Universität, Kaiserstraße 12, W-7500 Karlsruhe
Hübner, M.	Sektion Physik der Universität, Karl-Marx-Platz 10/11, O-7010 Leipzig
Kaufmann, Markus	Solmsstraße 52, W-Berlin 61
Kliem, Albrecht	Physikalisches Institut der Universität, Nußallee 12, W-5300 Bonn
Kreimer, Dirk	Institut für Physik der Universität, Staudinger Weg 5, W-6500 Mainz
Kubitza, Markus	Trockener Kamp 88, W-3200 Hildesheim
Laartz, Jürgen	Metzstraße 21, W-2300 Kiel
Lemke, Horst	Ludwigstraße 10, W-5000 Köln 50
Liebscher, Dierck E.	Zentralinstitut für Astrophysik der ehem. Akademie der Wissenschaften der DDR, Rosa-Luxemburg-Straße 17a, O-1561 Potsdam
Loll, Renate	Physikalisches Institut der Universität, Nußallee 12, D-5300 Bonn
Luce, Thomas	Institut für Theoretische Physik der Universität, Arnimallee 14, W-1000 Berlin 33
Marquardt, Uli	Institut für Theoretische Physik der Universität, Philosophenweg 16, W-6900 Heidelberg

Matthes, R.	Sektion Physik der Universität, Karl-Marx-Platz 10/11, O-7010 Leipzig
Meier, Wolfgang	Sektion Physik der Universität, Max-Wien-Platz 1, O-6900 Jena
Meinhardt, Wolfram	Sektion Physik der Universität, Max-Wien-Platz 1, O-6900 Jena
Meinrenken, Eckhard	Sternwaldstraße 13, W-7800 Freiburg
Moonen, Boudewijn	Max-Planck-Institut für Mathematik, Gottfried-Claren-Straße 26, W-5300 Bonn 3
Müller, Wolfgang	Max-Planck-Institut für Mathematik, Gottfried-Claren-Straße 26, W-5300 Bonn 3
Nahm, Werner	Physikalisches Institut der Universität, Nußallee 12, W-5300 Bonn
Nierste, Ulrich	Institut für Theoretische Physik der Universität, Am Hubland, W-8700 Würzburg
Papadopoulos, Nikolas A.	Institut für Physik der Universität, Staudinger Weg 7, W-6500 Mainz
Pflaum, Markus	Helmtrudenstraße 4, W-8000 München 40
Recknagel, Andreas	Karl-Frowein-Straße 19, W-5300 Bonn
Rehren, Karl-Henning	II. Institut für Theoretische Physik der Universität, Luruper Chaussee 149, W-2000 Hamburg 50
Reifenhäuser, Bernd	Institut für Physik der Universität, Staudinger Weg 7, W-6500 Mainz
Reuter, Martin	Institut für Theoretische Physik der Universität, Appelstraße 2, W-3000 Hannover
Richter, O.	Sektion Physik der Universität, Karl-Marx-Platz 10/11, O-7010 Leipzig
Römer, Hartmann	Physikalisches Institut der Universität, Hermann-Herder-Straße 3, W-7800 Freiburg
Ruder, Dieter	Sektion Physik der Universität, Max-Wien-Platz 1, O-6900 Jena
Rudolph, G.	Sektion Physik der Universität, Karl-Marx-Platz 10/11, O-7010 Leipzig
Schirmer, Thomas	Wiedemannstraße 33, W-5300 Bonn 2
Schmidt, Hans-Jürgen	Zentralinstitut für Astrophysik der ehem. Akademie der Wissenschaften der DDR, Rosa-Luxemburg-Straße 17a, O-1561 Potsdam
Schmidt, Joachim	Am Waldesrand 2, W-3410 Northeim
Schücker, Thomas	Institut für Theoretische Physik der Universität, Philosophenweg 16, W-6900 Heidelberg

Seifert, Karsten	Johann-Sigismund-Straße 12, W-1000 Berlin 31
Staszkiewicz, Carl-Philipp	Institut für Theoretische Physik der Freien Universität, Arnimallee 14, W-1000 Berlin 33
Striker, Timothy	Weender Landstraße 57a, W-3400 Göttingen
Thienel, Hans-Peter	Weidenweg 11, W-7150 Backnang
Tolksdorf, Jürgen	Institut für Physik der Universität, Philosophenweg 16, W-6900 Heidelberg
Varnhagen, Raimund	Physikalisches Institut der Universität, Nußallee 12, W-5300 Bonn

Index